112
# Advances in Biochemical Engineering/Biotechnology

**Series Editor: T. Scheper**

**Editorial Board:**
W. Babel · I. Endo · S.-O. Enfors · M. Hoare ·W.-S. Hu
B.Mattiasson · J. Nielsen · G. Stephanopoulos
U. von Stockar · G. T. Tsao · R. Ulber · J.-J. Zhong

# Advances in Biochemical Engineering/Biotechnology
Series Editor: T. Scheper

Recently Published and Forthcoming Volumes

**Bioreactor Systems for Tissue Engineering**
Volume Editors: Kasper, C., van Griensven, M.,
Poertner, R.
Vol. 112, 2008

**Food Biotechnology**
Volume Editors: Stahl, U., Donalies, U. E. B.,
Nevoigt, E.
Vol. 111, 2008

**Protein – Protein Interaction**
Volume Editors: Seitz, H., Werther, M.
Vol. 110, 2008

**Biosensing for the 21st Century**
Volume Editors: Renneberg, R., Lisdat, F.
Vol. 109, 2007

**Biofuels**
Volume Editor: Olsson, L.
Vol. 108, 2007

**Green Gene Technology**
Research in an Area of Social Conflict
Volume Editors: Fiechter, A., Sautter, C.
Vol. 107, 2007

**White Biotechnology**
Volume Editors: Ulber, R., Sell, D.
Vol. 105, 2007

**Analytics of Protein-DNA Interactions**
Volume Editor: Seitz, H.
Vol. 104, 2007

**Tissue Engineering II**
Basics of Tissue Engineering and Tissue
Applications
Volume Editors: Lee, K., Kaplan, D.
Vol. 103, 2007

**Tissue Engineering I**
Scaffold Systems for Tissue Engineering
Volume Editors: Lee, K., Kaplan, D.
Vol. 102, 2006

**Cell Culture Engineering**
Volume Editor: Hu, W.-S.
Vol. 101, 2006

**Biotechnology for the Future**
Volume Editor: Nielsen, J.
Vol. 100, 2005

**Gene Therapy and Gene Delivery
Systems**
Volume Editors: Schaffer, D.V., Zhou, W.
Vol. 99, 2005

**Sterile Filtration**
Volume Editor: Jornitz, M.W.
Vol. 98, 2006

**Marine Biotechnology II**
Volume Editors: Le Gal, Y., Ulber, R.
Vol. 97, 2005

**Marine Biotechnology I**
Volume Editors: Le Gal, Y., Ulber, R.
Vol. 96, 2005

**Microscopy Techniques**
Volume Editor: Rietdorf, J.
Vol. 95, 2005

**Regenerative Medicine II**
Clinical and Preclinical Applications
Volume Editor: Yannas, I. V.
Vol. 94, 2005

**Regenerative Medicine I**
Theories, Models and Methods
Volume Editor: Yannas, I. V.
Vol. 93, 2005

# Bioreactor Systems for Tissue Engineering

Volume Editors:
Cornelia Kasper · Martijn van Griensven · Ralf Pörtner

With contributions by

M. Al-Rubeai · K. Baar · S. Boehm · M.H. Chowdhury
M. Cioffi · S. Cancaro · R.G. Dennis · S. Diederichs
K. Donnelly · D. Eibl · R. Eibl · A.J. El Haj · E. Eisenbarth
Y. Fu · P. Gatenholm · C. Goepfert · G. Gogniat · M. van Griensven
F. Gustavson · K. Hampson J. Hansmann · D.W. Hutmacher
E. Ilinich · M. Israelowitz · R. Janssen · C. Kasper · I. Martin
J.M. Melero-Martin ·H. Mertsching · M. Morlock · H. Paethaold
A. Peterbauer · A. Philp · R. Pörtner · S.A. Riboldi · D. Riechers
S. Roeker · S. Santhalingam · H.P. von Schroeder · H. Singh
B. Smith · F. Stahl · P.M. Vogt · R. Wendt · B. Weyand
K. Wiegandt · S. Wolbank

*Advances in Biochemical Engineering/Biotechnology* reviews actual trends in modern biotechnology. Its aim is to cover all aspects of this interdisciplinary technology where knowledge, methods and expertise are required for chemistry, biochemistry, micro-biology, genetics, chemical engineering and computer science. Special volumes are dedicated to selected topicswhich focus on newbiotechnological products and new processes for their synthesis and purification. They give the state-of-the-art of a topic in a comprehensive way thus being a valuable source for the next 3–5 years. It also discusses new discoveries and applications. Special volumes are edited by well known guest editors who invite reputed authors for the review articles in their volumes.
In references *Advances in Biochemical Engineering/Biotechnology is abbeviated Adv Biochem Engin/ Biotechnol* and is cited as a journal.

Springer WWW home page: springer.com
Visit the ABE content at springerlink.com

ISBN 978-3-540-69356-7    e-ISBN 978-3-540-69357-4
DOI: 10.1007/978-3-540-69357-4

Advances in Biochemical Engineering/Biotechnology ISSN 0724-6145

Library of Congress Control Number: 2008939436

© 2009 Springer-Verlag Berlin Heidelberg

This work is subject to copyright. All rights are reserved, whether the whole or part of the material is concerned, specifically the rights of translation, reprinting, reuse of illustrations, recitation, roadcasting, reproduction on microfilm or in any other way, and storage in data banks. Duplication of this publication or parts thereof is permitted only under the provisions of the German Copyright Law of September 9, 1965, in its current version, and permission for use must always be obtained from Springer. Violations are liable to prosecution under the German Copyright Law.

The use of general descriptive names, registered names, trademarks, etc. in this publication does not imply, even in the absence of a specific statement, that such names are exempt from the relevant protective laws and regulations and therefore free for general use.

*Cover design*: WMXDesign GmbH, Heidelberg, Germany

Printed on acid-free paper

9 8 7 6 5 4 3 2 1

springer.com

# Series Editor

Prof. Dr. T. Scheper

Institute of Technical Chemistry
University of Hannover
Callinstraße 3
30167 Hannover, Germany
scheper@iftc.uni-hannover.de

# Volume Editors

Dr. Cornelia Kasper

Institute of Technical Chemistry
University of Hannover
Callinstraße 3
30167 Hannover, Germany
kasper@iftc.uni-hannover.de

Prof. Martijn van Griensven

Ludwig Boltzmann Institut für
Klinische und Experimentelle
Traumatologie
Donaueschingenstr. 13
1200 Wien, Austria
Martijn.van.Griensven@LBITRAUMA.ORG

Dr. Ralf Pörtner

TU Hamburg-Harburg
Institut für Biotechnologie und
Verfahrenstechnik Denickestr. 15
21073 Hamburg, Germany

# Editorial Board

Prof. Dr.W. Babel

Section of Environmental Microbiology
Leipzig-Halle GmbH
Permoserstraße 15
04318 Leipzig, Germany
*babel@umb.ufz.de*

Prof. Dr. I. Endo

Saitama Industrial Technology Center
3-12-18, Kamiaoki Kawaguchi-shi
Saitama, 333-0844, Japan
*a1102091@pref.saitama.lg.jp*

Prof. Dr. S.-O. Enfors

Department of Biochemistry`
and Biotechnology
Royal Institute of Technology
Teknikringen 34,
100 44 Stockholm, Sweden
*enfors@biotech.kth.se*

Prof. Dr. M. Hoare

Department of Biochemical Engineering
University College London
Torrington Place
London, WC1E 7JE, UK
*mhoare@ucl.ac.uk*

Prof. Dr.W.-S. Hu

Chemical Engineering
and Materials Science
University of Minnesota
421Washington Avenue SE
Minneapolis, MN 55455-0132, USA
wshu@cems.umn.edu

Prof. Dr. B. Mattiasson

Department of Biotechnology
Chemical Center, Lund University
P.O. Box 124, 221 00 Lund, Sweden
bo.mattiasson@biotek.lu.se

Prof. Dr. J. Nielsen

Center for Process Biotechnology
Technical University of Denmark
Building 223
2800 Lyngby, Denmark
jn@biocentrum.dtu.dk

Prof. Dr. G. Stephanopoulos

Department of Chemical Engineering
Massachusetts Institute of Technology
Cambridge, MA 02139-4307, USA
gregstep@mit.edu

Prof. Dr. U. von Stockar

Laboratoire de Génie Chimique et
Biologique (LGCB)
Swiss Federal Institute of Technology
Station 6
1015 Lausanne, Switzerland
urs.vonstockar@epfl.ch

Prof. Dr. G. T. Tsao

Professor Emeritus
Purdue University
West Lafayette, IN 47907, USA
tsaogt@ecn.purdue.edu
tsaogt2@yahoo.com

Prof. Dr. Roland Ulber

FB Maschinenbau und Verfahrenstechnik
Technische Universität Kaiserslautern
Gottlieb-Daimler-Straße
67663 Kaiserslautern, Germany
ulber@mv.uni-kl.de

Prof. Dr. C.Wandrey

Institute of Biotechnology
Forschungszentrum Jülich GmbH
52425 Jülich, Germany
c.wandrey@fz-juelich.de

Prof. Dr. J.-J. Zhong

Bio-Building #3-311
College of Life Science & Biotechnology
Key Laboratory of Microbial Metabolism,
Ministry of Education
Shanghai Jiao Tong University
800 Dong-Chuan Road
Minhang, Shanghai 200240, China
jjzhong@sjtu.edu.cn

# Honorary Editors

Prof. Dr. A. Fiechter

Institute of Biotechnology
Eidgenössische Technische Hochschule
ETH-Hönggerberg
8093 Zürich, Switzerland
ae.fiechter@bluewin.ch

Prof. Dr. K. Schügerl

Institute of Technical Chemistry
University of Hannover, Callinstraße 3
30167 Hannover, Germany
schuegerl@iftc.uni-hannover.de

# Advances in Biochemical Engineering/ Biotechnology Also Available Electronically

For all customers who have a standing order to Advances in Biochemical Engineering/ Biotechnology, we offer the electronic version via SpringerLink free of charge. Please contact your librarian who can receive a password or free access to the full articles by registering at:

springerlink.com

If you do not have a subscription, you can still view the tables of contents of the volumes and the abstract of each article by going to the SpringerLink Homepage, clicking on "Browse by Online Libraries", then "Chemical Sciences", and finally choose Advances in Biochemical Engineering/Biotechnology.

You will find information about the

– Editorial Board
– Aims and Scope
– Instructions for Authors
– Sample Contribution

at springer.com using the search function.

*Color figures* are published in full color within the electronic version on SpringerLink.

# Attention all Users
# of the "Springer Handbook of Enzymes"

Information on this handbook can be found on the internet at springeronline.com

A complete list of all enzyme entries either as an alphabetical Name Index or as the EC-Number Index is available at the above mentioned URL. You can download and print them free of charge.

A complete list of all synonyms (more than 25,000 entries) usedfor the enzymes is available in print form (ISBN 3-540-41830-X).

# Save 15%

We recommend a standing order for the series to ensure you automatically receive all volumes and all supplements and save 15% on the list price.

# Preface

The editors of this special volume would first like to thank all authors for their excellent contributions. We would also like to thank Prof. Dr. Thomas Scheper, Dr. Marion Hertel and Ulrike Kreusel for providing the opportunity to compose this volume and Springer for organizational and technical support.

Tissue engineering represents one of the major emerging fields in modern biotechnology; it combines different subjects ranging from biological and material sciences to engineering and clinical disciplines. The aim of tissue engineering is the development of therapeutic approaches to substitute diseased organs or tissues or improve their function. Therefore, three dimensional biocompatible materials are seeded with cells and cultivated in suitable systems to generate functional tissues.

Many different aspects play a role in the formation of 3D tissue structures. In the first place the source of the used cells is of the utmost importance. To prevent tissue rejection or immune response, preferentially autologous cells are now used. In particular, stem cells from different sources are gaining exceptional importance as they can be differentiated into different tissues by using special media and supplements. In the field of biomaterials, numerous scaffold materials already exist but new composites are also being developed based on polymeric, natural or xenogenic sources. Moreover, a very important issue in tissue engineering is the formation of tissues under well defined, controlled and reproducible conditions. Therefore, a substantial number of new bioreactors have been developed. Depending on the target tissue, different concepts have already been realized for dynamic cultivations. For the generation of functional tissue it is often necessary or beneficial to apply mechanical forces during or prior to cultivation.

This book comprises contributions of researchers active in the field of bioreactor design and optimization for the controlled cultivation of cells for tissue engineering purposes. The knowledge and expertise of the authors cover disciplines like engineering, biotechnology and clinical sciences. Recent developments in bioreactor developments for the use of cartilage, bone, and cardiovascular, muscle and connective tissue are presented in separate chapters as well as the current status of disposable bioreactors. Furthermore, contributions are included on the considerations and requirements of fluid dynamics for bioreactor optimization.

We hope that this state-of-the-art book is helpful to your research. Please enjoy reading it as much as we enjoyed preparing it.

Hannover, Summer 2008
Cornelia Kasper
Martijn van Griensven
Ralf Pörtner

# Contents

**Bioreactors in Tissue Engineering: Scientific Challenges and Clinical Perspectives** .................................................................. 1
D. Wendt, S.A. Riboldi, M. Cioffi, and I. Martin

**Bioreactor Technology in Cardiovascular Tissue Engineering** ................. 29
H. Mertsching and J. Hansmann

**Bioreactors for Guiding Muscle Tissue Growth and Development** .......... 39
R.G. Dennis, B. Smith, A. Philp, K. Donnelly, and K. Baar

**Bioreactors for Connective Tissue Engineering: Design and Monitoring Innovations** .................................................. 81
A.J. El Haj, K. Hampson, and G. Gogniat

**Mechanical Strain Using 2D and 3D Bioreactors Induces Osteogenesis: Implications for Bone Tissue Engineering** ............ 95
M. van Griensven, S. Diederichs, S. Roeker, S. Boehm, A. Peterbauer, S. Wolbank, D. Riechers, F. Stahl, and C. Kasper

**Bioreactors for Tissue Engineering of Cartilage** ........................................ 125
S. Concaro, F. Gustavson, and P. Gatenholm

**Technical Strategies to Improve Tissue Engineering of Cartilage-Carrier-Constructs** ................................................................ 145
R. Pörtner, C. Goepfert, K. Wiegandt, R. Janssen, E. Ilinich, H. Paetzold, E. Eisenbarth, and M. Morlock

**Application of Disposable Bag Bioreactors in Tissue Engineering and for the Production of Therapeutic Agents** ..................... 183
R. Eibl and D. Eibl

**Methodology for Optimal In Vitro Cell Expansion in Tissue Engineering** .................................................................................... 209
J.M. Melero-Martin, S. Santhalingam, and M. Al-Rubeai

**Bioreactor Studies and Computational Fluid Dynamics** ........................... 231
H. Singh and D.W. Hutmacher

**Fluid Dynamics in Bioreactor Design: Considerations for the Theoretical and Practical Approach** .............................................. 251
B. Weyand, M. Israelowitz, H.P. von Schroeder, and P.M. Vogt

**Index** ............................................................................................................... 269

# Bioreactors in Tissue Engineering: Scientific Challenges and Clinical Perspectives

**D. Wendt, S.A. Riboldi, M. Cioffi, and I. Martin**

**Abstract** In this Chapter we discuss the role of bioreactors in the translational paradigm of Tissue Engineering approaches from basic research to streamlined tissue manufacturing. In particular, we will highlight their functions as: (1) *Pragmatic tools for tissue engineers*, making up for limitations of conventional manual and static techniques, enabling automation and allowing physical conditioning of the developing tissues; (2) *3D culture model systems*, enabling us to recapitulate specific aspects of the actual in vivo milieu and, when properly integrated with *computational modeling* efforts and *sensing and control* techniques, to address challenging scientific questions; (3) *Tissue manufacturing devices*, implementing bioprocesses so as to support safe, standardized, scaleable, traceable and possibly cost-effective production of grafts for clinical use. We will provide evidences that fundamental knowledge gained through the use of well-defined and controlled bioreactor systems at the research level will be essential to define, optimize, and moreover, *streamline* the key processes required for efficient manufacturing models.

**Keywords** Bioreactor, Computational modeling, 3D model system, Sensing, Tissue manufacturing.

---

D. Wendt (✉), S.A. Riboldi, and I. Martin
Institute for Surgical Research and Hospital Management, University Hospital Basel, Hebelstrasse 20, 4031 Basel, Switzerland

M. Cioffi
Laboratory of Biological Structure Mechanics, Department of Structural Engineering, Politecnico di Milano, Piazza Leonardo Da Vinci 32, 20133 Milano, Italy

## Contents

1 Introduction ........................................................................................................ 2
2 Bioreactors: Pragmatic Tools for Tissue Engineers ....................................... 3
   2.1 Cell Seeding on Three-Dimensional Matrices ........................................ 3
   2.2 Maintenance of a Controlled Culture Environment ............................... 5
   2.3 Physical Conditioning of Developing Tissues ........................................ 6
3 Bioreactors as 3D In vitro Model Systems ..................................................... 7
   3.1 Open Challenges ........................................................................................ 10
   3.2 Computational Modeling in Bioreactor Systems .................................... 10
   3.3 Sensing in Tissue Engineering Bioreactors ............................................ 16
4 Bioreactors: The Clinical Perspective ............................................................. 20
   4.1 Streamlining Graft Manufacturing Processes ......................................... 21
   4.2 Centralized Versus De-Centralized Production Facilities ..................... 23
   4.3 "Intraoperative Engineering" Approaches .............................................. 23
5 Conclusions and Outlook .................................................................................. 24
References ................................................................................................................ 25

## Abbreviations

| | |
|---|---|
| BMSC | Bone marrow stromal cells |
| CFD | Computational fluid dynamics |
| 2D | Bi-dimensional |
| 3D | Three-dimensional |
| DOCT | Doppler optical coherence tomography |
| GAG | Glycosaminoglycan |
| GMP | Good manufacturing practice |
| HA | Hyaluronan |
| μCT | Micro-computed tomography |
| OCT | Optical coherence tomography |
| PFC | Perfluorocarbon |
| PIV | Particle image velocimetry |
| QC | Quality control |
| SZP | Superficial zone protein |
| TE | Tissue engineering |

# 1 Introduction

"Bioreactors", a term generally associated with classical industrial bioprocesses such as fermentation, was initially used in Tissue Engineering (TE) applications to describe little more than simple mixing of a Petri dish. Over the last two decades, the concept of a bioreactor has evolved not only in complexity, but also in the field of use. It is now clear that bioreactors represent not only powerful technical tools to support and direct the in vitro development of living, functional tissues, but also dynamic culture model systems to study fundamental mechanisms of cell function under physiologically relevant conditions.

Of primary importance in the field of tissue engineering and regenerative medicine is the consideration that despite the impressive scientific progress achieved, the need for safe and clinically effective autologous tissue substitutes still remains unsatisfied. In order to successfully translate TE technologies from bench to bedside while competing with alternative therapeutic options, the clinical efficacy of a tissue-engineered product needs to be accompanied by a cost-effective manufacturing process and compliance to the evolving regulatory framework in terms of Quality Control (QC) and Good Manufacturing Practice (GMP) requirements. In this context, bioreactors as a means to generate and maintain a controlled culture environment and enable directed tissue growth could represent the key element for the development of automated, standardized, traceable, cost-effective, and safe manufacturing processes for engineered tissues for clinical applications.

In this Chapter we discuss the role of bioreactors in the translational paradigm of TE approaches from basic research to streamlined tissue manufacturing. To this purpose, we first review the key functions of bioreactors traditionally employed in research applications, both as pragmatic tools overcoming limitations of conventional cell/tissue culture techniques (Sect. 2) and as 3D model systems recapitulating aspects of the actual in vivo milieu of specific tissues (Sect. 3). In particular, having identified the necessity of predictive tools and technological platforms to peek inside the "black box" bioreactor, we briefly review the state of the art of computational modeling in bioreactor systems and give an overview of the basic sensing techniques employed in the engineering of biological tissues. Finally, we describe and critically discuss examples, potentials and challenges for bioreactor-based manufacturing of tissue-engineered products (Sect. 4).

## 2 Bioreactors: Pragmatic Tools for Tissue Engineers

In the past years, bioreactors have proved to be crucial tools to initiate, maintain and direct cell cultures and tissue development in a three-dimensional (3D), physico-chemically defined, tightly controlled, aseptic environment. Specifically, state-of-the-art devices may offer the possibility to (1) dynamically seed cells within 3D matrices, (2) overcome the constraints of a static culture environment and (3) physically stimulate the developing constructs. In this Section, we briefly describe these features, which are key for bioreactors commonly used for research purposes (Fig. 1) [1–3].

### 2.1 Cell Seeding on Three-Dimensional Matrices

Traditionally, the delivery of a cell suspension within a three-dimensional scaffold is manually performed by means of pipettes and relying on gravity as a leading principle for cell settlement and subsequent adhesion to the scaffold pores. Such a

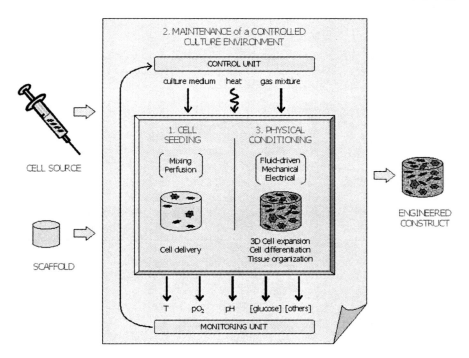

**Fig. 1** Schematic representation of the key functions of bioreactors used in research applications for tissue engineering, described in Sect. 2. (1) Cell seeding of three-dimensional matrices: bioreactors can maximize the cell utilization, control the cell distribution, and improve the reproducibility of the cell seeding process. (2) Maintenance of a controlled culture environment: bioreactors that monitor and control culture parameters can provide well-defined model systems to investigate fundamental aspects of cell function and can be used to enhance the reproducibility and overall quality of engineered tissues. (3) Physical conditioning of cell/scaffold constructs: bioreactors that apply physiological regimes of physical stimulation can improve the structural and functional properties of engineered tissues (T, temperature; $pO_2$, oxygen partial pressure; [], concentrations)

seeding method, besides being scarcely reproducible due to marked intra- and interoperator variability, is inevitably characterized by poor efficiency and nonuniformity of the resulting cell distribution within the scaffold [4]. The usual "static" seeding method may yield particularly inhomogeneous results when *thick and/or low-porosity scaffolds* are used, since gravity may not suffice for the cells to penetrate throughout the scaffold pores. Especially when dealing with *human cell sources*, optimizing the efficiency of seeding could be crucial in order to maximize the utilization of cells that can be obtained from the rather limited tissue biopsies.

Hence a variety of "dynamic" cell-seeding techniques, relying on the use of bioreactors, have been recently developed with the aim to increase quality, reproducibility, efficiency, and uniformity of the seeding process as compared to conventional static methods. Spinner flasks [5], wavy-walled reactors [6], and rotating wall vessels [7] are only examples of the numerous devices found in the literature. However, the most promising approach, enabling efficient and uniform seeding of different cell types in scaffolds of various morphologies and porosities, proved to be "perfusion seeding," consisting of direct—unidirectional or alternating—perfusion of a cell suspension through the pores of a 3D scaffold [4, 8–16]. Such an efficient method, relying on active driving forces rather than on gravity for the fluid to penetrate the scaffold pores, was revealed to be particularly suitable when seeding cells into thick scaffolds of low porosity [4]. Interestingly, the principle of perfusion has been recently used in the field of heart valve tissue engineering also for in vitro transformation of porcine valves into human valves, enabling decellularization of valve grafts of xenogenic origin and subsequent re-cellularization with human cells [17].

When defining and optimizing seeding protocols (i.e., the selection of parameters such as cell concentration in the seeding suspension, medium flow rate, flow directions, and timing of the perfusion pattern), most of the studies found in the literature rely upon experimental, application-specific, trial and error investigations, rather than turning to the support of theoretical models. Definition of the cell seeding parameters based on computational models (as extensively described in the following Section) could allow for a more rational design of experiments, ultimately leading to more efficient optimization strategies. However, the inherent complexity of dynamic seeding systems represents a major challenge for modeling, because of high dependence on the specific cell type and scaffold implemented (i.e., complex pore architecture and related fluid-dynamics, kinetics of cell adhesion, molecular mechanics, biomaterial properties, etc.). A notable effort in this direction was described by Li and co-authors, who developed and validated a mathematical model allowing predictive evaluation of the maximum seeding density achievable within matrices of different porosities, in a system enabling filtration seeding at controlled flow rates [18].

## 2.2 *Maintenance of a Controlled Culture Environment*

The high degree of structure heterogeneity usually noticed in 3D-engineered constructs cultured in static conditions (i.e., presence of a necrotic central region, surrounded by a dense layer of viable cells) suggests that diffusional transport does not properly assure uniform and efficient mass transfer within the constructs [19]. On the contrary, convective media flow around the construct and, even to a greater extent, direct medium perfusion through its pores, can aid in overcoming diffusional transport limitations (specifically via oxygen and metabolite supply and waste product removal). Bioreactors that perfuse culture medium directly through the pores of a scaffold have therefore been employed in the engineering of various tissues, demonstrating that perfusion enhances calcified matrix deposition by marrow-derived osteoblasts [9, 10, 20, 21], viability, proliferative capacities and expression of cardiac-specific markers of cardiomyocytes [22, 23], cell proliferation in engineered blood vessels [11] and extra-cellular matrix deposition, accumulation and uniform distribution by chondrocytes [14, 24].

In this context, a deep understanding of the basic mechanisms underlying perfusion-associated cell proliferation/differentiation and matrix production will be challenging to achieve, since the relative effects of perfusion-induced mechanical stresses acting on cells and enhanced mass transfer of chemical species, or a combination of the factors, cannot be easily discerned. As a result, similar to perfusion seeding parameters, optimization of perfusion culture conditions is commonly obtained by means of an experimental, trial-and-error approach. In future applications, both the design of new perfusion bioreactors and the optimization of their operating conditions will derive significant benefits from computational fluid dynamics (CFD) modeling aimed at estimating fluid velocity and shear profiles [25–27], as well as biochemical species concentrations within the pores of 3D scaffolds. A more comprehensive strategy that could help to elucidate and decouple the effects of mechanical stimuli and specific species should (1) combine theoretical and experimental approaches, i.e., validate simplified models with experimental data [28] and (2) make use of sensing and control technologies to monitor key parameters indicative of the culture progression. The key contribution of both these strategies in the optimization of bioreactor-based tissue-engineering procedures will be discussed in Sect. 3.

Here it is worth mentioning that bioreactors may represent crucial tools in the maintenance of *globally* well-balanced environmental conditions, together with the above cited *local* homeostasis (i.e., at the level of the engineered construct). Early bioreactors, developed for research purposes in the 1980s and 1990s, were in fact generally meant to be positioned inside cell culture incubators while in use. In such configurations, monitoring and control of key environmental parameters for homeostatic maintenance of cell cultures (such as temperature, atmosphere composition, and relative humidity) were supplied by the incubators themselves. More recently, a spreading demand for automated, user-friendly, and operationally simple bioreactor systems for cell and tissue culture catalyzed research towards the development of stand-alone devices integrating the key function of traditional cell culture incubators, namely environmental control.

Another noteworthy factor heavily hindering homeostatic control in cell-culture systems is the abrupt change in the concentration of metabolites/catabolites, signal molecules, as well as pH, when culture medium is exchanged in periodic batches. In traditional static culture procedures, the smoothening of these step-shaped variations can be achieved by performing partial medium changes, however requiring additional repeated manpower involvement. Bioreactor technology offers a better solution by enabling either semi-continuous automatic replenishment of exhausted media at defined time-points or feedback-controlled addition of fresh media, aimed at re-establishing a homeostatic parameter to a pre-defined set point (e.g., pH) [12, 29, 30].

## 2.3 *Physical Conditioning of Developing Tissues*

A number of in vivo and ex vivo studies over centuries contributed to demonstrate that physical forces (i.e., hydrodynamic/hydrostatic, mechanical, and electrical) play a key role in the development of tissues and organs during embryogenesis, as well as their remodeling and growth in postnatal life. On the basis of these findings, and in an attempt to induce the development of biological constructs that resemble the structure and function of native tissues, tissue engineers have aimed to recreate in vitro a physical environment similar to the one experienced by tissues in vivo. For this purpose, numerous bioreactors have been developed, enabling controlled and reproducible dynamic conditioning of three-dimensional constructs for the generation of functional tissues.

Bioreactors applying *fluid-driven mechanical stimulation*, for example, were employed to establish shear stress acting directly on cells (e.g., in the case of cartilage [28], bone [20], cardiac tissue [31]), create a differential pressure (e.g., for blood vessels [32] and heart valves [33]) or combine these two mechanisms (again with vessels [34–37] and heart valves [38]). Furthermore, coherently with what was expected on the basis of in vivo findings, the development of tissues natively experiencing relevant mechanical cues was enhanced by means of bioreactors enabling *mechanical conditioning*, namely direct tension (e.g., tendons, ligaments, skeletal muscle tissue [39–41], cardiac tissue [42]), compression (e.g., cartilage [43, 44]) and bending (e.g., bone [45]). Similarly, interesting findings on the effect of *electrical stimulation* on the development of excitable tissues were derived by conditioning skeletal muscle [41, 46] and cardiac constructs [47]. Moreover, a promotion of neural gene expression by activation of calcium channels was observed as a result of the application of physiological electrical patterns to primary sensory neurons [48].

Consistent with the tight correlation existing in nature between the *structure* and *function* of biological tissues (the spatial arrangement of load-bearing structures in long bones and the presence of tightly parallel arrays of fibers in skeletal muscles being just two examples of this principle), appropriate tissue structural arrangements have been induced in vitro via the dynamic conditioning of engineered tissues. Physical conditioning was shown to be an effective means to improve cell/tissue structural organization, mainly entering the mechanism of mutual influence that cells and extracellular proteins reciprocally exert via integrin binding [16, 49–51].

As previously discussed with respect to flow-associated effects in perfusion bioreactors, it is imperative to underline that current scientific knowledge is far from allowing a deep understanding of the mechano-responsive dynamics lying behind cell function. As a consequence, the idea of precisely directing tissue development in vitro by means of specific physical cues still remains an immense challenge. Necessarily, a rational design of dynamic culture protocols, intended to actively modulate the growth of engineered tissues, will be conditional on gaining a more comprehensive insight into fundamental cell functions and tissue development mechanisms; bioreactors, representing 3D culture model systems recapitulating specific aspects of the actual in vivo milieu, can enable a step forward in this direction, as will be discussed in the next Section.

## 3 Bioreactors as 3D In vitro Model Systems

Over the last two decades, both in basic biology and tissue engineering studies, we have been witnessing the constant inadequacy of conventional 2D culture systems (i.e., Petri dishes, culture flasks) in resembling the in vivo developmental microenvironment.

At the same time, as it has become evident that cell differentiation and tissue development in vivo are strongly dependent on cell spatial arrangement and directional cues, it has been recognized that 3D culture model systems will be vital to gain a greater understanding of basic cell and tissue function within its native microenvironment.

Urged by a rising curiosity in the mechanisms regulating cell metabolism, proliferation and differentiation, as well as cell/cell and cell/material interactions and mechano-transduction dynamics, 3D model systems have been developed to recapitulate specific aspects of the actual in vivo milieu of defined tissues (e.g. cartilage, bone, skeletal muscle, bone marrow). Such culture systems, besides comprising the appropriate differentiated/undifferentiated cellular source and the proper 3D microenvironment (relying on scaffolding materials or on alternative solutions such as micro-mass culture or cell sheet technology), generally encompass the application of suitable *biochemical* and *physical* cues, resembling the ones sensed by the corresponding tissue in vivo, by making use of bioreactors. To better illustrate the potentialities of bioreactors as in vitro tools for quantitative biological research, we provide examples of bioreactor-based 3D culture model systems, allowing investigation of chondrocyte mechano-responsiveness and oxygen transport in cardiac muscle tissue.

- Functionality of engineered human cartilage

   One of the most compelling questions in the engineering of grafts for clinical use is related to the minimal stage of development required for safe and successful implantation. To answer the question "how good is good enough?," bioreactor-based in vitro model systems could be implemented to test and predict the behavior of engineered grafts upon implantation and exposure to the associated physiological forces. With the ultimate goal to define the effective functionality of engineered human cartilage, Demarteau et al. exposed engineered constructs at different stages of development to a loading regime resembling a mild post-operative rehabilitation using a bioreactor applying dynamic compression [44]. Results clearly indicated that the response of engineered tissues to dynamic compression was correlated with the amount of glycosaminoglycans in the constructs prior to loading. Despite the limitation to a specific scaffold type and loading regime, the study suggests a potential role for bioreactors in defining the specific criteria for a graft prior to implantation, or as a potential functional quality control for engineered tissues. Conversely, the same experimental setup could be exploited to identify potential regimes of physical rehabilitation which are most appropriate for a specific graft.

- Mechano-responsiveness of nasal chondrocytes

   The same concept and paradigm described above can be used to address fundamental questions relevant to the selection of appropriate cell sources. For example, the use of nasal chondrocytes for the repair of articular cartilage defects has long been proposed, mostly due to the higher and more reproducible chondrogenic capacity as compared to articular chondrocytes [52]. However, the possible use of nasal chondrocytes in a joint critically depends on their capacity to respond to physical forces similarly to articular chondrocytes. Before more complex and costly animal models are introduced, bioreactors can provide a technical solution for addressing the raised question. In fact, Candrian et al. used different types of bioreactor systems to demonstrate that nasal chondrocytes not only can increase the synthesis and accumulation of extracellular matrix molecules in response to dynamic compression, but also can upregulate the expression of lubricating molecules, typically expressed by cells in the surface zone of articular cartilage, in response to surface motion (Fig. 2) [53].

- Oxygen transport in cardiac muscle tissue

In cardiac muscle tissue, a high cell density is supported by the flow of oxygen-rich blood through a dense capillary network, as oxygen diffuses from the blood into the tissue surrounding each capillary. While oxygen has a rather low solubility in blood plasma alone, hemoglobin, a natural oxygen carrier, increases the total amount of oxygen in the blood by 65-fold. In order to better understand the influence of specific factors on the development of thick, synchronously contracting cardiac constructs, a bioreactor-based in vitro model system was developed to recapitulate aspects of the native cardiac tissue environment [54].

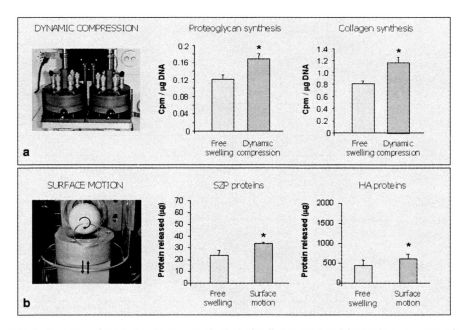

**Fig. 2** Response of nasal chondrocytes to physical stimuli that simulate joint loading. Amounts of newly synthesized proteoglycans and collagen, measured by the incorporation of [$^{35}$S]SO$_4$ and [$^3$H] proline, respectively, were significantly higher in constructs subjected to a single application of *dynamic compression* as compared to those maintained under free swelling. Proteins involved in joint-lubrication (superficial zone protein "SZP" and hyaluronan "HA") were released into the culture medium in significantly higher amounts when constructs were subjected to intermittent applications of *surface motion* as compared to those maintained under free swelling conditions [53]

To mimic the capillary network, cardiomyocytes and fibroblasts were cultured on scaffolds fabricated with a parallel array of channels that were perfused with culture medium. To mimic the oxygen supply in blood, including the role of hemoglobin, culture medium was supplemented with a perfluorocarbon (PFC) emulsion as an oxygen carrier. Engineered constructs cultivated in the presence of the PFC oxygen carrier had higher amounts of DNA, troponin I and Cx-43, and significantly better contractile properties as compared to control constructs cultured in medium without the PFC emulsion. In conjunction with the experimental tests, a mathematical model was developed to simulate the oxygen profiles and cell distributions within the constructs based on diffusive-convective oxygen transport and its utilization by the cells. The model was first used to assess the effects of perfusion rate, oxygen carrier and scaffold geometry on viable cell density, and subsequently to define the scaffold geometry and flow conditions necessary to cultivate cardiac constructs with clinically relevant thicknesses.

## 3.1 Open Challenges

While bioreactor-based 3D model systems have clearly begun to play a crucial role in answering fundamental scientific questions about cell and tissue function in a more physiological 3D environment, their potential value is far from being fully exploited. However, as highlighted in Sect. 2 of this Chapter, the development of bioreactors and the definition of their operating parameters are far too often based on inefficient trial and error investigations rather than on a rational design strategy. Moreover, once implemented, bioreactors have often been treated as black boxes, which rely on qualitative, endpoint, destructive, offline observations of the culture outcomes.

The necessity of *predictive tools*, aiding a sound and rational experimental design, together with the integration of *technological platform*s to noninvasively and nondestructively monitor the culture progression in real-time, remain critical challenges to be faced in order to allow an efficient use of bioreactor-based 3D model systems in both scientific research and clinically compliant tissue manufacturing. In the following Sections, we will provide an overview, including current limitations and potential future directions, of computational modeling and sensing technologies that will be central to the establishment of a controlled and well-defined bioreactor system.

## 3.2 Computational Modeling in Bioreactor Systems

Perhaps due to a lack of true integration between the engineering and biological fields, computational modeling still remains underutilized in tissue engineering and basic biological research. However, in recent years, it has become more apparent that computational methods can not only be a powerful yet cost-effective tool for the design and optimization of the bioreactor systems, but moreover, can be applied to gain a better understanding of fundamental aspects of cell responses in complex 3D environments. In this Section, we begin by describing computational models as they relate to rather classical engineering applications: to aid in the basic design and the optimization of bioreactors for tissue engineering applications. We then review and discuss innovative micro-scale models, aimed at revealing the influence of the scaffold microstructure on transport phenomena. Finally, we give a critical overview of a well-needed emerging area of computational methods: macro-scale and micro-scale tissue growth models for understanding and predicting the effects of physicochemical culture parameters on cell responses and tissue development.

### 3.2.1 Macro-Scale Computational Models for Bioreactor Design

As with bioreactor systems developed previously in other fields of biotechnology, the fundamental design of a bioreactor for tissue engineering applications should be founded on a rational design process that is based in part on computational modeling. The use of computational methods to predict and understand the flow-dependent processes in the bioreactor can not only improve the overall performance of the system, but will likely reduce both the time and cost of development. Nevertheless, to date, only a handful of research groups have implemented modeling approaches for TE bioreactor design and optimization.

Macro-scale computational models have been developed in recent years, which simulated the fluid-dynamics and mass transport environment at the level of the bioreactor system. CFD simulations, validated through imaging techniques such as particle image velocimetry (PIV), showed that flow in spinner flasks and wavy-walled bioreactors was unsteady, periodic and fully turbulent, resulting in heterogeneous fluid-induced shear stress distributions over the outer surface of the scaffolds [55, 56]. Computational modeling of these systems helped to tune construct location and agitation rate in order to provide a more homogeneous shear stress distribution over the scaffolds [55]. Similar approaches, based on macro-scale models, have been carried out to evaluate the effect of the scaffold shape (spheroid versus cylinder) on the wall shear stress distributions within rotating-wall vessel bioreactors [57] and to simulate oxygen transport and velocity/shear profiles over the surfaces of constructs placed at various locations within a concentric cylinder bioreactor [58]. Computational modeling has also been used to optimize the geometry of a direct perfusion bioreactor to achieve a self de-bubbling device in order to overcome the common problem of bubble accumulation [59].

### 3.2.2 Micro-Scale Models for Scaffold Design and Bioreactor Optimization

In an attempt to better characterize the local hydrodynamic environment seen by the cells (i.e., within the scaffold *pores*), micro-scale CFD models have been developed based on idealized scaffold structures with well-defined simple pore architectures. Raimondi et al. first developed a simplified two-dimensional (2D) micro-scale model of a mesh scaffold subject to direct perfusion [60]. Simulations showed that the scaffold's random fiber architecture generated highly variable shear stresses, indicating that a scaffold with a homogeneous distribution of pores would allow for more precise control over the shear. A 3D micro-scale model of a polymeric foam scaffold was later developed, which idealized the pore micro-geometry as a honeycomb-like pattern [61], and served as a powerful tool for defining scaffold design criteria [62]. Simulations showed that the foam's pore sizes strongly influenced the predicted average shear stress, whereas the porosity strongly affected the statistical distribution of the shear stresses but not the average value.

While these simplified models provided for the first time estimates of the shear stress acting on the surface of the internal walls of a 3D porous scaffold, the idealized scaffold structures are not realistic and may not fully capture the influence of the complex and tortuous microstructure of typical porous scaffolds. Recently, micro-scale CFD models have been developed, based on micro-computed tomography ($\mu$CT) reconstructions of the 3D scaffolds, to predict local velocity and shear profiles throughout the actual pore microarchitecture of perfused scaffolds [25, 27, 28]. Interestingly, the globally averaged shear stresses that were predicted within a foam scaffold based on a $\mu$CT model [25] and a simplified model [62] were quite similar; however, the $\mu$CT-based model could reveal a more variable shear distribution within individual pores and among different pores throughout the foam (Fig. 3). From these studies we can speculate that for foams with highly interconnected pores perfused at low Reynolds numbers (low flow rates), the actual micro-geometry does not significantly affect the average shear stress acting at scaffold walls.

However, $\mu$CT-based models should not be replaced by simplified models when an accurate map of the flow fields and shear stresses may be necessary. For example, the $\mu$CT-based models were later extended not only to quantify the hydrodynamic shear stress throughout a perfused porous scaffold, but also, to predict the oxygen profiles within a cell-seeded construct during the initial stage of bioreactor culture [63]. Since low levels of oxygen were to be supplied to the inlet of the construct during culture (to replicate oxygen levels in native articular cartilage), an accurate map of *local* oxygen concentrations within the pores was required to predict whether cells in specific regions of the scaffold would suffer from anoxic oxygen levels.

A significant limitation of the previously described micro-scale computational approaches, which were based on a defined cell population and/or fixed scaffold architecture, is that the models are generally relevant only at the initial stage of the culture. At later time points, CFD models would need to account for modified velocity profiles due to changes in the scaffold's effective pore microstructure as cells proliferate/migrate and extracellular matrix is deposited. Moreover, models of mass transport would not only have to consider these altered flow profiles, but may also need to consider changes in nutrient consumption and waste production due to cell proliferation and possible changes in cell metabolism due to cell differentiation. To be relevant throughout the culture time, computational models will need to correlate specific cell responses and tissue development with physicochemical parameters, as discussed in the following Section.

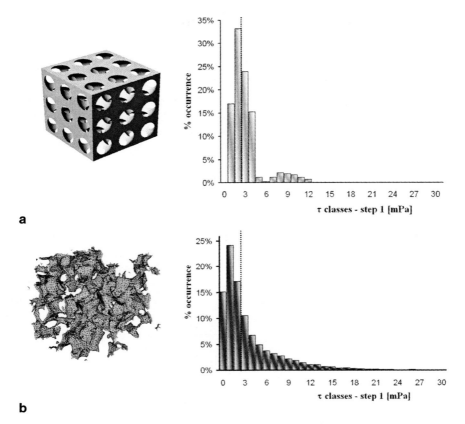

Fig. 3 Distribution of wall shear stresses, τ, within the pores of 3D scaffolds (average pore diameter of 100 μm, 77% porous) subjected to direct perfusion at rates of 0.5 cm$^3$ min$^{-1}$. (**a**) Simulations based on a simplified pore architecture (*left panel*) showed a relatively uniform distribution of shear stress with an average of 2.7 mPa (indicated by *vertical dashed line*). (**b**) Simulations based on a μCT reconstruction of the actual porous microarchitecture (*left panel*) showed an *average* shear stress (3.0 mPa) similar to that based on the simplified geometry, however, the μCT-based model revealed a more heterogeneous distribution of shear within the pores [25]

### 3.2.3 Computational Models of Cell/Tissue Development

Historically, the first tissue-growth models simulated macroscopically the culture system, mostly aiming at a correlation between experimental findings in terms of construct development and the predicted distribution of specific nutrients. In a series of pioneering studies, Galban and Locke [64–66] modeled cell growth kinetics within engineered 3D constructs, ultimately simulating the effects of spatial variations of cells and local nutrient and product concentrations on cell growth. The models highlighted the influence of external mass transfer limitations and internal diffusional limitations on the heterogeneous cell growth generally observed experimentally at the periphery of statically cultured constructs.

Obradovic et al. soon after developed a partial differential equations model to rationalize the spatial glycosaminoglycan (GAG) distributions within engineered cartilage as a function of the local oxygen concentrations, which were governed by diffusional transport and cellular consumption [67]. Despite the gross approximations of the model, predicted GAG distributions were in close agreement to experimentally derived data [68], supporting the authors' hypothesis of a dependence of GAG synthesis on oxygen in the studied bioreactor system. Similarly, Lewis et al. [69] developed a simple mathematical model to describe the interaction between the spatial and temporal development of oxygen and cell density profiles within engineered cartilage constructs. Despite neglecting the influence of the developing extracellular matrix, model predictions were in close agreement with experimental data [70, 71], and led to the conclusion that engineered constructs relying upon diffusional transport of oxygen would inevitably develop heterogeneously, with a proliferation-dominated region at the periphery of the scaffold.

While the aforementioned models helped to establish the existence of relationships between mass transport phenomena and tissue development, and fitted specific sets of existing experimental data, they have relatively limited predictive capabilities. Future models will certainly benefit not only by taking into consideration multiple nutrients and product species, but moreover, by considering additional phenomena which are likely to have a significant impact on tissue development, including cell migration and extra cellular matrix synthesis within the scaffold pores.

To predict cell motility, cell–cell collisions, and cell proliferation, Lee et al. developed a preliminary 2D cell automata model and simulated cell movement based on random walks inhibited by cell–cell collision [72]. To simulate cell migration within 3D constructs, the models of Galban and Locke [66] were further developed by Chung et al. by introducing a cell diffusion term to describe the effects of cell random walks [73]. Although the simulations tended to fit previous experimental data [74], cell motility based on the concept of diffusion may not be described as comprehensively as by cellular automata models, such as the one recently applied to 3D-engineered constructs by Cheng et al., taking into account random walks, cell division, and contact inhibition.

Galbusera et al. took cellular automata models one step further, going beyond the conventional static culture systems with diffusive mass transport simulated by Chung and Cheng, and presented a computational model which accounted both for cell population dynamics and for the hydrodynamic microenvironment imposed by a perfusion bioreactor (i.e., velocity field and oxygen transport) [26]. The model simulated cell proliferation, migration and oxygen consumption within a perfused porous scaffold of simplified geometry previously described by Boschetti et al. [62]. Although the results of this work were not correlated to experimental data, the models showed the correspondence between cell location and local oxygen depletion and the contribution of convective transport in reducing the decreased oxygen concentrations. A key component which still remains absent in these models is the influence of the developing extracellular matrix, influencing the cellular microenvironment not only physically (e.g., altered velocity profiles and mass transport) but biologically as well (e.g., cell-matrix mediated responses to biochemical and physical factors).

Development of models which accurately simulate the development of an engineered tissue, based on the regulation of cell proliferation and differentiation by both biochemical factors and mechanical stimuli, remains a monumental challenge still to be addressed. However, mechano-regulation models of tissue differentiation have previously been developed to model and predict patterns of fracture healing [75, 76], and for the repair of osteochondral defects [77]. In particular, a mechano-regulation model for tissue differentiation was recently used to determine the optimal mechanical properties of a construct to properly recruit, proliferate and differentiate mesenchymal progenitor cells from the bone marrow within osteochondral defects, in order to induce orderly formation of hyaline and bone tissues rather than of fibrous tissue [78]. Prendergast recently continued to further develop this later model by combining a random walk algorithm to account for cell proliferation and migration, with the mechano-regulation model for tissue differentiation [79] (Fig. 4). While the developed models include several and sometimes arbitrary assumptions, these studies are a key milestone in the field, since they represent serious engineering efforts to develop quantitative principles of design

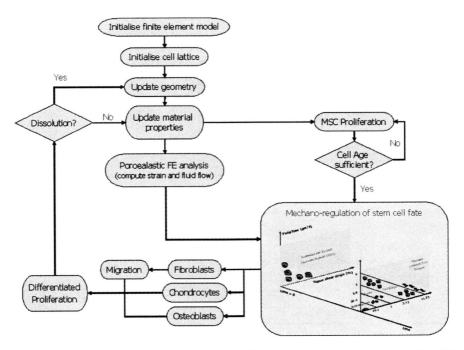

**Fig. 4** Computational model of temporal tissue differentiation and bone regeneration in a 3D printed scaffold during fracture healing. The model accounted for cell proliferation and migration based on a three-dimensional random-walk approach and for tissue differentiation based on a mechano-regulation algorithm both in terms of the prevailing biophysical stimulus and number of precursor cells. Simulations were a function of the scaffold porosity, Young's modulus, and dissolution rate, under both low and high loading conditions. Figure adapted from [79]

for engineered constructs, which should direct the setting up of hypothesis-driven experimental studies and controlled in vitro model systems.

## 3.3 Sensing in Tissue Engineering Bioreactors

Sensing in tissue culture bioreactors represents a major premise to clarify still unknown aspects of the cellular response in dynamic culture conditions, as well as a step forward towards the automation and in-process control of tissue manufacturing processes. Monitoring the partial pressure of $O_2$ and $CO_2$ in the culture medium, or detecting the concentrations of glucose and lactate, for instance, allows quantitative evaluation of the metabolic behavior of cultured cells, thus supporting/substituting subjective, qualitative conclusions traditionally derived by simply observing the color of the medium.

Borrowing from the nomenclature of Mason et al. [80] and Starly and Choubey [81], the noteworthy parameters which should be adequately monitored and controlled during in vitro organogenesis can be classified in two main categories, namely the *milieu parameters* and the *construct parameters*. Milieu parameters are then physical (e.g., temperature, pressure, flow rate), chemical (e.g., pH, dissolved $O_2$ and $CO_2$, chemical contaminants, concentration of significant metabolites/catabolites such as glucose, lactate or secreted proteins) and biological (e.g., sterility). Similarly, the construct parameters can be different in nature: physical (e.g., stiffness, strength, and permeability), chemical (e.g., composition of the scaffold and of the developing extracellular matrix) and biological (e.g., cell number and proliferation rate, concentration of intracellular proteins, cell viability).

### 3.3.1 Monitoring of the Milieu

Several technological solutions were developed in recent years to monitor the *milieu parameters*. Beside the general requirements that sensors need to meet in the common practice (i.e., accuracy, sensitivity, specificity), probes employed in cell and tissue culture are required to fulfill peculiar specifications. In particular, they might need to be rather small in size compared to many commercially available sensors, have a lifetime of several weeks, unless they are sufficiently low cost to be disposed of and replaced during culture, and must ensure stable response over time, since repeated calibration might be difficult to carry out due to accessibility of the sensors during culture. Regardless of technical details related to the specific parameter under investigation (reviewed in [82]), sensors for milieu monitoring can be generally classified in different categories, according to the position of the sensing probe relative to the culture chamber of the reactor (Fig. 5) [80].

1. *Invasive* (embedded) are those sensors whose probes are placed directly inside the culture chamber of the bioreactor, either immersed in the culture fluid or in direct contact with the engineered construct. Clearly, invasive sensors must be

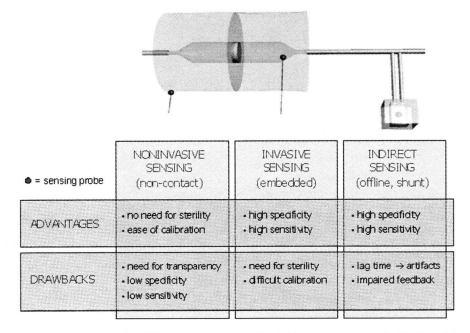

**Fig. 5** The three main modalities to monitor the milieu in bioreactors, as described in Sect. 3.3. (1) *"Invasive sensing"* implies that the sensor probe is placed directly inside the culture chamber of the bioreactor, either immersed in the culture fluid or in direct contact with the engineered construct. (2) *"Noninvasive sensing"* involves sensors that do not come in contact with the interior of the culture chamber, but are capable of measuring via interrogation through the bioreactor wall. (3) *"Indirect sensing"* is performed directly on the culture media, but via sampling means, either by offline analysis or shunt sensing. Figure adapted from [82] by Rosaria Santoro

sterile and therefore either disposable or capable of withstanding repeated sterilization protocols. Integrating these sensors within a bioreactor for clinical applications may have limitations considering the potential high cost of single-use disposable sensors. Clear advantages of invasive sensing are the high precision and accuracy achieved with the measurement. The most common invasive sensors currently in use are based on *optical* and *electrochemical* principles: fiber optic fluorescence-based sensors, for example, have been manufactured for the measurement of dissolved oxygen and of pH, while typical electrochemical sensors are membranes functionalized with appropriate enzymes for the detection of glucose or urea.

2. *Noninvasive* (noncontact) sensors are sensors which do not come in contact with the interior of the culture chamber, and are capable of measuring via interrogation through the bioreactor wall, for example by using *ultrasounds* or *optical methods* such as spectrophotometry or fluorimetry. This approach, while avoiding the sterility issues associated with invasive sensors, implies that the bioreactor wall

must be either entirely or locally transparent to the investigating wave. Moreover, due to the presence of an intermediate material between the probe and the object of interest, it is more challenging to achieve high specificity and high sensitivity of the measurement with these sensors. Typically, the flow rate of medium in perfusion bioreactors is detected via noninvasive techniques, mainly based on Doppler velocimetry. DOCT (Doppler Optical Coherence Tomography), for example, is a novel technique allowing noninvasive imaging of the fluid flow at micron-level scales, in highly light scattering media or biological tissues. Derived from clinical applications, DOCT has been adapted to characterize the flow of culture medium through a developing engineered vascular construct within a bioreactor chamber [80].

3. *Indirect* sensing is performed directly on the culture media, but via sampling means. The two main options included in this category are "offline analysis" and "shunt sensing." In the first case, manual or automated online medium sampling is performed (with possible negative implications for the sterility of the closed system) and analyses are conducted with common instruments for bioanalytics (e.g., blood-gas analyzers). In the latter, the measurement is directly carried out within the fluid, driven through a sensorized shunting loop and later either returned to the body of the bioreactor or discarded. Since probes do not need to be placed inside the culture chamber, indirect sensing can be performed by means of advanced, accurate instruments, with clear advantages in terms of specificity and sensitivity with respect to the invasive method. On the other hand, the lag-time introduced by sampling can heavily hinder the significance of the measurement itself (with possible introduction of artifacts) and impair the efficacy of feedback control strategies. With this method, $pO_2$, $pCO_2$, pH, and glucose concentration in the culture medium are typically measured [83]; protein and peptide analysis can be also conducted via spectrophotometric and fluorimetric assays within shunting chambers.

### 3.3.2 Monitoring of the Construct

While monitoring of the milieu is gradually entering the practice of bioreactor-based tissue engineering, monitoring the function and structure of developing engineered constructs still remains a relatively uncharted area and a highly challenging field of research [84]. In this application, it would be limiting to use the term "sensor" in the traditional sense, since the techniques currently under study are based on highly sophisticated cutting edge technology, often inherited from rather unrelated fields (e.g., clinics, telecommunications). Systems for the nondestructive online monitoring of the construct developmental state would allow continuous and immediate optimization of the culture protocol to the actual needs of the construct itself, thus overcoming the drawbacks traditionally related to the use of endpoint detection methods or fixed time point analyses. Typically, research is being driven by the need for real-time characterization of (1) *functional* and (2) *morphological* properties of engineered constructs, both at the micro- and at the macro-scale. The following paragraphs give a brief overview of the most recent techniques applied in bioreactors for these cited purposes.

1. *Monitoring of the functional properties* comprises characterization of the *construct*'s overall physical properties (e.g., strength, elastic modulus, permeability), but also monitoring of *cell* function within the engineered construct itself, e.g., in terms of proliferation, viability, metabolism, phenotype, biosynthetic activity, and adhesive forces. In this context, Stephens and collaborators [85] recently proposed a method to image real-time cell/material interactions in a perfusion bioreactor based on the use of an upright microscope. The kinetics of cell aggregation and organoid assembly in rotating-wall vessel bioreactors, instead, could be performed according to the method developed by Botta et al. [86], relying on a diode pumped solid state laser and on a CCD video camera. Boubriak and co-authors [87] recently proposed the use of micro-dialysis for detecting local changes in cellular metabolism (i.e., glucose and lactate concentrations) within a tissue-engineered construct. By means of this method, concentration gradients could be monitored within the construct, with the highest lactate concentrations in the construct center, thus allowing early detection of inappropriate local metabolic changes.

2. *Monitoring of the morphological properties* of engineered constructs essentially encompasses assessing the amount, composition, and distribution of the extracellular matrix which is being deposited throughout the scaffold during bioreactor culture. Monitoring morphological properties is particularly pertinent when engineering tissues whose function is strictly dependent on the structural organization of their extracellular matrix, e.g., bone and tendons. In this context, OCT (Optical Coherence Tomography) has been successfully employed as a real-time, nondestructive, noninvasive tool to monitor the production of extracellular matrix within engineered tendinous constructs in a perfusion bioreactor [88]. OCT is analogous to conventional clinical ultrasound scanning, but using near infrared light sources instead of sound, it enables higher resolution images (1–15 μm vs. 100–200 μm); the technique is compact and flexible in nature, as well as relatively low cost since it can be implemented by commercially available optic fibers [89]. However, the most promising technique in the field of real-time imaging is undoubtedly μ-CT (micro-computed tomography). Using this technique, the mineralization within a three-dimensional construct cultured in a perfusion bioreactor was monitored over time, allowing quantification of the number, size, and distribution of mineralized particles within the construct [90].

We have underlined how progress made in the in vitro generation of 3D tissues starting from isolated cells is slowed down by the complexity of the process and of the interplay among different culture parameters. The establishment of well-defined and controlled bioreactor-based 3D culture model systems, supported by modeling efforts and sensing technologies, will be key to gain a deep insight into the mechanisms of tissue development at the research level and, consequently, may provide an advanced technological platform for the achievement of more applicative, high-throughput objectives, e.g. enabling drug screening and toxicology studies, fostering the development of new, rational design criteria for advanced biomaterials/implants, as well as allowing functional quality control of engineered tissues. Besides limiting the recourse to complex, costly and ethically questionable in vivo experiments in animal models, such an approach would found the basis for safe and standardized manufacture of grafts, which will be the subject of the last Section of this Chapter.

## 4 Bioreactors: The Clinical Perspective

During the initial phase of the emerging TE field, we have been mainly consumed by the new biological and engineering challenges posed in establishing and maintaining three-dimensional cell and tissue cultures. After nearly two decades, with exciting and promising research advancements, tissue engineering is now at the stage where it must begin to translate this research-based technology into large-scale and commercially successful products. However, just as other biotech and pharmaceutical industries came to realize in the past, we are ultimately faced with the fact that even the most clinically successful products will need to demonstrate: (1) cost-effectiveness and cost-benefits over existing therapies, (2) absolute safety for patients, manufacturers, and the environment, and (3) compliance to the current regulations.

But what has been hindering cell-based engineered products from reaching the market and what can be done to increase their potential for clinical and commercial success? As described in Sect. 2, the basic procedures for generating engineered tissues have traditionally been based around conventional manual benchtop cell and tissue culture techniques. It is therefore quite natural that these manual techniques, due to their simplicity and wide-spread use, were included in the initial phases of product development, and ultimately in the final manufacturing processes of early cell-based products. Manual techniques still remain particularly appealing for start-up companies since the simple level of technology minimizes initial development time and investment costs, allowing for more rapid entry to clinical trials and into the market. An example of the straightforward benchtop-based manufacturing process is that employed by Genzyme Tissue Repair (Cambridge, MA, USA) for the production of Carticel®, an autologous cell transplantation product for the repair of articular cartilage defects currently used in the clinic. Hyalograft C™, marketed by Fidia Advanced Biopolymers (Abano Terme, Italy), is an alternative autologous cell-based product for the treatment of articular cartilage defects, also manufactured through conventional benchtop techniques. However, these manufacturing processes require a large number of manual and labor-intensive manipulations; as a result, the production costs of the resulting products are rather high, and the process would be difficult to scale as product demands increase. It is therefore becoming more and more evident that tissue engineering firms will inevitably have to follow in the footsteps of other biotechnology fields and begin to introduce process engineering into their manufacturing strategies.

As an alternative to essentially mimicking established manual procedures, bioreactor systems that implement novel concepts and techniques that streamline the conventional engineering processes will likely have the greatest impact on manufacturing. In the following, we comment on the potential of bioreactor-based manufacturing approaches to improve the clinical and commercial success of engineered products by controlling, standardizing, and automating cell and tissue culture procedures in a cost-effective and regulatory-compliant manner. In particular, brief insight will be given in regards to (1) strategies to automate and streamline tissue manufacturing processes (i.e., comprising automation of the three-dimensional culture phase), (2) centralized and de-centralized manufacturing approaches, and (3) the rising interest in the "intraoperative engineering" approaches.

## 4.1 Streamlining Graft Manufacturing Processes

As discussed in Sects. 2 and 3, the structure, function, and reproducibility of engineered constructs can be dramatically enhanced by employing bioreactor-based strategies to establish, maintain, and possibly physically condition cells within the 3D environment. Therefore, efficient manufacturing processes should ideally rely on bioreactor systems to automate and control the *entire* graft generation, from cell isolation to the obtainment of a semi-mature tissue. The advantages of this comprehensive approach would be manifold. A closed, standardized, and operator-independent system would possess great benefits in terms of safety and regulatory compliance, and despite incurring high product development costs initially, these systems would have great potential to improve the cost-effectiveness of a manufacturing process, maximizing the potential for large-scale production in the long-term.

Advanced Tissue Sciences was the first tissue engineering firm to address the issues of automation and scale-up for their production of Dermagraft®, an *allogenic* product manufactured with dermal fibroblasts grown on a scaffold for the treatment of chronic wounds such as diabetic foot ulcers [91] (currently manufactured by Smith and Nephew, London, UK). Skin grafts were generated in a closed manufacturing system within bioreactor bags inside which cells were seeded onto a scaffold, cell-scaffold constructs were cultivated, cryopreservation was performed, and finally that also served as the transport container in which the generated grafts were shipped to the clinic [92]. Eight grafts could be manufactured within compartments of a single bioreactor bag, and up to 12 bags could be cultured together with automated medium perfusion using a manifold, allowing the scaling of a single production run to 96 tissue grafts. Nevertheless, despite this early effort to automate the tissue engineering process, the production system was not highly controlled and resulted in many batches that were defective, ultimately contributing to the overall high production costs [2]. Considering that significant problems were encountered in the manufacturing of this *allogenic* product, tremendous challenges clearly lie ahead in order to automate and scale the production of *autologous* grafts (technically, biologically, and in terms of regulatory issues), particularly since cells from each patient will be highly variable and cells must be processed as completely independent batches.

A particularly appealing approach to automate the production of autologous cell-based products would be based on a modular design, where the bioprocesses for each single cell source are performed in individual, dedicated, closed system sub-units. In this strategy, a manufacturing process can be scaled-up, or perhaps more appropriately considered "scaled-out" [93], simply by adding more units to the production as product demand increases. This strategy is exemplified by the concept of ACTES™ (Autologous Clinical Tissue Engineering System), previously under development by Millenium Biologix. As a compact, modular, fully automated and closed bioreactor system, ACTES would digest a patient's cartilage biopsy, expand the chondrocytes, and provide either (1) an autologous cell suspension, or (2) an osteochondral graft (CartiGraft™) generated by seeding and culturing the cells onto the surface of an osteoconductive porous scaffold. Clearly, full automation of an entire tissue engineering process possesses the greatest risks upfront, requiring considerable investment costs and significant time to develop a highly technical and complex bioreactor system such as ACTES. In fact, Millenium Biologix was forced to file for bankruptcy in late 2006, and the ACTES system never reached the production stage.

Bioreactor designs could be dramatically simplified, and related development costs significantly reduced, if we re-evaluate the conventional tissue engineering paradigms and could streamline the numerous individual processing steps. A bioreactor-based concept was recently described by Braccini et al. for the engineering of osteoinductive bone grafts [8], which would be particularly appealing to implement in a simple and streamlined manufacturing process. In this approach, cells from a bone marrow aspirate, introduced into a perfusion bioreactor, were seeded, expanded and differentiated directly within the pores of a ceramic scaffold, completely bypassing the conventional phase of selection and cell expansion on plastic dishes. The approach resulted in the engineering of a highly osteogenic stromal-like tissue, within a single culture system and minimal operator handling (e.g., no serial passaging in plastic flasks by enzymatic treatment and transfer into a 3D culture system). Simplified tissue engineering processes could be key to future manufacturing strategies by requiring a minimal number of bioprocesses and unit operations, facilitating simplified bioreactor designs with reduced automation requirements, permitting compact designs, with the likely result of reduced product development and operating costs.

For the enthusiastic engineer, developing a fully automated and controlled system would probably necessitate state-of-the-art technologies to monitor and control a full range of culture parameters, and when possible, to monitor cell behavior and tissue development throughout the production process [2]. Significant benefits would derive from implementing sensing and monitoring devices within the manufacturing system in terms of *traceability* and *safety* of the process itself, features that are crucial to compliantly face current GMP guidelines. However, sensors and control systems will add significant costs to the bioreactor system. Keeping in mind low-cost bioreactor systems will be required for a cost-effective manufacturing process, it will be imperative to identify the essential process and construct parameters to monitor and control to standardize production and which can provide meaningful quality control and traceability data. In this context, the monitoring and control of bioreactor systems as discussed in Sect. 3 will be crucial at the research stage of product development in order to identify these key parameters and to establish standardized production methods.

## *4.2 Centralized Versus De-Centralized Production Facilities*

To date, all TE products currently on the market have been and continue to be manufactured within *centralized* production facilities. While manufacturing a product at central locations has the clear advantage of enabling close supervision over the entire production process, this requires establishing and maintaining large and expensive GMP facilities. But unlike in the production of other biotech products such as pharmaceuticals, critical processes and complicated logistical issues (e.g., packaging, shipping, and tracking of living biopsies and engineered grafts), and the considerable associated expenses, must be considered for the centralized production of engineered tissue grafts.

As an alternative to manufacturing engineered products within main centralized production facilities, a *de-centralized* production system, such as a fully automated closed-bioreactor system (e.g., ACTES), could be located on-site within the confines of a hospital. This would eliminate complex logistical issues of transferring biopsies and engineered products between locations, eliminate the need for large and expensive GMP tissue engineering facilities, facilitate scale-up, and minimize labor-intensive operator handling. On the other hand, as previously mentioned in the context of fully automated closed bioreactor systems, a de-centralized manufacturing strategy will clearly involve the greatest upfront risks in terms of development time and costs.

## *4.3 "Intraoperative Engineering" Approaches*

During the first two decades of the tissue engineering field, most research was aimed at the in vitro generation of tissue grafts that resemble the composition and function of native tissues. Trends may be changing. Perhaps due in part to the realization of the current high costs to engineer mature tissue grafts, there is now great emphasis on determining the *minimal* maturation stage of the graft (i.e., only cells seeded onto a scaffold, cells primed for (re-)differentiation within a scaffold, or a functional graft) that will promote defect repair in vivo (capitalizing on the in vivo "bioreactor"), with the ultimate goal of developing intraoperative therapies. In spite of a potential future paradigm shift, bioprocess engineering will continue to serve numerous vital roles in the tissue engineering/regenerative medicine field. Bioreactors could be used to automate the isolation of cells from a biopsy for intra-operative cell therapies (e.g., Biosafe from Sepax, Eysins, Switzerland), or to rapidly seed the isolated cells into a 3D scaffold for immediate implantation. Moreover, bioreactors will continue to be critical for in vitro research applications to identify the requirements for the "in vivo bioreactors" [94], and supporting the shift from tissue engineering approaches to the more challenging field of regenerative medicine.

## 5 Conclusions and Outlook

The ex vivo generation of living tissue grafts has presented new biological and engineering challenges for establishing and maintaining cells in three-dimensional cultures, therefore necessitating the development of new biological models as compared to those long established for traditional cell culture. In this context, bioreactors represent a key tool in the tissue engineering field, from the initial phases of basic research through the final manufacturing of a product for clinical applications.

As we have seen from past and present tissue engineering manufacturing strategies, manual benchtop-based production systems allowed engineered products to reach the clinic, despite their rather high cost and limitations for potential scale-up. Higher-level technology involves longer development time, increased costs, and the risk of technical difficulties, but on the other hand, maximizes the potential for a safe, standardized, scaleable, and cost-effective manufacturing process. Therefore, fundamental knowledge gained through the use of well-defined and controlled bioreactor systems at the research level will be essential to define, optimize, and moreover, streamline the key processes required for efficient manufacturing models.

The translation of bioreactors initially developed for research applications into controlled and cost-effective commercial manufacturing systems would benefit from collaborations between tissue engineering firms, academic institutions, and industrial partners with expertise in commercial bioreactor and automation systems. Academic partners would be key to provide the fundamental aspects of the system, while industrial partners could provide essential elements of automation, as well as making the system user-friendly and compliant with regulatory criteria. Working towards this ambitious goal, a number of multi-disciplinary consortia have already been established within Europe (e.g., REMEDI, AUTOBONE, STEPS) to develop automated and scaleable systems and processes to streamline and control the engineering of autologous cell-based grafts, such that the resulting products meet specific regulations and criteria regarding efficacy, safety, and quality, in addition to being cost-effective. Efforts in this direction will help to make tissue-engineered products more clinically accessible and will help the translational paradigm of TE approaches from research-based technology to a competitive commercial field.

# References

1. Chen HC, Hu YC (2006) Biotechnol Lett 28:1415
2. Martin I, Wendt D, Heberer M (2004) Trends Biotechnol 22:80
3. Portner R, Nagel-Heyer S, Goepfert C, Adamietz P, Meenen NM (2005) J Biosci Bioeng 100:235
4. Wendt D, Marsano A, Jakob M, Heberer M, Martin I (2003) Biotechnol Bioeng 84:205
5. Vunjak-Novakovic G, Martin I, Obradovic B, Treppo S, Grodzinsky AJ, Langer R, Freed LE (1999) J Orthop Res 17:130
6. Bueno EM, Laevsky G, Barabino GA (2007) J Biotechnol 129:516
7. Freed LE, Vunjak-Novakovic G (1997) In Vitro Cell Dev Biol Anim 33:381
8. Braccini A, Wendt D, Jaquiery C, Jakob M, Heberer M, Kenins L, Wodnar-Filipowicz A, Quarto R, Martin I (2005) Stem Cells 23:1066
9. Janssen FW, Hofland I, van OA, Oostra J, Peters H, van Blitterswijk CA (2006) J Biomed Mater Res A 79:338
10. Janssen FW, Oostra J, Oorschot A, van Blitterswijk CA (2006) Biomaterials 27:315
11. Kitagawa T, Yamaoka T, Iwase R, Murakami A (2006) Biotechnol Bioeng 93:947
12. Sun T, Norton D, Haycock JW, Ryan AJ, MacNeil S (2005) Tissue Eng 11:1824
13. Timmins NE, Scherberich A, Fruh JA, Heberer M, Martin I, Jakob M (2007) Tissue Eng 13:2021
14. Wendt D, Stroebel S, Jakob M, John GT, Martin I (2006) Biorheology 43:481
15. Zhao F, Ma T (2005) Biotechnol Bioeng 91:482
16. Chen JP, Lin CT (2006) J Biosci Bioeng 102:41
17. Karim N, Golz K, Bader A (2006) Artif Organs 30:809
18. Li Y, Ma T, Kniss DA, Lasky LC, Yang ST (2001) Biotechnol Prog 17:935
19. Fassnacht D, Portner R (1999) J Biotechnol 72:169
20. Bancroft GN, Sikavitsas VI, van den DJ, Sheffield TL, Ambrose CG, Jansen JA, Mikos AG (2002) Proc Natl Acad Sci U S A 99:12600
21. Sikavitsas VI, Bancroft GN, Lemoine JJ, Liebschner MA, Dauner M, Mikos AG (2005) Ann Biomed Eng 33:63
22. Dvir T, Benishti N, Shachar M, Cohen S (2006) Tissue Eng 12:2843
23. Radisic M, Yang L, Boublik J, Cohen RJ, Langer R, Freed LE, Vunjak-Novakovic G (2004) Am J Physiol Heart Circ Physiol 286:H507
24. Davisson T, Sah RL, Ratcliffe A (2002) Tissue Eng 8:807
25. Cioffi M, Boschetti F, Raimondi MT, Dubini G (2006) Biotechnol Bioeng 93:500
26. Galbusera F, Cioffi M, Raimondi MT, Pietrabissa R (2007) Comput Methods Biomech Biomed Engin 10:279
27. Porter B, Zauel R, Stockman H, Guldberg R, Fyhrie D (2005) J Biomech 38:543
28. Raimondi MT, Moretti M, Cioffi M, Giordano C, Boschetti F, Lagana K, Pietrabissa R (2006) Biorheology 43:215
29. Kino-Oka M, Ogawa N, Umegaki R, Taya M (2005) Tissue Eng 11:535
30. Prenosil JE, Kino-Oka M (1999) Ann N Y Acad Sci 875:386
31. Radisic M, Euloth M, Yang L, Langer R, Freed LE, Vunjak-Novakovic G (2003) Biotechnol Bioeng 82:403
32. Thompson CA, Colon-Hernandez P, Pomerantseva I, MacNeil BD, Nasseri B, Vacanti JP, Oesterle SN (2002) Tissue Eng 8:1083
33. Mol A, Driessen NJ, Rutten MC, Hoerstrup SP, Bouten CV, Baaijens FP (2005) Ann Biomed Eng 33:1778
34. Hahn MS, McHale MK, Wang E, Schmedlen RH, West JL (2007) Ann Biomed Eng 35:190
35. Hoerstrup SP, Zund G, Sodian R, Schnell AM, Grunenfelder J, Turina MI (2001) Eur J Cardiothorac Surg 20:164
36. Niklason LE, Gao J, Abbott WM, Hirschi KK, Houser S, Marini R, Langer R (1999) Science 284:489

37. Sodian R, Lemke T, Fritsche C, Hoerstrup SP, Fu P, Potapov EV, Hausmann H, Hetzer R (2002) Tissue Eng 8:863
38. Flanagan TC, Cornelissen C, Koch S, Tschoeke B, Sachweh JS, Schmitz-Rode T, Jockenhoevel S (2007) Biomaterials 28:3388
39. Altman GH, Lu HH, Horan RL, Calabro T, Ryder D, Kaplan DL, Stark P, Martin I, Richmond JC, Vunjak-Novakovic G (2002) J Biomech Eng 124:742
40. Mantero S, Sadr N, Riboldi SA, Lorenzoni S, Montevecchi FM (2007) JABB 5:107
41. Powell CA, Smiley BL, Mills J, Vandenburgh HH (2002) Am J Physiol Cell Physiol 283:C1557
42. Fink C, Ergun S, Kralisch D, Remmers U, Weil J, Eschenhagen T (2000) FASEB J 14:669
43. Davisson T, Kunig S, Chen A, Sah R, Ratcliffe A (2002) J Orthop Res 20:842
44. Demarteau O, Wendt D, Braccini A, Jakob M, Schafer D, Heberer M, Martin I (2003) Biochem Biophys Res Commun 310:580
45. Mauney JR, Sjostorm S, Blumberg J, Horan R, O'Leary JP, Vunjak-Novakovic G, Volloch V, Kaplan DL (2004) Calcif Tissue Int 74:458
46. Pedrotty DM, Koh J, Davis BH, Taylor DA, Wolf P, Niklason LE (2005) Am J Physiol Heart Circ Physiol 288:H1620
47. Radisic M, Park H, Shing H, Consi T, Schoen FJ, Langer R, Freed LE, Vunjak-Novakovic G (2004) Proc Natl Acad Sci U S A 101:18129
48. Brosenitsch TA, Katz DM (2001) J Neurosci 21:2571
49. Grad S, Lee CR, Gorna K, Gogolewski S, Wimmer MA, Alini M (2005) Tissue Eng 11:249
50. Shangkai C, Naohide T, Koji Y, Yasuji H, Masaaki N, Tomohiro T, Yasushi T (2007) Tissue Eng 13:483
51. Wernike E, Li Z, Alini M, Grad S (2008) Cell Tissue Res 331:473
52. Kafienah W, Jakob M, Demarteau O, Frazer A, Barker MD, Martin I, Hollander AP (2002) Tissue Eng 8:817
53. Candrian C, Vonwil D, Barbero A, Bonacina E, Miot S, Farhadi J, Wirz D, Dickinson S, Hollander A, Jakob M, Li Z, Alini M, Heberer M, Martin I (2007) Arthritis Rheum 58:197
54. Radisic M, Deen W, Langer R, Vunjak-Novakovic G (2005) Am J Physiol Heart Circ Physiol 288:H1278
55. Bilgen B, Barabino GA (2007) Biotechnol Bioeng 98:282
56. Sucosky P, Osorio DF, Brown JB, Neitzel GP (2004) Biotechnol Bioeng 85:34
57. Gutierrez RA, Crumpler ET (2008) Ann Biomed Eng 36:77
58. Williams KA, Saini S, Wick TM (2002) Biotechnol Prog 18:951
59. Moretti M (2005) PhD Dissertation Politecnico di Milano, Italy
60. Raimondi MT, Boschetti F, Falcone L, Fiore GB, Remuzzi A, Marinoni E, Marazzi M, Pietrabissa R (2002) Biomech Model Mechanobiol 1:69
61. Raimondi MT, Boschetti F, Falcone L, Migliavacca F, Remuzzi A, Dubini G (2004) Biorheology 41:401
62. Boschetti F, Raimondi MT, Migliavacca F, Dubini G (2006) J Biomech 39:418
63. Cioffi M, Kueffer J, Stroebel S, Dubini G, Martin I, Wendt D (2006) J Biomech 39:S578
64. Galban CJ, Locke BR (1997) Biotechnol Bioeng 56:422
65. Galban CJ, Locke BR (1999) Biotechnol Bioeng 65:121
66. Galban CJ, Locke BR (1999) Biotechnol Bioeng 64:633
67. Obradovic B, Meldon JH, Freed LE, Vunjak-Novakovic G (2000) AICHE J 46:1860
68. Martin I, Obradovic B, Freed LE, Vunjak-Novakovic G (1999) Ann Biomed Eng 27:656
69. Lewis MC, Macarthur BD, Malda J, Pettet G, Please CP (2005) Biotechnol Bioeng 91:607
70. Malda J, Martens DE, Tramper J, van Blitterswijk CA, Riesle J (2003) Crit Rev Biotechnol 23:175
71. Malda J, Woodfield TB, Van d V, Kooy FK, Martens DE, Tramper J, van Blitterswijk CA, Riesle J (2004) Biomaterials 25:5773
72. Lee Y, Kouvroukoglou S, McIntire LV, Zygourakis K (1995) Biophys J 69:1284
73. Chung CA, Yang CW, Chen CW (2006) Biotechnol Bioeng 94:1138
74. Freed LE, Vunjak-Novakovic G, Marquis JC, Langer R (1994) Biotechnol Bioeng 43:597

75. Carter DR, Blenman PR, Beaupre GS (1988) J Orthop Res 6:736
76. Lacroix D, Prendergast PJ (2002) J Biomech 35:1163
77. Duda GN, Maldonado ZM, Klein P, Heller MO, Burns J, Bail H (2005) J Biomech 38:843
78. Kelly DJ, Prendergast PJ (2006) Tissue Eng 12:2509
79. Byrne DP, Lacroix D, Planell JA, Kelly DJ, Prendergast PJ (2007) Biomaterials 28:5544
80. Mason C, Markusen JF, Town MA, Dunnill P, Wang RK (2004) Biosens Bioelectron 20:414
81. Starly B, Choubey A (2008) Ann Biomed Eng 36:30
82. Rolfe P (2006) Meas Sci Technol 17:578
83. Sud D, Mehta G, Mehta K, Linderman J, Takayama S, Mycek MA (2006) J Biomed Opt 11:050504
84. Pancrazio JJ, Wang F, Kelley CA (2007) Biosens Bioelectron 22:2803
85. Stephens JS, Cooper JA, Phelan FR, Jr., Dunkers JP (2007) Biotechnol Bioeng 97:952
86. Botta GP, Manley P, Miller S, Lelkes PI (2006) Nat Protoc 1:2116
87. Boubriak OA, Urban JP, Cui Z (2006) J R Soc Interface 3:637
88. Bagnaninchi PO, Yang Y, Zghoul N, Maffulli N, Wang RK, Haj AJ (2007) Tissue Eng 13:323
89. Mason C, Markusen JF, Town MA, Dunnill P, Wang RK (2004) Phys Med Biol 49:1097
90. Porter BD, Lin AS, Peister A, Hutmacher D, Guldberg RE (2007) Biomaterials 28:2525
91. Marston WA, Hanft J, Norwood P, Pollak R (2003) Diabetes Care 26:1701
92. Naughton GK (2002) Ann N Y Acad Sci 961:372
93. Mason C, Hoare M (2006) Regen Med 1:615
94. Stevens MM, Marini RP, Schaefer D, Aronson J, Langer R, Shastri VP (2005) Proc Natl Acad Sci USA 102:11450

# Bioreactor Technology in Cardiovascular Tissue Engineering

**H. Mertsching and J. Hansmann**

**Abstract** Cardiovascular tissue engineering is a fast evolving field of biomedical science and technology to manufacture viable blood vessels, heart valves, myocardial substitutes and vascularised complex tissues. In consideration of the specific role of the haemodynamics of human circulation, bioreactors are a fundamental of this field. The development of perfusion bioreactor technology is a consequence of successes in extracorporeal circulation techniques, to provide an in vitro environment mimicking in vivo conditions. The bioreactor system should enable an automatic hydrodynamic regime control. Furthermore, the systematic studies regarding the cellular responses to various mechanical and biochemical cues guarantee the viability, bio-monitoring, testing, storage and transportation of the growing tissue.

The basic principles of a bioreactor used for cardiovascular tissue engineering are summarised in this chapter.

**Keywords** Cardiac muscle tissue, Cardiovascular tissue engineering, Dynamic mechanical stimulation, Heart valve, In vitro vascularised tissue, Vascular grafts.

## Contents

1 Cardiovascular Tissue Engineering .................................................. 30
2 Vascular Grafts ..................................................................... 30
3 Heart Valves ........................................................................ 33
4 In Vitro Vascularised Tissue ....................................................... 34
5 Cardiac Muscle Tissue .............................................................. 36
References ........................................................................... 36

H. Mertsching (✉), J. Hansmann
Fraunhofer Institute Interfacial Engineering and Biotechnology (IGB), Nobelstr. 12, 70569 Stuttgart, Germany

# 1 Cardiovascular Tissue Engineering

Cardiovascular diseases are a major cause of death in the Western world. In the past three decades there have been a number of improvements in artificial devices and surgical techniques for cardiovascular diseases; however, there is still a need for novel therapies. The major disadvantage of current artificial devices is that they cannot grow, remodel, or repair in vivo. Tissue engineering offers the possibility to develop autologous substitutes in vitro with the essential morphological, biological, chemical and mechanical properties [1]. In general tissue-engineering bioreactors should enable the culture of bioartificial tissue under sterile, physiological (37°C and 5% $CO_2$) conditions and should be easy to clean and maintain. In order to engineer functional cardiovascular substitutes bioreactor systems have to be developed that simulate in vitro the dynamic physiological, biomechanical and biochemical conditions of the human in vivo circulation. The effects of the simulated physiological conditions of the bioreactor on the growing tissue have to be examined with respect to the material properties of vascular grafts (Sect. 2), heart valves (Sect. 3) and cardiac muscle tissue (Sect. 5).

The aim of tissue engineering is the creation of complex tissues or organs. This requires the development of vascularised scaffolds (Sect. 4) or the induction of angiogenesis [2]. The overall concept is the in vitro generation of a bioartificial blood vessel network that can be connected to the host's vasculature following implantation in order to maintain graft viability [3]. These approaches include the use of biological extracellular matrices such as collagen, hydrogels, porous biodegradable polymeric scaffolds with macro- and micro-lumens and micro-channels, the co-culture of different cell types, the incorporation of growth factors, and last but not least culture in dynamic bioreactor systems [2, 3].

All bioreactors for cardiovascular tissue engineering create dynamic mechanical stimulations of the biological substitutes during in vitro culture. These devices typically rely on pulsatile flow and have been demonstrated to promote both the development of mechanical strength [4–6], and the modulation of cellular function [7] within the growing bioartificial tissues.

Although these devices are promising in terms of development of functional biological vascular grafts [1, 2, 4, 6], they still present several drawbacks regarding biomechanical properties, the anatomical sample geometry and a restricted sample capacity.

A couple of bioreactors have been designed to evaluate systematically the intrinsic effect of specific mechanical and biochemical stimuli on engineered tissues [8–10]. These devices offer a user-defined mode of mechanical stimulation, provide a sufficient sample capacity for statistically significant comparisons at multiple time points and accommodate basic sample geometry (Fig. 1).

# 2 Vascular Grafts

Bioartificial vessels must mimic the responsive nature of native arteries based on the communication of endothelial cells and smooth muscle cells in the vessel wall. Endothelial cells lining the luminal surface of native blood are exposed to pulsatile

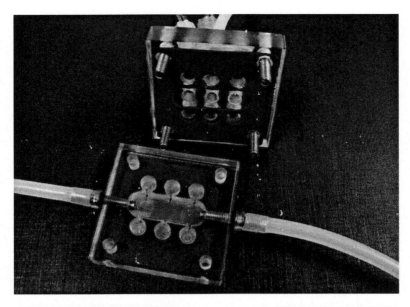

**Fig. 1** Bioreactors to evaluate systematically the angiogenetic effect. The lumen of the chamber can be filled with a tubular scaffold, collagen or hydrogels. The six notches, three on the left and three on the right can be filled up with different growth factors in diverse concentrations. The lumen of the scaffold will be seeded with endothelial cells and cultured under mechanical stimulation. The influence on the angiogenetic behaviour of the endothelial cells can be followed under the microscope or in a histological or molecular follow up. This bioreactor provides a sufficient sample capacity for statistically significant comparisons at multiple time points (developed by Jan Hansmann, Fraunhofer IGB)

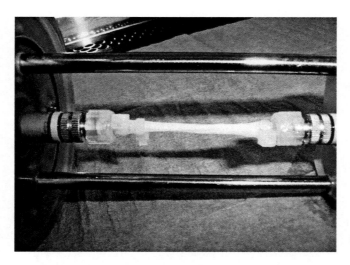

**Fig. 2** Carotid scaffold inserted in a typical bioreactor for vascular tissue engineering. A scaffold for the generation of a carotid was inserted into a typical two-chamber vascular bioreactor. The whole construct can be inserted in a glass tube to guaranty sterility during the culture period. The whole construct will be placed in an apparatus for pulsatile perfusion, to mimic the cardiovascular conditions in vitro. The perfusion system provides intra-luminal pulsatile flow ensuring the necessary shear stress and culture medium for the endothelial cells. The surrounding container can be filled up with smooth muscle cell specific culture medium

physical forces during vasculogenesis and throughout life. One consequence is that all bioreactors for vessel engineering consist of two chambers (Fig. 2) and are developed as pulsatile perfusion bioreactors to mimic the cardiovascular conditions in vitro [4].

The perfusion systems provide intra-luminal pulsatile flow ensuring the necessary shear stress for the endothelial cells. The pumps in all bioreactor systems can produce pressure values of 270 ± 30 mmHg, with variable stroke volumes of 0–10 ml per stroke and the bioreactor can be operated at a range of pulse rates from 60 to 165 beats per min [11]. Mathematical modelling of the fluid flow regime in the perfusion bioreactor showed that integrating a flow-distributing mesh 1.5 cm upstream from the construct compartment imposed an equal medium flow and shear stress of 0.6 dynes cm$^{-2}$ along the entire cell construct cross-sectional area [12].

The scaffold of smooth muscle cells of a functional blood vessel substitute is remodelled in response to mechanical stimulation. It has been demonstrated that cyclic mechanical strain enhances the function and remodelling activity of smooth muscle cells. Additionally, cyclic strains significantly increased the mechanical strength and material modulus, as indicated by an increase in circumferential tensile properties of the constructs [7]. Some bioreactor systems additionally allowed pulsatile perfusion cyclic waveforms of bandwidth between 0 and 20 Hz with a mean displacement error of 0.26% of the full-scale output. The maximum overshoot is 0.700% [4, 11].

Current studies demonstrate that the appropriate choice of a scaffold for vascular tissues in a mechanically dynamic environment depends on the time frame of the mechanical stimulation, and on elastic scaffolds allowing mechanically directed control of cellular function [11, 13]. The engineered vessels display improved mechanical strength, enhanced collagen deposition, contain endothelial cell and smooth muscle cell expression markers of differentiation, and show contractile responses to pharmaceutical agents. Nevertheless, to date the luminal diameter of engineered autologous blood vessels is restricted to 2 mm.

## 3 Heart Valves

In simple terms, a heart valve is a blood vessel with leaflet structures, which control the blood flow by opening and closing the lumen of the vessel. Thus, the characteristics of the bioreactor technology for heart valve tissue engineering are very similar to the above-described vascular bioreactor technology. Additionally, the pulsatile flow of the luminal cell-culture medium is generated by periodic expansion of an elastic membrane that is inflated and deflated by an air pump. Under these dynamic conditions average shear rate, systolic and diastolic pressures and pressure waveforms comparable to the conditions in the human carotid artery could be obtained. At the moment the culture period for growing bioartificial heart valves is on average 21 days. Culture conditions vary amongst the different groups, including both steady and pulsatile flow of between 3 and 20 l min$^{-1}$ and a shear stress from 1 up to 22 dyne cm$^{-2}$ [6].

Dynamic flexure is a major mode of deformation in the native heart valve cusp. Essential for functional heart valve engineering is the knowledge of normal and pathological heart valve function based on the quantification of leaflet deformation during the cardiac cycle. Because of the technical complexities an experimental method to investigate dynamic leaflet motion using a non-contacting structured laser-light projection technique has been developed. "Using a simulated circulatory loop, a matrix of 150–200 laser light points were projected over the entire leaflet surface. To obtain unobstructed views of the leaflet surface, a stereo system of high-resolution boroscopes was used to track the light points at discrete temporal points during the cardiac cycle. The leaflet surface at each temporal point was reconstructed in three dimensions. This method has high spatial and temporal resolution and can reconstruct the entire surface of the cusp simultaneously" [14]. This completely non-contacting method is applicable to studying fatigue and optimising heart valve bioreactor technology.

An alternative non-destructive in vitro method to investigate interaction between scaffolds and seeded cells prior to implantation is the near-infrared multiphoton technology. It allows the visualization of deep tissue cells and extracellular matrix (ECM) compartments with submicron resolution. The reduced fluorescent coenzyme NAD(P)H, flavoproteins, keratin, melanin and elastin are detected by two-photon excited autofluorescence, whereas myosin, tubulin and the ECM protein collagen can be imaged additionally by second harmonic generation (SHG) [15].

Tissue-engineered heart valves represent a promising strategy especially for paediatric applications, because the growth and remodelling potential is necessary for long-term function. The ideal graft design has not yet been clarified. A systematic evaluation of candidate scaffolds, constructs and bioreactors for the incubation of biomaterial samples under conditions of cyclic stretching has to be carried out in the future.

## 4 In Vitro Vascularised Tissue

Tissue engineering of complex tissues and organs is limited by the need for a vascular supply to guarantee graft survival and render bioartificial organ function. To overcome this hurdle numerous strategies and bioreactor technologies have been developed. To induce vessel sprouting and endothelial cell differentiation in vitro and in vivo a pulsatile haemodynamics flow pattern mimicking physiological conditions is essential. We recently introduced a unique biological capillarised matrix (BioCaM) and a PC-controlled bioreactor system for the generation of complex vascularised tissues [16].

The matrix represents an acellular collagenous network generated from porcine tissue by removal of all cellular components, similar to the small intestine scaffold (SIS) that has been characterised as a complete absorbable biocompatible framework enabling cellular migration and differentiation showing early capillary ingrowths and endothelialisation as well as high infection resistance [17].

Engineering of complex tissues using this scaffold needs a specially designed computer-controlled bioreactor ensuring culture conditions as described in Chap. 1.2 (vascular tissue engineering). For our bioreactor system we incorporated systematic mathematical modelling and computer simulation based on computational fluid dynamics into the bioreactor design process, and development of automatic systems of hydrodynamic regime control. So we are able to simulate the natural environment of the body, from blood pressure to temperature and control and regulate the culture conditions during the whole culture period. In the bioreactor (Fig. 3), the tissue has separate connections: an arterial supply provides it with fresh nutrient solution and through a venous connection the exhausted solution and metabolites are removed.

The dynamic in vitro system was designed to provide medium flow by a roller pump (500 ml min$^{-1}$) in the same way as the heart pumps blood through the human circulatory system in pulses; in the bioreactor the pulse is adjusted in a range from 80 to 200 ± 30 mmHg, with a variable stroke volume of 0–10 ml per stroke and a pulse rate of 60 to 180 beats min$^{-1}$. The computer regulates the arterial oxygen and nutrient supply via parameters such as blood pressure, temperature and flow rate [2, 18]. This work was awarded in 2006 the 1st Hugo Geiger Prize for the life sciences of the Fraunhofer Gesellschaft and the Lewa prize of the University of Stuttgart (Fig. 4).

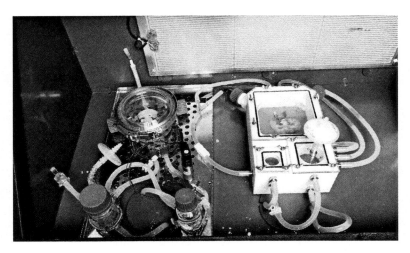

**Fig. 3** Bioreactor for the generation of vascularised tissue with an arterial inflow and a venous reflux system. In the bioreactor a scaffold for the in vitro generation of vascularised liver tissue is inserted and connected to an arterial supply to provide fresh nutrients to the culture medium. The metabolites are then removed with the pulsatile flowing medium through the venous reflux system into the smaller chamber. The small chamber and the larger chamber are connected, providing a closed circulation. Using this bioreactor we are able to simulate the natural environment of the body, from blood pressure to temperature and control and regulate the culture conditions during the whole culture period

**Fig. 4** A computer-based bioreactor system for vascular tissue engineering developed at the Fraunhofer IGB. The figure shows the whole bioreactor system. The bioreactor described in Fig. 3 is inserted into a climate chamber. A computer simulation forms the basis of the fluid dynamics of the bioreactor system to control the hydrodynamic regime and temperature. The medium flow through the growing vascularised tissue is provided by a roller pump in the same way as the heart pumps blood through the human circulatory system. The computer regulates the arterial oxygen and nutrient supply via parameters such as blood pressure, temperature and flow rate

This scaffold and bioreactor system may provide answers to the problem of bioartificial graft vascularisation, which is an inalienable prerequisite for the generation of clinically applicable bioartificial tissues and organs of sufficient strength and size. The first successful implantation of a tracheal patch has been published [19, 20].

## 5 Cardiac Muscle Tissue

Cardiac tissue engineering has emerged as a promising approach to replace or support an infracted cardiac tissue and thus may hold great potential to treat and save the lives of patients with heart disease. By its broad definition, tissue engineering involves the construction of tissue equivalents from donor cells. This is the major limitation in the field of cardiac tissue engineering, human cardiomyocytes could not proliferate in vitro. Several enabling technologies have been established over the past 10 years to create functional myocardium. None of the presently employed technologies yields a perfect match to natural heart muscle. So far only engineered myocardial patches in animals-mostly rats-exist. The suitability of engineered heart tissue (EHT) depends on the degree of syncytoid tissue formation and cardiac myocyte differentiation in vitro, contractile function and electrophysiological properties. State of the art is still the Zimmermann and Eschenhagen [24] developed technology. The cardiac myocytes from neonatal rats were mixed with collagen I scaffold, cast in circular moulds, and subjected to phasic mechanical stretch, which leads to intensively interconnected, longitudinally oriented, cardiac muscle bundles. The myocardial tissue displayed contractile characteristics of native myocardium with a high ratio of twitch (0.4–0.8 mN) to resting tension (0.1–0.3 mN) and a strong adrenergic inotropic response. Action potential recordings demonstrated stable resting membrane potentials of 66–78 mV, fast upstroke kinetics, and a prominent plateau phase [21].

A new development is a triple perfusion bioreactor for vascularised tubular tissue-engineered cardiac constructs. Two unique features integrated into the bioreactor provided a homogenous fluid flow along the bioreactor cross-section and maximal exposure of the cellular constructs to the perfusing medium. Mathematical modelling of the fluid flow regime documented along the entire cell construct an equal medium flow and shear stress of 0.6 dynes cm$^{-2}$ [12].

## References

1. Fuchs JR, Nasseri BA, Vacanti JP (2001) Tissue Engineering: a 21st century solution to surgical reconstruction. Ann Thorac Surg 72(2):577–591
2. Walles T et al. (2007) The potential of bioartificial tissues in oncology research and treatment. Onkologie 30(7):388–394

3. Ko HC, Milthorpe BK, McFarland CD (2007) Engineering thick tissues—the vascularisation problem. Eur Cell Mater 25 (14):1–18
4. Niklason LE et al (1999) Functional arteries grown in vitro. Science 284 (5413):489–493
5. Seliktar D et al. (2000) Dynamic mechanical conditioning of collagen-gel blood vessel constructs induces remodeling in vitro. Ann Biomed Eng 28(4):351–362
6. Schmidt D, Stock UA, Hoerstrup SP (2007) Tissue engineering of heart valves using decellularized xenogeneic or polymeric starter matrices. Philos Trans R Soc Lond B Biol Sci 29362(1484):1505–1512
7. Seliktar D, Nerem RM, Galis ZS (2001) The role of matrix metalloproteinase-2 in the remodeling of cell-seeded vascular constructs subjected to cyclic strain. Ann Biomed Eng 29(11):923–934
8. Minuth WW et al. (2000) Physiological and cell biological aspects of perfusion culture technique employed to generate differentiated tissues for long term biomaterial testing and tissue engineering. J Biomater Sci Polym Ed 11(5):495–522
9. Mitchell SB et al. (2001) A device to apply user-specified strains to biomaterials in culture. IEEE Trans Biomed Eng 48 (2):268–273
10. Jockenhoevel S et al. (2002) Cardiovascular tissue engineering: a new laminar flow chamber for in vitro improvement of mechanical tissue properties. ASAIO J 48 (1):8–11
11. Mironov V et al. (2006) Cardiovascular tissue engineering I. Perfusion bioreactors: a review. J Long Term Eff Med Implants 16(2):111–130
12. Dvir T et al. (2006) A novel perfusion bioreactor providing a homogenous milieu for tissue regeneration. Tissue Eng 12(10):2843–2852
13. Chen HC, Hu YC (2006) Bioreactors for tissue engineering. Biotechnol Lett 28(18):1415–1423
14. Iyengar A et al. (2001) Dynamic in-vitro 3D reconstruction of heart valve leaflets using structured light projection. Ann Biomed Eng 29:963–973
15. Schenke-Layland K et al. (2006) Two-photon microscopes and in vivo multiphoton tomographs—powerful diagnostic tools for tissue engineering and drug delivery. Adv Drug Deliv Rev 58(7):878–96
16. Mertsching H et al. (2005) Engineering of a vascularized scaffold for artificial tissue and organ generation. Biomaterials 26 (33):6610–6617
17. Hodde J (2002) Naturally occurring scaffolds for soft tissue repair and regeneration. Tissue Eng 8:295–308
18. Linke K et al. (2007) Engineered vascularised liver tissue for applied research. Tissue Eng 13:2699–2707
19. Walles T et al. (2005) Tissue remodelling in a bioartifical fibromuscular patch following transplantation in human. Transplantation 80(2):284–285
20. Macchiarini P et al. (2004) First human transplantation of a bioengineered airway tissue. J Thorac Cardiovasc Surg 128(4):638–641
21. Zimmermann W-H et al. (2002) Tissue engineering of a differentiated cardiac muscle construct. Circ Res 90(2):223–230
22. Walles T, Mertsching H (2007) Strategies for scaffold vascularization in tissue engineering. In: Pannone PJ (ed.) Trends in biomaterials research. Nova Science Publishers Inc, New York, ISBN 1–60021–361–8; (in press)

# Bioreactors for Guiding Muscle Tissue Growth and Development

R.G. Dennis, B. Smith, A. Philp, K. Donnelly, and K. Baar

**Abstract** Muscle tissue bioreactors are devices which are employed to guide and monitor the development of engineered muscle tissue. These devices have a modern history that can be traced back more than a century, because the key elements of muscle tissue bioreactors have been studied for a very long time. These include barrier isolation and culture of cells, tissues and organs after isolation from a host organism; the provision of various stimuli intended to promote growth and maintain the muscle, such as electrical and mechanical stimulation; and the provision of a perfusate such as culture media or blood derived substances. An accurate appraisal of our current progress in the development of muscle bioreactors can only be made in the context of the history of this endeavor. Modern efforts tend to focus more upon the use of computer control and the application of mechanical strain as a stimulus, as well as substrate surface modifications to induce cellular organization at the early stages of culture of isolated muscle cells.

**Keywords** Force production, Muscle development, Organ culture, Tissue culture

---

R.G. Dennis, B. Smith
Department of Biomedical Engineering, University of North Carolina at Chapel Hill, Chapel Hill, USA

A. Philp, K. Baar (✉)
Functional Molecular Biology Lab, Division of Molecular Physiology, University of Dundee, James Black Centre, Dow Street, Dundee, DD1 5EH, UK

K. Donnelly
Division of Mechanical Engineering and Mechatronics, University of Dundee, James Black Centre, Dow Street, Dundee, DD1 5EH UK

**Contents**

| | | |
|---|---|---|
| 1 | Introduction | 40 |
| 2 | A Muscle Tissue and Organ Bioreactor System | 41 |
| | 2.1 What is a Muscle | 41 |
| | 2.2 Development of a Muscle Bioreactor System | 42 |
| | 2.3 Design Specification for Muscle Bioreactors | 42 |
| | 2.4 Muscle Development and Dynamic Feedback Control | 43 |
| | 2.5 Historical Perspective | 44 |
| | 2.6 Basic Questions | 44 |
| 3 | Categories of Muscle Bioreactors | 45 |
| | 3.1 Category 1: In Vivo Bioreactors for Whole Muscle Organ Maintenance | 46 |
| | 3.2 Category 2: In Vitro Bioreactors for Whole Muscle Organ Maintenance | 50 |
| | 3.3 (Category 3) In Vivo Guided Muscle Tissue/Organ Development | 55 |
| | 3.4 (Category 4) In Vitro Guided Muscle Tissue/Organ Development | 56 |
| 4 | Modern Approaches | 56 |
| | 4.1 Discovery Versus Delivery | 57 |
| | 4.2 Mechanical Stimulation | 57 |
| | 4.3 Electrical Stimulation | 60 |
| | 4.4 Measurement of Forces Generated by Muscle Constructs | 61 |
| | 4.5 Perfusion | 62 |
| | 4.6 Chemical Stimulation | 63 |
| | 4.7 Multiple-Mode Stimulation | 64 |
| 5 | Failure Modes for Muscle Tissue Bioreactors | 65 |
| | 5.1 Septic Contamination | 65 |
| | 5.2 Mechanical Failure Within the Tissue | 66 |
| | 5.3 Mechanical Failure at the Tissue-to-Tissue Interfaces | 66 |
| | 5.4 Metabolic Failure | 67 |
| | 5.5 Cellular Necrosis and Cell Death | 68 |
| | 5.6 Toxic Contamination | 68 |
| | 5.7 Electrochemical Tissue Damage | 68 |
| | 5.8 Operator Error | 69 |
| 6 | Conclusions and Future Directions | 69 |
| 7 | Key Questions and Technical Requirements | 70 |
| | References | 71 |

# 1 Introduction

In the field of tissue engineering the term "bioreactor" has a distinct meaning from its use in other disciplines such as chemical engineering or pharmacology. Throughout this chapter the term bioreactor will exclusively mean a tissue engineering bioreactor, which is a device or system that has the specific function of providing a controlled environment to either maintain a whole muscle organ ex vivo indefinitely or promote the differentiation of cells and the maturation of both intra and extracellular structures resulting in the generation of an organized tissue. Generically, many tissue engineering bioreactors share the same basic functions: (1) maintaining asepsis; (2) controlling temperature and pH; (3) excluding harmful energy such as intense light or excessive mechanical vibration; (4) improving nutrient delivery; and (5) enabling controlled experimental interventions. More advanced bioreactors tend to be tissue specific, and for skeletal muscle the bioreactors generally include either some form of mechanical stimulation, some form of electrical stimulation, or less frequently both. Because of the architectural similarity of the bioreactor technologies that are generally employed, for the purposes of this chapter the term muscle will refer generally to all three types of muscle (skeletal, cardiac, and smooth) unless otherwise stated.

The development of bioreactors to guide the development of engineered human tissues has been extensively studied and recently reviewed [1]. For muscle tissues this interest is driven by the increasing scope of applications for engineered muscle tissue, including: (1) basic research into skeletal, cardiac, and smooth muscle development, injury, aging, and disease [2–9]; (2) in vitro screening of new drugs and gene therapies [10–20]; (3) developing mechanical actuators using muscle to drive hybrid robotic and prosthetic devices [21, 22]; (4) implanting genetically engineered muscle tissue as a source of therapeutic protein production [11–13, 15–17, 20, 23, 24]; (5) providing replacement tissues for surgical repair of injured or congenitally deformed muscle and (6) developing more humane technologies for the production of animal protein (i.e., meat) for human consumption as food [25, 26]. Some of the above listed application areas are already upon us at the time of writing, such as the use of engineered muscle in basic research and for drug screening, whereas others, those nearer the end of the list, present ever increasing technical challenges.

## 2 A Muscle Tissue and Organ Bioreactor System

### 2.1 What is a Muscle

A whole muscle organ is a complex structure including cells and tissues of many types in addition to myofibers. Other cell types include adipocytes, fibroblasts or tenocytes, vascular smooth muscle, endothelium, and peripheral nerve. A whole muscle organ is also a "smart material" that can rapidly change its biochemical and mechanical properties over a wide range. A smart material is a material with properties that can be significantly altered and controlled by one or more external stimuli. To achieve this, muscle organs use embedded sensors (proprioceptive structures) that allow the central nervous system to exert feedback control over the function of each muscle. The embedded sensors include intrafusal muscle fibers (muscle spindles), Golgi tendon organs, and their associated afferent nerves. Because muscle is a smart material that allows dynamic control of its mechanical properties with time constants on the order of tens of milliseconds, organisms can use muscle to dynamically react to their environment by employing individual muscles and muscle groups variously as brakes, motors, struts, or springs, depending upon the needs of the organism

[27]. Because of the immense complexity of muscle structure and function, all attempts to date to build a muscle tissue bioreactor have only incorporated a small subset of the full range of capabilities that would be needed to engineer and monitor a fully functional whole muscle organ.

## 2.2 Development of a Muscle Bioreactor System

It is generally believed that a sufficiently advanced tissue bioreactor system will allow the development, growth, and maintenance of living muscle in the form of complete muscle organs. These muscle organs would be guided in their development to achieve the final desired phenotype, including organ-level structures such as tissue to–tissue interfaces (muscle-nerve, muscle-tendon, and vascular). The technological design of such a bioreactor system is a complex engineering enterprise. A prerequisite to this goal is the development of a comprehensive design specification for the bioreactor system. This is essentially a systems biology problem which draws upon our understanding of the molecular biology, anatomy, development, physiology, aging, injury, functional adaptation and disease of muscle tissue. Such a design must also take into account the biological feedback and control mechanisms involved at all levels of muscle development, adaptation, and aging. Since our understanding of the biology of muscle is incomplete, the greatest engineering problem in the design of muscle tissue bioreactors is simply that it is unclear precisely what capabilities are required in the design specification. Those specific capabilities that are believed to be required will be discussed in detail in the following sections.

## 2.3 Design Specification for Muscle Bioreactors

The desired outcome of the field of skeletal muscle tissue engineering is the ability to engineer fully functional adult phenotype muscle organs of predefined architecture. This includes the proper tissue geometry, fiber type, and architecture, as well as functional tissue to tissue interfaces. The tissue-to-tissue interfaces include the vascular bed, neuromuscular interface, and the mechanical transition at each end in the form of a muscle-tendon or muscle-tendon-bone interface with transition zones of increasing or decreasing mechanical stiffness to achieve mechanical impedance matching, and to allow mechanical signal transduction to the appropriate muscle structures. To achieve this desired outcome, the design specification for a muscle bioreactor system should include specific details of the input and control signals that must be provided to the muscle. Key parameters to consider include: (1) potential failure modes and countermeasures; (2) means to probe and monitor muscle development nondestructively; and (3) algorithms that will be employed to effect a feedback loop that guides the developing muscle along a series of developmental milestones (states). System complexity and reliability must also be considered because as any system becomes more complex, it generally also becomes increasingly subject to failure. The basic elements that will be required for the success of future muscle tissue bioreactors include a fundamental change in the basic architecture that allows order-of-magnitude increases in system complexity while also increasing overall system reliability and ease of use.

## 2.4 Muscle Development and Dynamic Feedback Control

A developing muscle can be viewed as a "state machine." Each developmental stage is one "state" and the outputs from the state machine depend upon the history of the inputs as well as the current state. At each state transition, the muscle will require a different set of inputs to guide it to the next state (or stage) of development. A simple way to appreciate this is to note that muscular exercises appropriate for an adult are not necessarily appropriate for a newborn and vice versa. In engineering terms, the contractile performance of a muscle will depend both upon its current developmental state as well as its history of use. An advanced muscle bioreactor system will require some means by which the current status of the muscle can be determined automatically and frequently without harming the muscle. Then, the bioreactor system or the system's technician must make intelligent decisions about the quantity and duration of stimuli to be applied in the current tissue state, as well as when rest periods will be required. Importantly, the system will also need to be able to detect when the anticipated muscle development has not occurred so that corrective action can be taken. A simple set of important state transitions includes: (1) terminal differentiation of myoblasts to myocytes; (2) fusion of myoblasts to form primary (weak and with low excitability) myotubes; (3) development of secondary (stronger, more excitable) myotubes; and (4) expression of adult muscle proteins and formation of fast or slow muscle. Once the final state transition occurs, the bioreactor system will need to be able to maintain the phenotype of the engineered muscle indefinitely. These basic requirements for advanced muscle bioreactor systems are far beyond our current technological capabilities.

One of the key aspects in the creation of a finite state bioreactor is the identification of a number of biomarkers of development that can be determined nondestructively. Beginning with the state transitions described above, the nondestructively testable biomarkers (NDTB) that can be used to monitor the muscle state are the contractility—force and the rate of contraction and relaxation; excitability—the energy required to achieve stimulation, measured in terms of chronaxie and rheobase [28]; and metabolism, which has several proxies such as oxygen consumption. The most important measures for determining state transitions are contractility and excitability. The advent of active force indicates a transition to state 2, while the progression to state 3 can be observed as an increase in both contractility and excitability of the tissue (Fig. 1). The final state transition can be observed in the contractility with progressively faster rates of contraction, increases in both normalized contractile force and power, more rapid and complete relaxation and a divergence of contraction speed depending on the phenotype of the muscle (i.e., fast vs. slow).

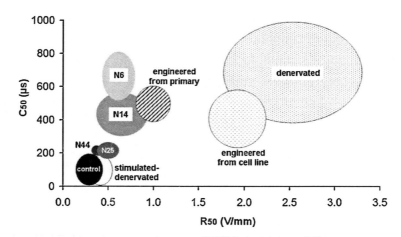

**Fig. 1** Excitability of muscle from control, 6, 14, 25, and 44 days after birth (N), denervated, denervated muscle stimulated in vivo for 5 weeks, as well as muscle engineered from primary muscle cells or cell lines. Note that as the muscle matures the chronaxie (C50) and rheobase (R50) values decrease indicative of improved bulk tissue excitability, from [29]

## 2.5 Historical Perspective

What may not be generally appreciated is the depth of the history of research in this area, and the breadth of technologies that will be required to achieve the ultimate goal. When the scientific literature is studied in detail, going back to original sources, one often finds that "new ideas" or "new technologies" are in fact not new at all, but rather are reinventions or rediscoveries of work that significantly predates current trends, often by many decades and in some cases by centuries. In some cases, the earlier work is both more careful and thorough than the more recent work, and important conclusions can be drawn from it. Thus, a simple discussion of the modern state of the technology and the recent developments in muscle tissue bioreactors in the past 5–10 years would be misleading, both in terms of what is currently known, and in terms of how far and how quickly we are likely to progress in the future. For this reason, the current technology for muscle tissue bioreactors will be treated within the context of the history of the research in such cases where this will make clear the significance of recent work.

## 2.6 Basic Questions

The basic questions at the core of muscle bioreactor research are:

1. Can a whole muscle organ be maintained by the artificial replacement of one, several, or all of the natural signals that a whole muscle normally receives within the body?

2. Can myogenic precursor cells be induced to organize and develop into whole muscle organs with adult phenotype by the artificial replacement of one, several, or all of the natural signals that a developing muscle normally receives within the body?

Buried within these two questions, one can find the many specific questions that drive much of the research in muscle development, functional adaptation, disease, and aging. For example, which signals are key, and which can be ignored or replaced? What combinations of signals are required? When during development, healing, or aging does a signal become more or less effective? Can a native signal be replicated by a very different mechanism and still be effective?

Some of these questions for muscle have been addressed directly and scientifically for at least 160 years [30]. For example, can electrical stimulation be used to maintain the mass and contractility of a muscle organ when innervation has been lost, or can mechanical vibrations be used to replace normal locomotion and exercise to maintain muscle mass and health? Bioreactors in one form or another have been used throughout this scientific enquiry into muscle development and maintenance. It is absolutely essential to appreciate the evolution of the key elements of modern muscle tissue bioreactors so that we can place into proper context the current embodiments of these devices. A failure to do so will result in the periodic and uninformed repetition of earlier work and an unrealistic view of recent progress and the time scale of likely future advances in the field.

## 3 Categories of Muscle Bioreactors

The overall scope of the field of muscle tissue bioreactors allows us to classify bioreactors based upon the following criteria: does the bioreactor primarily maintain the muscle tissue/organ or does it primarily promote tissue organization and development, and is the bioreactor implanted within or connected to an organism (in vivo) or is the muscle tissue/organ maintained in isolation from a host organism (in vitro or ex vivo). Thus, for the purposes or organizing the history and recent advances in muscle bioreactors, we organize the overall field according to the following 2 × 2 matrix (Table 1).

To use Table 1 effectively one should consider how each of the four categories relates to our basic scientific understanding of muscle development and maintenance. Then, consider how each contributes directly to a comprehensive design specification for muscle tissue bioreactor systems that would be capable of guiding the development of muscle precursor cells into adult phenotype whole muscle organs. It is also noteworthy that modern approaches to tissue engineering of muscle often employ two or more of these bioreactor classifications during the development of the muscle construct. An example would be the initial treatment of a muscle within an organism to induce satellite cell activation [29, 31–34], followed by ex-vivo tissue culture to form a muscle organ [3, 4, 8], followed by re implantation of the muscle into a host organism with or without further artificial stimulation [12, 35, 36].

**Table 1** Conceptual organization of the core technologies in muscle bioreactors into four categories

|  | Muscle maintenance | Muscle development de novo |
|---|---|---|
| In vivo | Whole muscle organs are maintained within a living organism, with or without alterations in position or activity pattern | Engineered muscle tissue or tissue precursor materials are implanted within a host organism to promote development toward adult phenotype; may or may not be within a synthetic chamber |
| In vitro | Whole intact muscle organs or large functional portions of organs are maintained in isolation from the host organism | This is archetypal muscle tissue engineering: myogenic cells are cultured ex vivo to produce functional muscle tissues and organs |

## 3.1 Category 1: In Vivo Bioreactors for Whole Muscle Organ Maintenance

On first pass this category of muscle tissue bioreactor technology might be considered by some to be trivial, but nothing could be further from the truth. As a first step, it must be appreciated that the host organism can be viewed as a bioreactor for the maintenance of all of its constituent organs, as well as for other "nonself" organisms, such as intestinal flora, parasites, and developing embryos in the case of females. To begin to develop a comprehensive design specification for a synthetic, in vitro muscle tissue bioreactor it is both logical and productive to study precisely how it is that the host organism maintains adult phenotype muscle organs, usually many hundreds of them, in a healthy state. Specific questions include: (1) what is the range of acceptable mechanical loading patterns? (2) what is the pattern of passive and active contractions? (3) what types of stimuli cause hypertrophy (growth) vs. atrophy (wasting)? (4) what soluble and extra cellular matrix signals are required at each developmental state? (5) what are the perfusion and mass transport requirements under differing metabolic loads? and (6) how do the internal feedback loops that control muscle organ maintenance respond to changing demands? The fundamental importance of this first category of bioreactors can not be overstated. One could easily argue that further progress in muscle tissue bioreactor development is impossible without better understanding of the basic science of how whole muscle organs are maintained within a living organism.

One way to view the importance of this problem is to consider an alteration of the in vivo conditions under which a muscle organ is maintained. For example, imagine an arm or leg immobilized in a cast. Despite decades of clinical research into this topic, the simple removal of one of the key mechanical stimuli for muscles (the ability to change length over the physiologic range of motion) is sufficient to produce muscle atrophy and degeneration of the muscles and associated tissues (tendons, vasculature, motor nerves, and bones) [37–42]. A suitable replacement stimulus for maintaining the health of immobilized but otherwise healthy muscles has not been identified, though some strategies do show promise [43]. This is despite the persistence of all of the other key mechanical and chemical signals that remain available to the immobilized muscle tissue within the otherwise intact organism. Similar problems arise for humans who remain inactive for prolonged periods, such as during long term bed rest. Contrast this with mammals that hibernate. These animals are able to remain dormant for many weeks/months with little or no loss in muscle mass and function [44]. What physiological strategies are used to circumvent the muscle atrophy associated with this form of inactivity, and can they be exploited for tissue engineering?

With the advent of prolonged weightlessness in space travel, and prolonged immobilization and bed rest while on life support following extreme trauma, the question of how to maintain healthy, functional adult skeletal muscle within the body is of growing importance. Our future success in the human exploration of space beyond our own moon, or the ability to extend healthy human aging beyond the first century of life may well be limited ultimately by our understanding of how the human body, as a bioreactor, maintains the health of its organs. And muscle, comprising approximately 50% of the total body mass of a human adult, is among the most poorly understood organs in this regard.

One productive approach to understanding the signals required to maintain healthy muscle tissue is to selectively and systematically remove one signal at a time while the muscle is maintained otherwise intact in vivo. Among all the experimental possibilities perhaps the most well studied and most relevant to the development of muscle tissue bioreactors is the effect of chronic denervation of skeletal muscle and the substitution of electrical stimulation to stimulate muscle activity.

The studies of the action of nerve impulses and exogenously generated electrical impulses on musculoskeletal tissues have a long scientific history and have left their imprint upon modern culture. The first reported use of electricity as a medical intervention is from 46 AD, when the Roman Scribonius Largus describes the use of the torpedo fish (an electric ray), applied directly to painful areas on the patient to alleviate musculoskeletal pain. Luigi Galvani carried out a series of investigations in 1783, later published in 1791 as a comprehensive book titled *de Viribus Electricitatis in Motu Musculari* [Effects of electricity on muscular motion] [45]. As early as 1841 it was suggested that electrical stimulation could be employed to replace stimulation from nerves to attenuate the rapid loss of both skeletal muscle mass and contractility following total denervation [30]. Two decades later in 1861 the use of electricity to study and treat paralyzed limbs was explored [46]. Since that time, experiments along these lines have been carried out more or less continuously until it was recently demonstrated that with the proper electrical stimulation protocol it is possible to maintain, or even improve upon, the contractility and mass of chronically denervated mammalian skeletal muscles in vivo [28, 47–51]. The scientific path from 1841 to the present includes over 400 published articles specifically on the topic of the use of electrical stimulation on muscle in vivo to restore lost function, many during the last century giving the impression along the way that the solution was near at hand.

Galvani's work involved mammals (lambs), birds (chickens), and amphibians (frogs and turtles), and he developed several technologies specifically related to the use of electricity to excite whole muscle organs in a bioreactor. Some of his work involved experiments with surgically isolated muscles while still in situ within the host organism, the remainder involved the explantation of whole muscles while remaining connected to a more extensive mass of support tissues including nerves and vasculature, as well as the surrounding musculoskeletal structures (bones, tendons, ligaments). The technology Galvani developed includes several types of simple whole muscle bioreactor systems to isolate and maintain the explanted tissues while the experiments were conducted (Figs. 2 and 3). Some of these experiments required the muscle to be maintained for long periods so that the effects of lightning and other natural electromagnetic phenomena could be observed [45]. These bioreactors, though primitive by modern standards, enabled Galvani to maintain viable muscle ex vivo for many hours, and as a practical matter even heroic modern attempts to significantly extend these time periods have met with only limited success [53, 54].

**Fig. 2** The various experimental apparati of Luigi Galvani [52]. In addition to several forms of electrical generating devices (a spinning disk electrostatic generator and a Leyden jar), the anatomical muscle preparation and two early forms of ex vivo muscle bioreactors are shown. These early whole muscle explant bioreactors contained whole limbs of frogs with intact nerves, as well as conducting electrodes (iron wires), conducting fluids (salt solutions) and the provision of resealable access to the tissue specimens

**Fig. 3** The use of a muscle tissue bioreactor was essential for the success of Galvani's tests involving atmospheric electricity [52]. The tissue preparations needed to be maintained in viable condition for many hours, and it was beneficial to provide electrical isolation of the explanted nerves and muscles so that the ground path for the "electrical fluid" could be defined. In this case, atmospheric electricity would pass from the lightening rod (*upper left*), through a long iron wire conductor, through the frog tissues, and then to Earth by way of an iron wire conductor extending to the bottom of a deep well (*right*)

### 3.1.1 Cultural Perspective on Tissue Bioreactors

The scientific results of Galvani were widely known outside of scientific circles and had lasting cultural impact. The young Mary Shelley, then aged 19, reading the work of Galvani during the dreary summer of 1816, a "volcanic winter" caused by the eruption of Mount Tambora the previous year, and further inspired by the French and Industrial Revolutions, both of which were viewed by many as having had unforeseen negative consequences for humanity, wrote her first draft of the gothic novel "Frankenstein, or, The Modern Prometheus" [55]. The book itself stands as a cultural icon and criticism of scientists who are ostensibly unconcerned by the potential consequences of their work. Frankenstein, the archetypal "mad scientist," is a term now commonly applied to scientists whose work is viewed by some as unnatural, socially irresponsible and potentially dangerous [56–58]. The recent use of the term "Frankenfood" to describe genetically altered agricultural products is just one such example [59–69]. The use of electricity to stimulate muscle tissues in bioreactors derives from the same intellectual source that has galvanized many generations of neo-Luddite anti-technology groups and activists, so it is advisable to consider the depth of meaning and symbolism when reporting or describing scientific progress in this area.

### 3.1.2 Additional Lessons from In Vivo Muscle Bioreactors

The plasticity of skeletal muscle tissue has been recognized for many decades. The ability of the enzyme profile of muscle to transform from fast to slow or vice versa was first suggested in 1960 through the use of cross-innervation, where motor nerves were transected and transposed, so that a predominantly "fast" muscle was now innervated by a predominantly slow twitch motor nerve bundle and vice versa. [32, 70–77]. Importantly, in 1976 Salmons and Sreter demonstrated that the plasticity of muscle was determined by the pattern of activity and not chemical signals from the nerve by preventing the slow-to-fast muscle transition following reinnervation using a tonic 10 Hz (slow nerve-like) electrical stimulus [75]. Together these experiments led to the understanding that muscle fiber type was determined by activity, not chemical or genetic factors determined during development. This led further to the use of implantable electrical stimulators to re engineer muscle in vivo for basic research [75, 78–82], and for clinical applications such as the surgical transposition and use of whole muscles in cardiomyoplasty, or the creation of skeletal muscle ventricular assist devices for the failing heart [83–87], and to replace the function of smooth muscle in the gut [5, 86, 88], as well as to the use of electrical stimulation of muscle cells in culture to control myosin isoform expression [89, 90]. This extensive body of work forms the basis for the belief that muscle tissue bioreactors can employ exogenous electrical stimulation to (1) guide and define the metabolic phenotype of adult muscle fibers as fast or slow twitch, (2) supplant natural innervation for the maintenance of muscle tissue mass, excitability, and phenotype, and (3) allow nondestructive computer control of skeletal muscle contractile activity during culture.

## 3.2 Category 2: In Vitro Bioreactors for Whole Muscle Organ Maintenance

The second category of muscle tissue bioreactors is defined by the explantation of whole muscle organs into a bioreactor system isolated entirely from the host organism. Many tissue engineers assume that whole tissues and organs are readily maintained outside of a host organism, and that the modern challenge is only to devise a technology to promote the ex vivo development of masses of cells into fully differentiated adult phenotype tissues. This, however, is not true. Another implicit assumption is that the field of tissue and organ bioreactor development is a new and emerging field, reaching back only to about the late 1980s. The general acceptance of both of these assumptions can be confirmed by careful study of the background and reference sections of nearly any recent peer-reviewed publication on this topic. When attempting to outline the current state of this technology, and where it will likely develop in the next 5–10 years, it is sobering to consider the facts concerning these two assumptions.

To begin with, although it is possible to maintain mammalian musculoskeletal cells in culture for many decades, it has proven impossible to maintain isolated whole mammalian organs within a bioreactor or on any other "life support" system, no matter how carefully the physiologic environment is maintained [53]. Specifically for muscle cells and tissues it was demonstrated in the first half of the twentieth century that isolated fibroblast cells could be maintained in culture for very long periods of time, often many decades as recently described by Boulay and Hardy [54], and isolated myocytes and myotubes could be maintained for weeks or months in culture [91–93], but a whole muscle as found in any animal is an organ, not just a mass of cells, and the maintenance of explanted whole organs for indefinitely long periods of time has never been accomplished [53]. This is true even though whole organs transplanted from one organism into another can in many cases be maintained for the normal lifespan of the host organism. So, therefore, the removal of an organ from a mammal does not mean that organ will have a reduced functional lifespan. However, the prolonged isolation of any whole mammalian organ from an organism results in rapid deterioration. The only exceptions are experiments that either maintain only a small mass of cells incorrectly termed an "organ" or those that employ a perfusate that includes fresh plasma from a live animal.

Research in this area has been ongoing since the early 1930s, and extraordinary progress was reported at that time [94, 95]. The early pioneers in this area were Charles Lindbergh, the famous aviator, and Alexis Carrel, the 1912 Nobel Prize winner in Physiology or Medicine for his pioneering work in organ transplantation. Together, they co-authored a classic text on this subject in 1938 titled "The Culture of Whole Organs" [94]. Lindbergh's contribution was in the capacity of "bioengineer," developing an alarmingly sophisticated and practical "bombeador de perfusión," the Carrel–Lindbergh Pulsatile Organ Perfusion Pump, shown in Fig. 4 [94]. These early ex vivo organ perfusion systems were so reliable in the hands of skilled technicians that about 20 identical systems were manufactured by skilled craftsmen, and as many as three could be set up simultaneously within a one-half height tissue culture incubator. Following this work, after about three decades of scientific inactivity, Lindbergh came out of retirement to take part in the development of pulsatile whole organ bioreactor systems for the U.S. Navy Medical Research Institute Tissue Bank [94, 96].

Using the Carrel–Lindbergh system the general procedure for the organ culture was essentially the same in all cases. The organ was dissected from a live donor and was placed within a cylinder of glass, the stoppered test-tube like chamber shown in Fig. 4, running diagonally across the top of the apparatus. The main artery of the organ was cannulated by means of the glass perfusate supply tube in the upper portion of the chamber. The organs were variously perfused with both natural (whole blood or blood-derived) and artificial media. Temperature and pH were monitored and controlled, and the perfusate was carefully filtered. Other than media immersion and pulsatile perfusion, no other means of active organ stimulation were reported.

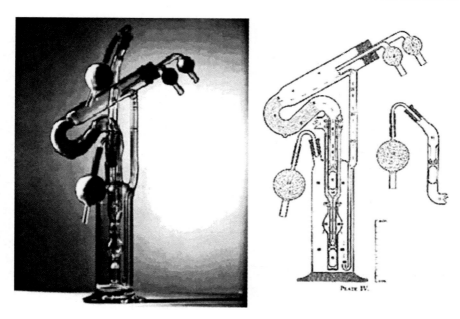

**Fig. 4** Carrel–Lindbergh Pulsatile Organ Perfusion Pump, a.k.a. the Lindbergh pump, 1935, showing a photograph of the actual device and a functional schematic of the perfusion system

Carrel and Lindbergh assert triumphantly in the preface to their 200+ page book that "… the techniques have been tested in over a thousand experiments during about one hundred thousand hours of perfusion," in at least eight independent laboratories across Europe and the United States [95]. The organs were subjected to pulsating perfusion to simulate normal arterial blood supply in the body, considered then and now to be important to the maintenance of healthy organs [94, 96–103].

This early organ bioreactor was extremely versatile. A wide range of whole organs and tumors were harvested from various species, including cat, guinea pig, chicken, and humans and perfused for various amounts of time. Bear in mind that this was all accomplished a few years before the general availability of modern antibiotics, though even today tissue and organ cultures are plagued by septic contamination. Extensive quantitative assessments of organ function and metabolism were reported, and the authors conclude that "A new era has opened… [permitting study of] how the organs form, how the organism grows, ages, heals its wounds, resists disease, and adapts itself with marvelous ease to changing environment"[95]. The work was so widely heralded that Alexis Carrel and Charles Lindbergh shared the cover of Time Magazine (June 13, 1938), and at the time it was believed that Lindbergh's contributions to medicine and biology would eclipse his accomplishments in aviation.

However, obscurity descended upon this important work and remains to this very day. Lindbergh's later work on these perfusion systems with the U.S. Naval Medical Research Institute in the 1960s also eventually ground to an anticlimax and was ultimately cancelled before it was carried through to successful completion. The "Lindbergh pump" though reliable was considered too difficult to use except by the most skilled operator, and fell out of use by about 1940; all but one or two of the systems were scrapped to recover the platinum in their filters. Researchers quickly found alternative approaches, either through the use of simple cell culture techniques or the use of simple organ sections which could be maintained ex vivo for a few hours, long enough for their experimental needs. The whole organs perfused by the Carrel–Lindbergh system only lasted for a few days and the smaller vascular tissue sections lasted at most several weeks, not indefinitely, as had been implied by their early reports in the popular press.

It is important to place modern advances in tissue engineering and tissue bioreactor development into this historical context, because many billions of dollars have been invested in these technologies [104–107], and it would be wise to learn from the past rather than to repeat it. Foremost, it is advisable to temper our claims and our expectations of the timeline for the development of these "new" technologies in light of the progress made since the first and often forgotten early successes in these areas.

The important lessons for tissue engineering and bioreactor design include:

1. Tissue culture bioreactor systems that are complex, difficult to use, expensive, or prone to failure will not have lasting scientific or commercial impact.
2. Modern successes often do not measure up to early successes in bioreactor systems and our technological progress in bioreactor design is not nearly as rapid as is generally presumed.
3. Fantastic claims about success with engineered tissues and bioreactor systems often have undesirable long-term consequences.

We remain faced with the assumption stated at the beginning of this section, restated as a question: is it now possible to maintain whole explanted mammalian organs ex vivo for indefinite periods of time? Briefly, the answer is "no." This alarming technological fact was recently reviewed by Dr. R.H. Bartlett [53], a widely regarded expert in extracorporeal life support systems (ECLS) who comprehensively analyzes the state of the art of our knowledge of ex vivo organ maintenance. Despite all attempts to the contrary, any organ isolated from a living mammal will experience an inexorable sequence of events leading to failure within ~48 h. Bartlett argues convincingly that it is specifically isolation from the midbrain that leads to accelerated organ failure. Many studies and a great deal of anecdotal evidence point to this conclusion [108–110]. Bartlett hypothesizes that the midbrain produces a very unstable and yet-to-be-identified hormone, which he terms "vitalin," the absence of which by any mechanism leads to accelerated organ failure [53]. However, experiments employing small groups of cells incorrectly termed "organs," or the use of fresh serum in the perfusate can appear to contradict this general observation, resulting in some confusion as to our state of understanding in this area.

Our failure to succeed at the ostensibly "easy" task of maintaining whole organs ex vivo (when compared with the heroic ask of engineering these same organs de novo from isolated cells) is by itself one of the greatest obstacles to a broad range of useful clinical technologies. Were it possible to maintain organs ex vivo indefinitely in a state of health, it would then be possible to: (1) bank organs for transplantation after precise matching to the recipient; (2) test and verify donor organ function and asepsis; (3) transplant a greater variety of organs including spleen and pancreas to treat diseases such as hemophilia and diabetes; (4) remove cancerous organs for more specific and aggressive treatment followed by reimplantation; and (5) use banked organs to produce needed cells and biochemicals (such as whole blood, hormones, etc.) [53, 95]. We lose much by our inability to maintain whole organs ex vivo, and the advancement of this technology should be a primary target of a major portion of our investment in this area of research [105].

Directing our discussion specifically to muscle tissue bioreactors, successful organ-level muscle bioreactor development will, at least for the time being, necessitate parallel development of in vivo and hybrid bioreactor systems [21, 22] that are integrated in some way with a living host organism. A number of models have been developed to provide muscle with a satisfactory environment for long term culture [21, 22, 53, 111, 112]. Many of these systems have attempted to recapitulate the mechanical and electrical environment of the organism (Fig. 5), but in general it is still impossible to maintain the health of a functioning whole muscle organ ex vivo for more than a few days.

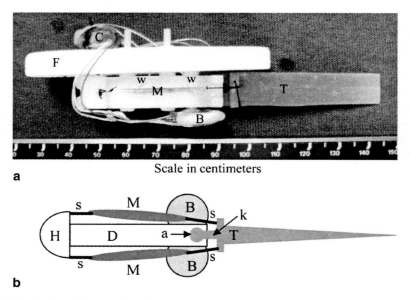

**Fig. 5** An in vitro bioreactor for whole muscle organ maintenance. (**a**) Muscle-powered robotic "fish" and (**b**) the schematic representation of the fish [60, 198]. The semitendinosus muscles of a frog are sutured near the base of the robot on each side (one is visible, sutured at each end). The robot swims in amphibian Ringer's solution, and muscle activity is controlled by an embedded microcontroller encapsulated in silicone, shown at the top of the biohybrid fish robot

## 3.3 (Category 3) In Vivo Guided Muscle Tissue/Organ Development

As described above, the body can be viewed as the ultimate bioreactor [35, 36, 113–117]. For many applications the use of the body as a bioreactor has many advantages over ex vivo or in vitro tissue engineering approaches, in particular it greatly simplifies the entire process if autologous cells and tissues are used so as to avoid tissue rejection. There are several examples of success with this approach in muscle tissue engineering. For example, de novo engineered skeletal and cardiac muscle have been implanted in vivo to promote vascularization and the formation of functional nerve-muscle synapses by surgical neurotization [35, 36, 115, 116, 118]. After 3–4 weeks of implantation engineered muscles have demonstrated ~eightfold increase in peak contractile force capacity [35], and neurotization resulted in an increase of ~fivefold in the contractility over non-neurotized controls [116]. Implantation also results in vascularization and a significant increase in the viable radius for cell survival of the engineered muscle constructs [36].

When using a host organism as the bioreactor it is possible to either isolate the muscle tissue construct within an implanted chamber, or to leave the construct exposed to the internal environment without barrier isolation (Fig. 6) [35, 36, 115]. In any case it is essential to control the length of the muscle construct to prevent hypercontraction or stretch-induced muscle injury and subsequent damage to the muscle construct. It is also important to position the engineered construct in a receptive area of the body so that it is in apposition with a vascular bed and nerve supply and is able to receive passive mechanical signals. Potential regions within the body that match these criteria include the inguinal cavity and the diaphragm.

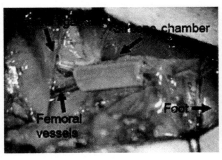 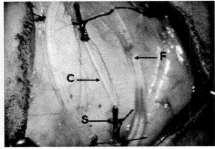

**a**  **b**

**Fig. 6 a** A skeletal muscle construct prepared for implantation within a silicone chamber, with axial vascular supply from the femoral vessels. **b** Implanted 3-dimensional cardiac muscle construct (C) first engineered in vitro over a period of 4 weeks then implanted subcutaneously in a rat. The muscle construct is not isolated from the host by use of a chamber, but rather supported by an open elliptical acrylic frame (F) using the suture anchors from the in vitro culture system (S). Three weeks after implantation, the cardioids were recovered and contractility was evaluated [23, 43, 227]

## 3.4 (Category 4) In Vitro Guided Muscle Tissue/Organ Development

The modern use of bioreactors to grow muscle tissue in vitro is heavily based upon an earlier and much more extensive scientific literature on bioreactors for bone and cartilage tissues. Often modern muscle tissue engineering bioreactors are just physical adaptations of existing systems already developed for these other tissue systems [119]. Interestingly, a recent review paper entitled "Bioreactors for tissues of the musculoskeletal system" [120] does not contain any information whatsoever on bioreactors for use with skeletal muscle tissue. This is not an isolated omission. The comprehensive and widely regarded book "Principles of Tissue Engineering, 3rd edition" has an entire section devoted to musculoskeletal tissue engineering (section XV), but there is not actually any material on skeletal muscle tissue engineering, only bone, ligament, tendon, and cartilage are covered [121]. Though skeletal muscle tissue engineering in vitro has a long history [91–93, 122–131], it is eclipsed to the point of near exclusion by the more dominant areas of skeletal tissue engineering. This is largely due to the complexity of the inputs required for skeletal muscle in relation to the other musculoskeletal tissues. Bone, cartilage, ligament, and tendon appear to require only passive mechanical input, whereas muscle requires both active and passive mechanical activity, often elicited by externally applied electrical interventions. Furthermore, the coordination of the mechanical and electrical inputs to produce the high force lengthening contractions that preferentially lead to greater increases in muscle mass are increasingly difficult to achieve without damaging the muscle construct. The important parameters to consider in this class of bioreactors (in vitro development) will be discussed in detail in the following sections.

## 4 Modern Approaches

Modern muscle bioreactors are classified according to whether they employ 2-dimensional cell cultures or 3-dimensional muscle constructs, the types of interventions to which the muscle is subjected (i.e., electrical, mechanical, or both), and whether or not functional measures of muscle contractility can be made. Bioreactors in this category can be further subdivided into various classes based upon their primary intended function, for example, whether the objective is discovery (basic science to study the response to a variety of stimuli) or delivery (uniform parameters for production of tissues for high-throughput screening) [132–134]. From the standpoint of understanding the technology of modern muscle tissue bioreactors it is best to consider primarily the types of stimulation employed and the basic architecture of the system that is used to achieve the intended stimulation.

## 4.1 Discovery Versus Delivery

As the name suggests, discovery bioreactors are intended for testing a number of different interventions to determine which is optimal for improving the parameter of muscle function that will form the basis of study and at what point a state transition has occurred within the tissue. As a result, they are generally small, independently controlled bioreactors, designed to hold up to six samples with ~four of the independent bioreactors fitting on one shelf of an average tissue culture incubator. In this orientation, four simultaneous experiments can be performed on one shelf and the number of samples in each group gives the investigator sufficient power to rapidly determine which intervention has had the optimal effect, if any. Once a paradigm of intervention has been determined, larger scale delivery bioreactors could be developed. These delivery bioreactors hold more tissues that undergo a uniform intervention program. These machines should have the capacity to perform nondestructive tests of tissue development and transition to new intervention programs once a state transition occurs.

## 4.2 Mechanical Stimulation

Mechanical stimulation has been widely employed in muscle cell culture and tissue engineering [11, 12, 14–17, 20, 28, 122–126, 128, 131, 135–176]. Many of these technologies have been reported only in abstract form, never quite achieving publication in a peer reviewed journal. Almost all of them have been hand built for use by a specific research group and their close collaborators. Mechanical stimulation of muscle can take various forms. For 2-dimensional muscle cultures, the forces are sometimes applied by the action of fluid flow to create shear, by mechanical vibration, or by the application of uni-or bi-directional mechanical strain applied to an elastic substrate on which the muscle cells are grown. When a 3-dimensional muscle organoid has been formed, it becomes possible to apply uniaxial strain directly to the tissue construct. The use of 3-dimensional tissues also permits direct measurement of tissue function, principally the contractility and excitability of the engineered muscle tissue. Three-dimensional muscle constructs can be passively stretched by an external mechanical actuator, or the culture conditions can be set up in such a way that the muscle tissue itself generates sufficient force to cause either active or passive length changes against a compliant anchor point. It is also important to realize that 3-dimensional muscle tissue constructs generate baseline axial stresses on the order of 3–5 kPa, similar to the stress generated during wound closure [8]. The reaction of the myotubes against fixed anchor points in culture has the important effect of causing myotube alignment along the axis formed by the two anchor points, even when no other external forces are applied [3, 4, 177]. This occurs in the static culture of engineered 3-dimensional muscle tissue, so it is correct to say that in general it is not possible to maintain engineered muscle in culture without some form of mechanical stimulation being present.

### 4.2.1 Mechanical Strain of Two Dimensional Muscle in Culture

The culture of layers of contracting myotubes in what amounts to two-dimensional sheets has been done for many decades using standard cell culture equipment [91–93], and this practice continues to the present for a very wide range of studies in the basic biology of muscle, not strictly within the field of tissue engineering. Tissue engineering properly enters the scope of this research when there is an attempt to organize the 2-dimensional layers of cells in some way to achieve a higher order or tissue organization. This may be through an attempt to achieve cellular alignment by micropatterning of surface topology [178] or adhesion molecules [179], by mechanical [180] or electrical stimulation, or by the controlled mechanical release of successive sheets of muscle cultures to build up 3-dimensional muscle tissue constructs [181–187]. The application of mechanical stimulation to 2-dimensional muscle cultures is relatively straight forward. It is generally not possible to mechanically affix to individual myotubes grown in culture because the myotubes grow in mechanically cohesive sheets with adhesion points to the substrate all along the length of each myotube. Therefore, the method most commonly employed is to culture the sheets of myocytes and myotubes on thin elastic substrates [135, 188–190] which are then stretched by any number of mechanical means.

Elastic substrates for 2-dimensional muscle cell culture are usually made from thin sheets of medical-grade silicone or from cast sheets of PDMS (polydimethylsiloxane). Though on first pass it would seem there are a virtually limitless variety of mechanical stimulation systems of this general category, in practice the number of basic system architectures is reasonably limited. To begin with, the elastic substrate can be deformed to provide either uniaxial or biaxial mechanical strain [191], and the strain fields produced can be either uniform [192] or nonuniform. Uniform strain fields simplify analysis, whereas nonuniform strain fields appear to have a more realistic basis in biological muscle function [193–195].

Uniaxial strain is accomplished by growing the muscle on a strip of elastic substrate and applying a mechanical displacement along a single axis [180, 196, 197]. Generally, if the substrate is uniform in geometry and material constitutive properties, the resulting strain field is approximately uniform. A less commonly employed alternative to this is to grow the muscle on a thin-walled tube fashioned from an elastomer, then inflating the tube while the cells are in culture. For thin-walled tubes of relatively small diameter the circumferential strain dominates, thus providing essentially uniaxial mechanical strain to the muscle tissue growing on the outside of the tube. This method, though relatively simple, is not widely employed because it is difficult to visualize the layer of muscle cells on the tube in culture.

Biaxial strain is readily accomplished by growing the muscle on an elastomer sheet that is typically constrained around its periphery, and then applying a mechanical displacement to the sheet by any number of means, including pneumatic distension or retraction of the sheet (which acts as a diaphragm) [119], or by mechanical distension of the elastomer sheet from the bottom (the opposite side from which the cells are being cultured) by means of a mechanical actuator of some kind. This could be a moveable platen, a braille reading machine, an electromagnetic solenoid, a voice coil actuator, a stepper-motor driven rod, hydrostatic pressure, or any other means that could be imagined. The resulting biaxial strain field can be measured directly by optical methods or calculated in a computer model [119, 198, 199]. The methods that have been reported for applying mechanical strain to 2-dimensional cultures are many, but only a very few systems have become commercially available for this purpose such as those marketed by Flexcell®, and none of the major commercially available systems are specifically designed for use only with muscle tissue [197, 200, 201].

Recent work has demonstrated that muscular thin films can be constructed on very thin PDMS substrates that are capable of generating active contractions to produce movement and mechanical work [202]. Although many new systems are reported to apply 2-D strain to cells in culture, it is unclear that any of them is a significant improvement over any of the previous systems. Since no 2-D system has been widely adopted for use to the exclusion of others, it is likely that all such systems have equivalent limitations.

### 4.2.2 Mechanical Strain of Three Dimensional Muscle Constructs

The culture of 3-dimensional muscle constructs is a very different technology than the culture of 2-dimensional sheets of muscle. Three dimensional muscle tissue constructs can be either self-organizing or they can be constructed using myocytes cast into biodegradable gels or grown on natural [203] or synthetic scaffolds [188, 204–207]. An alternative method for growing 3-D muscles and other tissues is the use of the NASA rotating bioreactor, commercially available from Synthecon [208–210]. Using microcarriers which are suspended by the rotation of a cylindrical culture vessel on a horizontal axis, it is possible to grow relatively large 3-D clusters of all types of tissue, including muscle [145, 211–213] which in some cases appear to be organized into tissue-like structures. Because of the differences in structure of skeletal muscle organs when compared with hollow organs containing smooth or cardiac muscle, the rotating bioreactor systems appear to be more suitable for the growth of smooth and cardiac muscle organs than for skeletal muscle organs.

With current technology, most self-organized muscle constructs tend to be small cylinders comprised predominantly of parallel myotubes, ranging in size up to several cm in length and up to about 1 mm in diameter [3]. Gel-cast muscle constructs can be of any shape, but usually are initially rectangular in cross section for ease of casting [36, 136, 151, 188, 190, 204, 207, 214]. It is with the three-dimensional muscle constructs that we begin to see significant tissue organization, especially in the self-organizing constructs that involve nonmuscle cell co-cultures [3, 4, 7, 8, 12, 131, 215–220]. These often have tissue-like structural features both along the axis and the radius of the constructs (Fig. 7), such as a fascicular arrangement of myotube bundles [3], or myotendinous junctions near the mechanical attachment points at each end [219, 221].

Three-dimensional muscle constructs offer important design opportunities and challenges for muscle tissue bioreactor design. These constructs can generally be excited electrically to generate active force, they spontaneously contract in culture, and they also generate baseline passive (elastic) forces. Because of force shunting in the scaffold material, the measurement of contractility in scaffold-based engineered muscle is often quite difficult unless the scaffold has a stiffness of approximately the same or lower magnitude than the muscle tissue itself. This is rarely achieved unless highly dense arrangements of parallel myotubes can be organized within low-stiffness gels [141, 142, 151, 152, 175, 176].

Unlike 2-D mechanical loading, the 3-D tissues do not contain adhesion points to a substrate along their length. The 3-D tissue constructs have only one adhesion point (actually a region) at each end which means that, once formed, the 3-D muscle constructs may be mechanically stimulated by connecting one end of the tissue to a mechanical actuator and performing uniaxial shortening and/or lengthening of a known distance and frequency. While having no adhesion points along its length increases the ease of bioreactor design, it also means that there are high strain concentrations at the interface between the engineered tissue and the points of attachment to the bioreactor. A durable mechanical interface is essential for the successful application of any mechanical intervention to engineered muscle. If the interface is weak, the tissue will fail at the interface under strain. A number of techniques and materials have been used to address this problem including making circular constructs to eliminate the interface [141, 143, 174–176, 222–224], using a highly porous plastic material [Flexcell(r)], using a material with a high surface area such as a woven suture [3, 4, 7, 8, 216, 217], or replacing the anchor material with engineered or native tendon [215, 221]. To date, only the last technique has produced a tissue interface that does not fail under high loads approaching physiologic stresses, ~280 kPa.

## 4.3 Electrical Stimulation

Electrical stimulation has been used for many years in muscle 2-dimensional cell culture to study its effect on myotube formation [225], myocyte alignment [226], fiber type [90, 152, 226, 226–230], metabolic function [231–234], transcription [227, 229, 231, 233–238], and protein synthesis [239, 240]. These studies are limited largely by the fact that 2-D cultures can only be stimulated for short periods (generally up to 4 days) before the myocytes detach from the culture dish [233]. In 3-dimensional muscle constructs, electrical stimulation can be used both to evaluate the contractility and excitability of the engineered tissues as well as to attempt to guide the development of the engineered tissues during culture [3, 4, 6]. Further, unlike 2-dimensional culture, 3-D constructs can be stimulated for a period in excess of 2 weeks without failure. Indeed, chronic low-frequency stimulation, 20 Hz train of five pulses with a rest phase of 3.75 s between pulse trains, has been shown to slow the contractility of a predominantly fast engineered muscle [235] demonstrating that electrical stimulation in vitro can alter the phenotype and function of engineered muscle. It appears to be more difficult to model a fast nerve/muscle system in vitro since engineered muscle is prone to electrochemical damage, however, few studies have tried to accomplish this task in engineered tissues. There is a much larger literature on the requirements for muscle stimulation in vivo, which should aid in the development of future in vitro electrical stimulation paradigms.

Chronic low frequency electrical stimulation of 10–20 Hz for 8–24 h a day has been successfully used in vivo to transform the fiber type of whole muscle organs from predominantly fast to predominantly slow twitch [32, 75, 81, 226, 241–265]. Many fewer studies have attempted to maintain the mass and contractility of chronically denervated fast muscle or shift the phenotype of muscle from slow to fast. This is a difficult task since fast muscle has a much lower work-to-rest ratio and a far more random firing pattern than slow muscle. Therefore, varying rest periods and innovative stimulation patterns may be essential to the development of the fast phenotype. In general, because electrical stimulation does more damage as the intensity, frequency, or duration of the stimulus is increased, some researchers report that for both in culture and in vivo electrical stimulation, optimal results are obtained by utilizing stimulation protocols that mimic the minimal activity patterns that are thought to prevail in vivo [28, 47–51, 266]. With increased levels of electrical stimulation it is thought that the damaging effects of the stimulation outweigh the beneficial effects of the increased activity. This is almost certainly due to limitations in the electrode and electrical interface technology currently available for tissue bioreactors.

## 4.4 Measurement of Forces Generated by Muscle Constructs

With the ability to measure both active and passive contractility of muscle, it becomes possible to devise muscle bioreactor systems that can both elicit muscle contractions by electrical stimulation or chemical additives to the media, as well as measure the resulting contractions. Many important parameters of the contractility of muscle can be measured, including the peak twitch and tetanus, rate of increase of force, half relaxation time, excitability, the delay period between stimulus and response, and dynamic measures of contraction velocity and power. These measures can be used as quantitative biomarkers for the state of the muscle tissue in culture, allowing nondestructive assessment of the developmental stage and general health of the engineered muscle [2–4, 8, 21, 22, 28, 216].

Force measurements also require some means by which the muscle constructs may be attached at two or more points so that the muscle tissue can be fastened at one end to a force transducer and held under tension while suspended in the culture media. As discussed above, this remains one of the greatest challenges in muscle bioreactor design. Several general approaches have been employed, including attachment to steel pins [177], steel screens [131, 156], Velcro [137, 212, 267], silk suture [3, 4, 7, 8, 216, 217], engineered tendons [215, 221, 268], or, alternatively, muscle geometries that naturally allow two point attachment such as a thin ring, which can be held in tension by placing two posts or hooks at opposite poles inside the ring and applying tension [5, 176, 269].

To assure accurate measures of contractility, it is necessary to exclude any force shunts, that is, it is necessary to make sure that all of the force generated by the muscle is transmitted to the ends of the construct and not to an internal scaffold or laterally to an external point of mechanical contact. A sensitive force transducer is required, generally capable of resolving down to 1 µN of force, and up to several mN [270–273]. The force transducer must pass through the wall of the bioreactor in some way such that it does not mechanically contact the wall, so usually this requires a wire or hook that passes over the open top of the bioreactor into the culture media to make a connection with the muscle construct. This arrangement results in a significant risk of contamination. To allow fully closed bioreactor systems it will be necessary to develop practical, immersible, high sensitivity force transducers which can be sealed into the bioreactor chamber. Some attempts have been made to use remote external camera optical methods with compliant attachment points inside the bioreactor chamber. The contracting muscle causes a detectable displacement of the compliant anchor point, usually the bending of a calibrated cantilever. The movement of the end of the cantilever is measured optically, and the force is calculated from the known stiffness of the cantilever. This method has been recently proposed for use in high-throughput drug screening devices in which measures of muscle contractility are to be used (Vandenburgh et al., 2008).

## 4.5 Perfusion

Perfusion bioreactors are sometimes employed with engineered muscle tissue for two reasons. The first is to provide improved mass transfer by circulating the culture media [111], the second is to use the fluid dynamic forces to induce tissue organization and development. Several perfusion bioreactor systems have been specifically developed for use with functional muscle tissue (Birla et al., in press) [138, 211, 274–276]. Unlike bone, muscle does not appear to respond to fluid shear stress as an anabolic stimulus.

This brings into question the need for dynamic fluid forces in the formation of mature skeletal muscle. Without the need for fluid dynamic forces, the main goal of muscle tissue perfusion is improved delivery of nutrients and removal of waste. In this case, a mass transfer system is warranted. Mass transfer systems can use either low shear stress mass perfusion or gyration and gentle shaking of the bioreactor. Both methods can be designed to increase the $pO_2$, decrease $pCO_2$, and maintain pH [138, 267, 275, 276]. Gyrating and shaking bioreactors are less well suited for 3-dimensional muscle constructs that are anchored to the underlying substrate. In these cases, a mass perfusion system is more feasible. Mass perfusion of scaffold-

Bioreactors for Guiding Muscle Tissue Growth and Development 63

based engineered tissues with flow rates from 0.6–3 mL min$^{-1}$ increases the number of cells deep within the matrix [138, 275]. A flow rate of 0.6 mL min$^{-1}$ can produce uniform cell density up to ~1 mm deep [275]. While the increase in functional cell depth is encouraging, neither the accumulation of protein nor the total DNA within the constructs was improved over static culture [138]. Furthermore, the rate of myosin heavy chain degradation was greater in the perfused muscle constructs than those grown in static conditions. One possible explanation for this finding is that mass perfusion often uses positive displacement (peristaltic) pumps to circulate the media. Peristaltic pumps compress a flexible tube to force the fluid through the perfusion loop. Unfortunately, mechanical friction and compression of cell culture media within silicon tubes often results in precipitation, aggregation, and loss of proteins within the serum. Many of the proteins lost in this manner are the chemical factors within the serum that promote muscle tissue growth and development. Other types of perfusion pumping systems, especially those that involve rapidly moving turbines or valve elements or sliding contact mechanism seals are also thought to cause damage to both whole blood and cell culture media [277]. Therefore, the development of more appropriate methods of perfusion pumping is required. Possible alternatives include magnetohydrodynamic pumping [278], passive gravity head-pumps [279], and positive pressure pumping systems with minimal valve closure cycles, similar to that employed by Lindbergh and Carrel. Positive pressure systems pump media from a positive gage pressure chamber to a slightly less positive gage pressure chamber on the other side of the perfusion loop. Since there is no mechanical compression of the media within the tube and fewer valve closure cycles per volume pumped, there are fewer problems with protein loss from the serum. Damage to the perfusate, especially the dissolved proteins, is also thought to result at the liquid–gas interface present in most perfusion systems. This would require the use of compliant reservoirs which can expand and contract to accommodate changes in perfusate volume, as well as an overall zero head space bioreactor design [280–282], to eliminate all liquid–gas interfaces by interposing semipermeable membranes where this interface might normally occur.

## 4.6 Chemical Stimulation

An increasing number of chemical agents and hormones are known to have important physiological effects on skeletal muscle growth and development. Importantly, different chemical agents effect muscle at different developmental states in vitro. For instance, the hormone myostatin and its inhibitor follistatin [283, 284], as well as fibroblast growth factor [285, 286] alter the terminal differentiation of myoblasts resulting in a greater/lesser number of myoblasts within the engineered muscle construct capable of fusing to form myotubes. A variety of drugs and hormones promote the terminal differentiation and fusion of myoblasts. For example, creatine increases myoblast fusion by approximately 40% [287]. Other agents that affect fusion include l-arginine (via the production of nitric oxide), and the insulin like growth factors

(IGFs) [52, 288–296]. Some hormones effect muscle function during more than one developmental state. For example, thyroid hormone (T3) promotes the terminal differentiation of muscle cells by directly activating the transcription of the myogenic regulatory factors MyoD and myogenin [297–299] and also regulates the development of more excitable and more mature muscle constructs. T3 increases excitability through the regulation of the Na+- K+- ATPase [300], directly controls muscle relaxation through the expression of the sarcoplasmic reticulum calcium reuptake channel [301, 302], and affects muscle maturation by directly altering the expression of adult myosin heavy chain [303–305]. Functionally, the presence of T3 results in a muscle that is more excitable, produces more force, and has a shorter time to peak tension and half relaxation [306]. In dwarf mice, with normal innervation and loading patterns, adult MyHC isoforms are not expressed in either the heart or skeletal muscle without thyroid hormone treatment [304]. However, while thyroid hormone has direct transcriptional effects on muscle, in cultured myocytes it is not sufficient by itself to produce adult muscle fibers. These findings suggest that a permissive level of thyroid hormone is required for the maturation of skeletal muscle fibers from an embryonic to adult phenotype. In the final state of muscle development, drugs such as cyclosporin A can be used to promote myosin heavy chain expression and the production of fast muscle [235]. Other hormones, such as IGF 1 and testosterone, can also be used during this state to promote muscle hypertrophy and increased force production. Further, addition of growth hormone, ascorbic acid, and proline can be used at this time to promote the production of extracellular matrix proteins resulting in a more mature muscle with either a fast or slow phenotype that produces more force and is able to transfer that force effectively to its anchor points resulting in improved contractility with less risk of tissue injury. Therefore, just as with mechanical and electrical interventions, chemical interventions need to be administered within the bioreactor in a state-specific manner to promote the final muscle phenotype.

## 4.7 Multiple-Mode Stimulation

Bioreactors that permit multiple modes of stimulation of engineered muscle have recently become relatively common in the literature. For example, bioreactors that allow control of mechanical stimulation and perfusion or electrical and mechanical stimulation have been reported by several groups [124, 126, 127, 131, 136, 153, 157, 161, 163, 171, 307].

As any system becomes more complex, it generally also becomes increasingly subject to failure. This remains true unless there is a radical change in the architecture of the system to accommodate the increased complexity while vastly reducing the overall probability of failure. One of the major reasons that new muscle bioreactor systems do not gain wide acceptance and use is that they have become so complex that they are difficult to set up, and once in use they are highly prone to failure. A familiarity with the various failure modes in muscle bioreactors is an important aspect in bioreactor design because the system architecture must minimize these failure modes in order for increasingly complex bioreactor systems to be practical.

## 5 Failure Modes for Muscle Tissue Bioreactors

Even though a bioreactor may function precisely as designed, it may still fail to elicit the desired response from the tissue specimen. Usually this is the type of "failure" that is reported in the scientific literature. But bioreactor system failures that cause unanticipated experiment termination are exceedingly common, and generally are only discussed briefly, if at all, within the methods sections of publications. Typically the discussion of bioreactor system failures are not categorized and analyzed and reported in sufficient detail to inform the reader fully. Thus, as a result, fundamental errors in the design of muscle bioreactor systems are undoubtedly repeated many times independently. As muscle tissue bioreactors become more complex, and as the culture periods ex vivo extend into many months, there are an increasing number of failure modes that become more likely to occur. The design of a muscle tissue bioreactor should include a careful consideration of all possible failure modes and provide countermeasures and some means for monitoring for these failures [21, 22, 136]. Briefly, the major failure modes include:

### 5.1 Septic Contamination

This mode of failure affects all types of muscle bioreactors and at room temperature typically results in rapid functional deterioration of the engineered tissues within 24 h in the absence of countermeasures. Barrier asepsis is prone to failure in complex bioreactor systems, especially those with intrusive sensors that come into contact with the culture media, and for any type of perfusion system [21]. Chemical countermeasures using broad spectrum antibiotic/antimycotic formulations in the culture media are effective [3, 4, 8, 216, 217], however the use of streptomycin can interfere with contractility in developing muscle because it is an aminoglycoside antibiotic which blocks stretch-activated channels (Birla et al. 2008) [308]. For long term culture of engineered muscle in bioreactor systems, it may be necessary to initially use, then taper off the concentration of certain chemical antibiotics/antimycotics. With increasing system complexity, the inclusion of media perfusion, intrusive sensors, mechanical actuators, and multichambered bioreactors with complex fluid manifolds, this failure mode has increased in practical importance to the point where it may be stated with some confidence that it sets the practical limit for muscle tissue bioreactor system complexity using the current methods and technologies. Until radical new system architectures are developed which mitigate this problem, further progress toward increasingly complex yet practical and usable muscle bioreactors is unlikely.

## 5.2 Mechanical Failure Within the Tissue

Also known as contraction induced injury, this mode of failure is prevalent in muscle tissue subjected to maximal contractions during forced lengthening, and affects engineered muscles, as well as aged and dystrophic muscle in vivo [309–311]. The effective countermeasure for tissue bioreactors will almost certainly involve employing control algorithms that prevent repeated eccentric contraction of fully-activated muscle actuators [21]. Living muscle can functionally adapt to tolerate lengthening contractions if the proper maintenance protocols are employed. An attempt can be made to implement such protocols in the muscle actuator bioreactors using feedback control.

## 5.3 Mechanical Failure at the Tissue-to-Tissue Interfaces

Less common for muscle in vivo, this is a major failure mode for explanted whole muscles and in vitro engineered muscle tissues in general. For whole explanted muscles, the interface typically involves suture or adhesive applied to the pre-existing tendons. Lack of process control in this tissue/synthetic junction leads to unpredictable mechanical failures over time. In engineered tissues the problem is more serious, as tissue failure frequently occurs at the tissue/synthetic interface under relatively mild mechanical interventions. The failure is hypothesized to be due to stress concentration at the tissue/synthetic interface resulting from a mismatch of the mechanical impedance between the muscle tissue and the material to which it is attached [215, 221, 268], compounded by inadequate force transduction from the appropriate intracellular force generating machinery through the myofiber cell membrane to the extracellular matrix, leading to cell membrane damage at the interface, with subsequent rapid tissue degradation and necrosis. This is a serious technological limitation to further advancement in muscle tissue engineering, and it is probable that fully-functional in vitro engineered skeletal muscles can not be realized until the technical challenge of engineering functional myotendinous junctions has first been mastered. This remains a major objective of current research in muscle tissue engineering [215, 219–221, 268]. The ECM of skeletal muscle is contiguous with the tendon tissue and provides both axial and transverse mechanical connectivity between the contractile proteins in muscle and the tendon [312–314], and thus the tendon at each end of an adult phenotype skeletal muscle is an important part of the whole muscle organ. Unlike muscle tissue, tendon tissue is 80–90% extracellular matrix, composed chiefly of parallel arrays of collagen fibers. Tendon is a much less fragile tissue than muscle. This reduces the difficulty of attaching engineered tendon to a synthetic material. Still, the tendon-to-synthetic interface, where biology interfaces with machine in the bioreactor system, is a separate yet equally important technical challenge.

## 5.4 Metabolic Failure

This failure mode results most frequently from inadequate delivery of metabolic substrates and inadequate clearance of metabolic byproducts, and is exacerbated at elevated temperatures. This failure mode can thus be equated at some level with the current inability to adequately engineer and perfuse vascular structures within in vitro engineered skeletal muscle. In the absence of vascular perfusion, the best countermeasure for this failure mode is to restrict the size of the muscle construct to remain below the $V_r$ (viable radius), where viable radius is defined as the maximum radius of a cylindrical muscle construct that can be achieved where living cells occupy the full depth of the tissue construct (Fig. 7). Though it is possible to generate much larger diameter tissue constructs, in these tissues the viable cells (fibroblasts, myocytes, myotubes, and fibers) will all be located in an annulus on the periphery of the tissue construct, surrounding a necrotic, noncontractile core of cellular debris and extracellular matrix [11, 148]. In static culture of 3-dimensional muscle in vitro, the viable radius is generally limited to about 40–50 µm for mammalian cardiac [23] and ~150 µm for mammalian skeletal muscle constructs [11, 83, 133, 148]. For this reason, muscle constructs larger than these grown in static culture in vitro will suffer metabolic failure at the core and have reduced peak contractile specific force (sPo), which is the peak contractile force normalized by dividing by the total cross sectional area of the tissue construct, in units of $kN/m^2$ or kPa [11]. This mode of failure typically initiates at the axial core of cylindrical muscle actuators [3]. For this reason, sustained angiogenesis and perfusion is a major technical objective in current tissue engineering research.

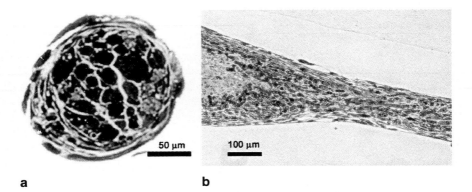

**Fig. 7** Cross section of an engineered skeletal muscle (*left*) showing an annulus of fibroblasts surrounding a core of viable, contractile skeletal muscle myotubes (adapted from [3]). In this case the tissue radius is <$V_r$ for skeletal muscle, ~150 µm. On the *right* is shown a longitudinal section of engineered cardiac muscle tissue through the middle of the tissue construct. The radius of the roughly cylindrical construct decreases from left to right. At the *extreme left*, the core of the tissue is disorganized and filled with nonviable, noncontractile cellular debris. As the radius of the construct approaches ~50 µm (to the right) the necrotic core disappears and viable cells are present throughout the tissue cross section

## 5.5 Cellular Necrosis and Cell Death

Other than metabolic failure induced necrosis, the second most common reason for this failure type is cellular hypercontraction and hyperextension in the muscle construct resulting in contraction-induced injury and in extreme cases in rapid necrosis [310, 315–317]. This mechanism can occur more or less uniformly across the muscle cross section, but will theoretically occur more frequently in areas with reduced physiologic cross-sectional area or inhibited sarcomeric function. These mechanisms are not necessarily due to dynamic mechanical interventions. Muscle maintained at an inappropriate length, either too short or too long, will deteriorate more rapidly, even if the muscle is quiescent. In explanted muscles, maintenance at lengths greater than the plateau of the length-tension curve appears to be the most damaging over time [318]. The occurrence of these failure mechanisms can be minimized in muscle bioreactors by use of nondestructive biomarkers of development, also known as nondestructible tissue biomarkers and the automated monitoring of the health of each muscle.

## 5.6 Toxic Contamination

A serious problem for all in vitro engineered muscle tissues is exposure to toxic agents. The best countermeasure for each of the three major sources of toxins is barrier exclusion of external toxic agents, the use of biocompatible materials in the media fluid space of the bioreactor, and the clearance of toxic metabolic byproducts via a perfusion and filtration system integrated with the bioreactor. The identification of which materials are biocompatible and which are not is somewhat problematic because this depends upon many variables including the cell types in question, the required level of compatibility, the amount and purity of the materials themselves, and several other factors.

There are many examples of where problems arising from toxic contamination of the bioreactor can occur. One example is the residual presence of sterilizing agents such as ethanol. Another is outgassing or leaching of chemicals from otherwise biocompatible materials [319]. This is sometimes the case with Delrin (polyoxymethylene) a commonly used material in medical implants and bioreactors which contain trace amounts of formaldehyde which can interfere with cell cultures unless properly treated prior to use [319, 320]. Also, the biocompatibility of many widely used biocompatible materials changes with repeated use and exposure time to culture media [319], so with the increasing duration of culture times of engineered muscles these considerations gain importance.

## 5.7 Electrochemical Tissue Damage

This failure mode affects muscle tissue under any conditions (in vivo or ex vivo) where electrical stimulation is applied. The single best countermeasure in all cases is to promote and maintain tissue phenotype exhibiting very high excitability, thus minimizing the required electrical excitation energy by up to three orders of magnitude [28]. In addition to vastly improving the excitation efficiency of the tissue, adult muscle phenotype excitability can yield as much as a 99.9% reduction in electrical pulse energy requirements for any given level of muscle activation, when compared with chronically denervated or tissue-engineered muscle tissue at early developmental stages [28]. For this reason, the development of electro-mechanical muscle bioreactor systems and maintenance stimulation protocols form a core component of all current research on muscle tissue engineering. Additional countermeasures include the selection of appropriate electrode materials, the use of minimally energetic stimulation protocols [47–50, 266], the use of pure bipolar stimulation pulses with careful attention to charge balancing [47, 48], and the use of high impedance outputs to the electrodes when not stimulating [47].

## 5.8 Operator Error

Ideally, a large number of semi automated bioreactor systems could be monitored by one or only a few individuals with a modest level of skill. To the extent that the use of muscle tissue bioreactors remains a labor intensive task carried out only by highly skilled individuals it will remain impractical to consider industrial-scaleup of bioreactor capacity. For some applications, such as basic research, this limitation is less important, whereas for others such as large-scale drug screening or the production of animal-based agricultural meat products in bioreactors this limitation renders the application both logistically and financially infeasible. But in any application, the increased frequency of operator error will lead to increased inefficiency and cost of operation. Semi-or fully-automated systems will only partially alleviate this problem, because at present most of the errors seem to occur during system setup: cleaning and disinfecting, harvesting and introduction of primary myogenic precursor cells and scaffolds, etc. To address this failure mode effectively will almost certainly require a dramatic change in the basic techniques currently employed for cell isolation, purification, and cell culture.

## 6 Conclusions and Future Directions

If one studied only the "recent" advances in muscle tissue engineering bioreactors from about the last decade, one might conclude that the advances in this area have been rapid, as have been recent advances in the basic understanding of the molecular biology of muscle development. However, a careful study of the history of tissue and organ bioreactor development clearly shows this not to be the case. As new technologies have become available, such as microfabrication, new and better imaging technologies, exponential increases in computing power, improved techniques and materials for basic biological research, and the availability of new biocompatible materials, these have been employed in the design of new muscle tissue bioreactor systems. Certainly, as a result there has been some advancement in our scientific understanding of muscle tissue development, but it is fair to say that this advancement has been at best incremental. There is no indication that any recent attempts to develop a muscle tissue bioreactor system have been systematic, comprehensive and successful in terms of achieving a major and lasting advance in the technology.

## 7 Key Questions and Technical Requirements

*Whole organ maintenance ex vivo*—What essential element is missing from our understanding that would allow us to maintain whole organs ex vivo indefinitely?

*Signals required for muscle development in vivo*—What are the essential signals required to drive muscle development from isolated myogenic precursor cells through to adult phenotype whole muscle organs?

*Emulation of the essential developmental signals*—How can each of these essential signals be duplicated in a satisfactory manner ex vivo?

*Tissue-to-tissue interfaces*—How can the essential tissue-to-tissue interfaces be engineered? As a minimum these will include vascular, muscle-tendon, and perhaps also nerve-muscle.

*Tissue-to-synthetic interfaces*—How can the essential tissue-to-synthetic interfaces be engineered? These will include connection of the vascular network to a perfusion system, the mechanical connection between tendon and machine, and a means to elicit active contractions in the muscle, either directly by electrodes or via a functional nerve-muscle interface.

*Co-culture technology*—How can diverse co-cultures be maintained in vitro? How can co-cultures of myogenic precursor cells be independently guided to form complex tissues in the same bioreactor compartment?

*Nondestructively testable biomarkers of development*—What are the nondestructively testable biomarkers that can be used to monitor the phenotype of the developing muscle in vitro? These must include as a minimum contractility, excitability, and metabolism.

*Sensors*—What existing sensor and actuator technologies exist, and which new technologies must be developed, in order to allow feedback control of muscle development in vitro using NDTB as the plant output? The sensors must be nontoxic, durable under prolonged implantation or cell culture conditions, low power, very low cost (essentially disposable), mass producible, cell culture compatible, and preferably noninvasive.

*Feedback control*—How do we effectively employ the NDTB as markers of muscle development within a feedback-controlled bioreactor system? It is unlikely that one set of input stimulus parameters will guide a muscle from the early stages of development through to an adult phenotype. Muscle bioreactor systems must be capable of monitoring development nondestructively and applying dynamic changes to setpoints for several controlled variables in order to guide the developing muscle to the desired phenotype.

*Usability*—How can systems be built that are user friendly enough to gain general acceptance?

*Reliability and cost*—What manufacturing processes can be employed (and what new processes might need to be developed) that will allow a very high level of bioreactor system integration to vastly improve reliability while reducing cost?

*Radical change in basic bioreactor system architecture*—Many tissue culture bioreactor systems have reached a point at which increased complexity or extension of the period of use is no longer practical due to the many failure modes and the correspondingly high probability of system failure. This is particularly true for perfusion systems, bioreactors that contain multiple tissue specimens, and systems with many sensors and actuators that protrude into the aseptic spaces of the bioreactor.

*Logical development sequence*—What interim strategies are most advantageous? Should implantable or hybrid bioreactor systems be developed first?

**Acknowledgements** KB and AP are supported by a grant from the Engineering and Physical Sciences Research Council (EPSRC—EP/E008925/1). KB and KD are supported by a grant from the Biotechnology and Biological Sciences Research Council (BBSRC—BB/F002084/1).

# References

1. PC, Mikos AG, Fisher JP, Jansen JA (2007) Tissue Eng 13:2827
2. Baar K, Birla R, Boluyt MO, Borschel GH, Arruda EM, Dennis RG (2005) FASEB J 19:275
3. Dennis RG, Kosnik PE (2000) In Vitro Cell Dev Biol Anim 36:327
4. Dennis RG, Kosnik PE, Gilbert ME, Faulkner JA (2001) Am J Physiol Cell Physiol 280:288
5. Hecker L, Baar K, Dennis RG, Bitar KN (2005) Am J Physiol Gastrointest Liver Physiol 289(2):G188–G196
6. Huang YC, Dennis RG, Baar K (2006) Am J Physiol Cell Physiol 291:C11
7. Kosnik, Paul E., Ph.D. Doctoral thesis: Contractile Properties of Engineered Skeletal Muscle, completed 2000, University of Michigan
8. Kosnik PE, Ph.D. Faulkner JA, Dennis RG (2001) Tissue Eng 7:573
9. McFarland DC (1992) J Nutr 122:818
10. Cox GA, Cole NM, Matsumura K, Phelps SF, Hauschka SD., Campbell KP et al. (1993). Overexpression of Dystrophin in Transgenic Mdx Mice Eliminates Dystrophic Symptoms Without Toxicity. Nature 364:725–729
11. Del Tatto M, Ferland P, Shansky J, Vandenburgh H (2000) FASEB J 14:A445
12. Lu Y, Shansky J, Del Tatto M, Ferland P, McGuire S, Marszalkowski J, Maish M, Hopkins R, Wang X, Kosnik P, Nackman M, Lee A, Creswick B, Vandenburgh H (2002) Ann N Y Acad Sci 961:78
13. Lu YX, Shansky J, Del Tatto M, Ferland P, Wang XY, Vandenburgh H (2001) Circulation 104:594
14. Payumo FC, Kim HD, Sherling MA, Smith LP, Powell C, Wang X, Keeping HS, Valentini RF, Vandenburgh HH (2002) Clin Orthop Relat Res S228

15. Powell C, Shansky J, Del TM, Forman DE, Hennessey J, Sullivan K, Zielinski BA, Vandenburgh HH (1999) Hum Gene Ther 10:565
16. Powell C, Shansky J, Del TM, Vandenburgh HH (2002) Methods Mol Med 69:219
17. Powell C, Shansky J, DelTatto M, Vandenburgh H (1997) Mol Biol Cell 8:2003
18. Sakai T, Ling Y, Payne TR, Huard J (2002) Trends Cardiovasc Med 12:115
19. Vandenburgh H, Del Tatto M, Shansky J, Goldstein L, Russell K, Genes N, Chromiak J, Yamada S (1998) Hum Gene Ther 9:2555
20. Vandenburgh H, Del TM, Shansky J, Lemaire J, Chang A, Payumo F, Lee P, Goodyear A, Raven L (1996) Hum Gene Ther 7:2195
21. Dennis RG, Herr H (2005) Engineered muscle actuators: cells and tissues. In: Y. Bar Cohen (ed.) Biomimetics: biologically inspired technologies, chap. 9. CRC Press, New York, NY, p. 234
22. Herr H, Dennis RG (2004) J Neuroeng Rehabil 1:6
23. Bertrand A, Ngo Muller V, Hentzen D, Concordet JP, Daegelen D, Tuil D (2003) Am J Physiol Cell Physiol 285:C1071
24. Smith LC, Nordstrom JL (2000) Curr Opin Mol Ther 2:150
25. Edelman PD, McFarland DC, Mironov VA, Matheny JG (2005) Tissue Eng 11:659
26. Heselmans M. Cultivated meat: the Dutch cultivate minced meat in a petri dish. NRC Handelsblad. 9–10–2005
27. Dickinson MH, Farley CT, Full RJ, Koehl MAR, Kram R, Lehman S (2000) Science 288:100
28. Dennis RG, Dow DE (2007) Tissue Eng 13:2395
29. Crameri RM, Langberg H, Magnusson P, Jensen CH, Schroder HD, Olesen JL, Suetta C, Teisner B, Kjaer M (2004) J Physiol 558:333
30. Reid J (1841) Lond Edinb Mon J Med Sci 1:320
31. Crameri RM, Aagaard P, Qvortrup K, Langberg H, Olesen J, Kjaer M (2007) J Physiol 583:365
32. Putman CT, Sultan KR, Wassmer T, Bamford JA, Skorjanc D, Pette D (2001) J Gerontol A Biol Sci Med Sci 56:B510
33. Tatsumi R, Liu X, Pulido A, Morales M, Sakata T, Dial S, Hattori A, Ikeuchi Y, Allen RE (2006) Am J Physiol Cell Physiol 290:C1487
34. Wozniak AC, Kong J, Bock E, Pilipowicz O, Anderson JE (2005) Muscle Nerve 31:283
35. Birla RK, Borschel GH, Dennis RG (2005) Artif Organs 29:866
36. Birla RK, Borschel GH, Dennis RG, Brown DL (2005) Tissue Eng 11:803
37. Gamble JG, Edwards CC, Max SR (1984) Am J Sports Med 12:221
38. Jarvinen TAH, Jozsa L, Kannus P, Jarvinen TLN, Jarvinen M (2002) J Muscle Res Cell Motil 23:245
39. Kannus P, Jozsa L, Kvist M, Lehto M, Jarvinen M (1992) Acta Physiol Scand 144:387
40. Kitahara A, Hamaoka T, Murase N, Homma T, Kurosawa Y, Ueda C, Nagasawa T, Ichimura S, Motobe M, Yashiro K, Nakano S, Katsumura T (2003) Med Sci Sports Exerc 35:1697
41. Picquet F, Stevens L, Butler Browne GS, Mounier Y (1998) J Muscle Res Cell Motil 19:743
42. Veldhuizen JW, Verstappen FT, Vroemen JP, Kuipers H, Greep JM (1993) Int J Sports Med 14:283
43. Blottner D, Salanova M, Puttmann B, Schiffl G, Felsenberg D, Buehring B, Rittweger J (2006) Eur J Appl Physiol 97:261
44. Shavlakadze T, Grounds M (2006) Bioessays 28:994
45. Galvani L (1791) De Viribus Electricitatis in Motu Musculari Commentarius. Academy of Science, Bologna, Italy
46. Lobb H (1862) Proc R Soc Lond 12:650
47. Dennis RG, Dow DE, Faulkner JA (2003) Med Eng Phys 25:239
48. Dow DE, Carlson BM, Hassett CA, Dennis RG, Faulkner JA (2006) Restor Neurol Neurosci 24:41
49. Dow DE, Dennis RG, Faulkner JA (2005) J Gerontol A Biol Sci Med Sci 60:416
50. Dow DE, Faulkner JA, Dennis RG (2005) Artif Organs 29:432

51. Kostrominova TY, Dow DE, Dennis RG, Miller RA, Faulkner JA (2005) Physiol Genomics 22:227
52. Florini JR, Ewton DZ, Coolican SA (1996) Endocr Rev 17:481
53. Bartlett RH (2004) J Am Coll Surg 199:286
54. Boulay C, Hardy MA (2001) Curr Surg 58:303
55. Shelley M (originally published anonymously) (1818) Frankenstein, or, The Modern Prometheus, 1 edn. Harding, Mavor & Jones, London
56. Editorial column by "Caveman", contact information: caveman@biologists.com (2002) J Cell Sci 115:3223
57. Finger S, Law MB (1998) J Hist Med Allied Sci 53:161
58. Wolpert L (2005) Philos Trans R Soc Lond B Biol Sci 360:1253
59. Campbell CS (2003) Camb Q Healthc Ethics 12:342
60. Davies H (2004) Med Humanit 30:32
61. Friedman LD (1985) Med Herit 1:181
62. Guinan P (2002) J Relig Health 41:305
63. Holm RP (2006) Pharos Alpha Omega Alpha Honor Med Soc 69:30
64. Kaplan PW (2004) J Clin Neurophysiol 21:301
65. Lawton A (1998) KY Law J 87:277
66. Morowitz HJ (1979) Hosp Pract 14:175
67. Rollin BE (1986) Basic Life Sci 37:285
68. Roth N (1978) Med Instrum 12:248
69. Westra L, Dandekar N, Zlotkowski E (1992) Between Species 8:216
70. Buller AJ, ECCLES JC, ECCLES RM (1960) J Physiol 150:417
71. Kirschbaum BJ, Heilig A, Hartner KT, Pette D (1989) FEBS Lett 243:123
72. Kirschbaum BJ, Pette D (1989) J Muscle Res Cell Motil 10:160
73. Kirschbaum BJ, Schneider S, Izumo S, Mahdavi V, Nadalginard B, Pette D (1990) FEBS Lett 268:75
74. Kirschbaum BJ, Simoneau JA, Bar A, Barton PJR, Buckingham ME, Pette D (1989) Eur J Biochem 179:23
75. Salmons S, Sreter FA (1976) Nature 263:30
76. Sreter FA, Gergely J (1974) Biochem Biophys Res Commun 56:84
77. Sreter FA, Luff AR, Gergely J (1975) J Gen Physiol 66:811
78. Jolesz F, Sreter FA (1981) Annu Rev Physiol 43:531
79. Mabuchi K, Szvetko D, Pinter K, Sreter FA (1982) Am J Physiol 242:C373
80. Pluskal MG, Sreter FA (1983) Biochem Biophys Res Commun 113:325
81. Salmons S, Gale DR, Sreter FA (1978) J Anat 127:17
82. Sreter FA, Pinter K, Jolesz F, Mabuchi K (1982) Exp Neurol 75:95
83. Dennis RG (1998) Med Biol Eng Comput 36:225
84. Faulkner JA, Carlson BM, Kadhiresan VA (1994) Biotechnol Bioeng 43:757
85. Gustafson KJ, Sweeney JD, Gibney J, Fiebig-Mathine LA (2006) J Surg Res 134:198
86. Kadhiresan VA, Guelinckx PJ, Faulkner JA (1993) J Appl Physiol 75:1294
87. Timek T, Ihnken K (2007) Expert Rev Cardiovasc Ther 5:251
88. Konsten J, Geerdes B, Baeten CGMI, Heineman E, Arends JW, Pette D, Soeters PB (1995) J Pediatr Surg 30:580
89. Naumann K, Pette D (1993) J Muscle Res Cell Motil 14:250
90. Naumann K, Pette D (1994) Differentiation 55:203
91. Lewis MR (1915) Am J Physiol 38:153
92. Lewis MR (1920) Carnegie Inst Washington Contrib Embryol 9:192
93. Lewis WH, Lewis MR (1917) Am J Anat 22:169
94. Carrel A, Lindbergh CA (1935) Science 81:621
95. Carrel A, Lindbergh CA (1938) The culture of organs. P. Hoeber Inc., New York
96. Lindbergh CA, Perry VP, Malinin TI, Mouer GH (1966) Cryobiology 3:252
97. Chen HC, Hu YC (2006) Biotechnol Lett 28:1415

98. Cooper JA Jr, Li WJ, Bailey LO, Hudson SD, Lin-Gibson S, Anseth KS, Tuan RS, Washburn NR (2007) Acta Biomater 3:13
99. Hahn MS, McHale MK, Wang E, Schmedlen RH, West JL (2007) Ann Biomed Eng 35:190
100. McFetridge PS, Abe K, Horrocks M, Chaudhuri JB (2007) ASAIO J 53:623
101. Morsi YS, Yang WW, Owida A, Wong CS (2007) J Artif Organs 10:109
102. Thompson CA, Colon-Hernandez P, Pomerantseva I, MacNeil BD, Nasseri B, Vacanti JP, Oesterle SN (2002) Tissue Eng 8:1083
103. Xu ZC, Zhang WJ, Li H, Cui L, Cen L, Zhou GD, Liu W, Cao Y (2007) Biomaterials 29(2):1464–1472
104. Hunziker E, Spector M, Libera J, Gertzman A, Woo SL, Ratcliffe A, Lysaght M, Coury A, Kaplan D, Vunjak-Novakovic G (2006) Tissue Eng 12:3341
105. Lysaght MJ, Hazlehurst AL (2004) Tissue Eng 10:309
106. Lysaght MJ, Nguy NA, Sullivan K (1998) Tissue Eng 4:231
107. Lysaght MJ, Reyes J (2001) Tissue Eng 7:485
108. Huber TS, Groh MA, Gallagher KP, D, Alecy LG (1993) Crit Care Med 21:1731
109. Powner DJ, Hendrich A, Lagler RG, Ng RH, Madden RL (1990) Crit Care Med 18:702
110. Powner DJ, Hendrich A, Nyhuis A, Strate R (1992) J Heart Lung Transplant 11:1046
111. Portner R, Nagel-Heyer S, Goepfert C, Adamietz P, Meenen NM (2005) J Biosci Bioeng 100:235
112. Shachar M, Cohen S (2003) Heart Fail Rev 8:271
113. Bach AD, Beier JP, Stern-Staeter J, Horch RE (2004) J Cell Mol Med 8:413
114. Baumert H, Simon P, Hekmati M, Fromont G, Levy M, Balaton A, Molinie V, Malavaud B (2007) Eur Urol 52:884
115. Borschel GH, Dow DE, Dennis RG, Brown DL (2006) Plast Reconstr Surg 117:2235
116. Dhawan V, Lytle IF, Dow DE, Huang YC, Brown DL (2007) Tissue Eng 13(11):2813–2821
117. Warnke PH, Wiltfang J, Springer I, Acil Y, Bolte H, Kosmahl M, Russo PA, Sherry E, Lutzen U, Wolfart S, Terheyden H (2006) Biomaterials 27:3163
118. Okano T, Matsuda T (1998) Cell Transplant 7:435
119. Banes AJ, Gilbert J, Taylor D, Monbureau O (1985) J Cell Sci 75:35
120. Abousleiman RI, Sikavitsas VI (2006) Adv Exp Med Biol 585:243
121. Lanza R, Langer R, Vacanti J (2007) Principles of tissue engineering, 3 edn. Elsevier Academic Press, Burlington, MA
122. Samuel JL, Vandenburgh HH (1990) In Vitro Cell Dev Biol 26:905
123. Vandenburgh HH (1987) In Vitro Cell Dev Biol 23:A24
124. Vandenburgh HH (1983) J Cell Physiol 116:363
125. Vandenburgh HH (1992) Am J Physiol 262:R350
126. Vandenburgh HH (1982) Dev Biol 93:438
127. Vandenburgh HH (1988) In Vitro Cell Dev Biol 24:609
128. Vandenburgh HH, Hatfaludy S, Karlisch P, Shansky J (1989) Am J Physiol 256:C674
129. Vandenburgh HH, Karlisch P, Farr L (1987) In Vitro Cell Dev Biol 23:A24
130. Vandenburgh HH, Sheff ME, Zacks SI (1982) Fed Proc 41:384
131. Vandenburgh HH, Swasdison S, Karlisch P (1991) FASEB J 5:2860
132. Ge X, Hanson M, Shen H, Kostov Y, Brorson KA, Frey DD, Moreira AR, Rao G (2006) J Biotechnol 122:293
133. Harms P, Kostov Y, French JA, Soliman M, Anjanappa M, Ram A, Rao G (2006) Biotechnol Bioeng 93:6
134. Kostov Y, Harms P, Randers-Eichhorn L, Rao G (2001) Biotechnol Bioeng 72:346
135. Baar K, Torgan CE, Kraus WE, Esser K (2000) Mol Cell Biol Res Commun 4:76
136. Birla RK, Huang YC, Dennis RG (2007) Tissue Eng 13:2239
137. Bursac N, Papadaki M, Cohen RJ, Schoen FJ, Eisenberg SR, Carrier R, Vunjak-Novakovic G, Freed LE (1999) Am J Physiol 277:433
138. Chromiak JA, Shansky J, Perrone C, Vandenburgh HH (1998) In Vitro Cell Dev Biol Anim 34:694

139. Chromiak JA, Vandenburgh HH (1992) Am J Physiol 262:C1471
140. Chromiak JA, Vandenburgh HH (1994) J Cell Physiol 159:407
141. Eschenhagen T, Didie M, Heubach J, Ravens U, Zimmermann WH (2002) Transpl Immunol 9:315
142. Eschenhagen T, Didie M, Munzel F, Schubert P, Schneiderbanger K, Zimmermann WH (2002) Basic Res Cardiol 97 Suppl 1:146
143. Eschenhagen T, Fink C, Remmers U, Scholz H, Wattchow J, Weil J, Zimmerman W, Dohmen HH, Schafer H, Bishopric N, Wakatsuki T, Elson EL (1997) FASEB J 11:683
144. Fink C, Ergun S, Kralisch D, Remmers U, Weil J, Eschenhagen T (2000) FASEB J 14:669
145. Freed LE, Guilak F, Guo XE, Gray ML, Tranquillo R, Holmes JW, Radisic M, Sefton MV, Kaplan D, Vunjak-Novakovic G (2006) Tissue Eng 12:3285
146. Fuchs JR, Pomerantseva I, Ochoa ER, Vacanti JP, Fauza DO (2003) J Pediatr Surg 38:1348
147. Guilak F, Kapur R, Sefton MV, Vandenburgh HH, Koretsky AP, Kriete A, O'Keefe RJ (2002) Ann N Y Acad Sci 961:207
148. Hatfaludy S, Shansky J, Smiley B, Vandenburgh HH (1990) J Muscle Res Cell Motil 11:76
149. Hatfaludy S, Shansky J, Vandenburgh HH (1989) Am J Physiol 256:C175
150. Hecker L, Birla RK (2007) Regen Med 2:125
151. Huang YC, Dennis RG, Larkin L, Baar K (2005) J Appl Physiol 98:706
152. Kofidis T, Akhyari P, Boublik J, Theodorou P, Martin U, Ruhparwar A, Fischer S, Eschenhagen T, Kubis HP, Kraft T, Leyh R, Haverich A (2002) J Thorac Cardiovasc Surg 124:63
153. Powell CA, Smiley BL, Mills J, Vandenburgh HH (2002) Am J Physiol Cell Physiol 283:1557
154. Powell CA, Smiley BL, Vandenburgh HH (2000) FASEB J 14:A444
155. Shansky J, Chromiak J, Vandenburgh HH (1995) Mol Biol Cell 6:2045
156. Shansky J, Del TM, Chromiak J, Vandenburgh H (1997) In Vitro Cell Dev Biol Anim 33:659
157. Vandenburgh H, Kaufman S (1979) Science 203:265
158. Vandenburgh H, Kaufman S (1980) J Biol Chem 255:5826
159. Vandenburgh HH (2002) Ann N Y Acad Sci 961:201
160. Vandenburgh HH (1987) Med Sci Sports Exerc 19:S142
161. Vandenburgh HH, Hatfaludy S, Karlisch P, Shansky J (1991) J Biomech 24 Suppl 1:91
162. Vandenburgh HH, Hatfaludy S, Sohar I, Shansky J (1990) Am J Physiol 259:C232
163. Vandenburgh HH, Karlisch P (1989) In Vitro Cell Dev Biol 25:607
164. Vandenburgh HH, Karlisch P, Farr L (1988) In Vitro Cell Dev Biol 24:166
165. Vandenburgh HH, Kaufman S (1981) J Cell Physiol 109:205
166. Vandenburgh HH, Shansky J, Karlisch P, Solerssi RL (1993) J Cell Physiol 155:63
167. Vandenburgh HH, Shansky J, Solerssi R (1992) Mol Biol Cell 3:A244
168. Vandenburgh HH, Shansky J, Solerssi R, Chromiak J (1995) J Cell Physiol 163:285
169. Vandenburgh HH, Solerssi R, Shansky J, Adams JW, Henderson SA (1996) Am J Physiol 270:C1284
170. Vandenburgh HH, Solerssi R, Shansky J, Adams JW, Henderson SA, Lemaire J (1995) Ann N Y Acad Sci 752:19
171. Vandenburgh HH, Solerssi R, Shansky J, Adams JW, Henderson SA, Lemaire J (1995) Ann N Y Acad Sci 752:19
172. Vandenburgh HH, Solerssi R, Shansky J, Adams JW, Henderson SA, Lemaire J (1995) Cardiac Growth Regen 752:19
173. Zimmermann WH, Didie M, Weyand M, Eschenhagen T (2001) Circulation 104:600
174. Zimmermann WH, Fink C, Kralisch D, Remmers U, Weil J, Eschenhagen T (2000) Biotechnol Bioeng 68:106
175. Zimmermann WH, Schneiderbanger K, Schubert P, Didie M, El Armouche A, Eschenhagen T (2001) Circulation 104:129
176. Zimmermann WH, Schneiderbanger K, Schubert P, Didie M, Munzel F, Heubach JF, Kostin S, Neuhuber WL, Eschenhagen T (2002) Circ Res 90:223

177. Strohman RC, Bayne E, Spector D, Obinata T, Micou-Eastwood J, Maniotis A (1990) In Vitro Cell Dev Biol 26:201
178. Gopalan SM, Flaim C, Bhatia SN, Hoshijima M, Knoell R, Chien KR, Omens JH, McCulloch AD (2003) Biotechnol Bioeng 81:578
179. McDevitt TC, Angello JC, Whitney ML, Reinecke H, Hauschka SD, Murry CE, Stayton PS (2002) J Biomed Mater Res 60:472
180. Collinsworth AM, Torgan CE, Nagda SN, Rajalingam RJ, Kraus WE, Truskey GA (2000) Cell Tissue Res 302:243
181. Okano T, Matsuda T (1998) Cell Transplant 7:71
182. Okano T, Satoh S, Oka T, Matsuda T (1997) ASAIO J 43:M749
183. Okano T, Yamada N, Okuhara M, Sakai H, Sakurai Y (1995) Biomaterials 16:297
184. Okano T, Yamada N, Sakai H, Sakurai Y (1993) J Biomed Mater Res 27:1243
185. Shimizu T, Yamato M, Akutsu T, Shibata T, Isoi Y, Kikuchi A, Umezu M, Okano T (2002) J Biomed Mater Res 60:110
186. Shimizu T, Yamato M, Isoi Y, Akutsu T, Setomaru T, Abe K, Kikuchi A, Umezu M, Okano T (2002) Circ Res 90:e40
187. Shimizu T, Yamato M, Kikuchi A, Okano T (2001) Tissue Eng 7:141
188. Boontheekul T, Hill EE, Kong HJ, Mooney DJ (2007) Tissue Eng 13:1431
189. Cunningham JJ, Linderman JJ, Mooney DJ (2002) Ann Biomed Eng 30:927
190. Figallo E, Cannizzaro C, Gerecht S, Burdick JA, Langer R, Elvassore N, Vunjak-Novakovic G (2007) Lab Chip 7:710
191. Hornberger TA, Armstrong DD, Koh TJ, Burkholder TJ, Esser KA (2005) Am J Physiol Cell Physiol 288:C185
192. Schaffer JL, Rizen M, L'Italien GJ, Benbrahim A, Megerman J, Gerstenfeld LC, Gray ML (1994) J Orthop Res 12:709
193. Finni T, Hodgson JA, Lai AM, Edgerton VR, Sinha S (2003) J Appl Physiol 95:829
194. Finni T, Hodgson JA, Lai AM, Edgerton VR, Sinha S (2003) J Appl Physiol 95:2128
195. Hodgson JA, Finni T, Lai AM, Edgerton VR, Sinha S (2006) J Morphol 267:584
196. Clark CB, Burkholder T.J., Frangos J.A (2001) Rev Sci Instrum 72:2415
197. Clarke MS, Feeback DL (1996) FASEB J 10:502
198. Gilbert JA, Weinhold PS, Banes AJ, Link GW, Jones GL (1994) J Biomech 27:1169
199. Williams JL, Chen JH, Belloli DM (1992) J Biomech Eng 114:377
200. Wozniak AC, Kong J, Bock E, Pilipowicz O, Anderson JE (2005) Muscle Nerve 31:283
201. Wozniak AC, Pilipowicz O, Yablonka-Reuveni Z, Greenway S, Craven S, Scott E, Anderson JE (2003) J Histochem Cytochem 51:1437
202. Feinberg AW, Feigel A, Shevkoplyas SS, Sheehy S, Whitesides GM, Parker KK (2007) Science 317:1366
203. Borschel GH, Dennis RG, Kuzon WM Jr (2004) Plast Reconstr Surg 113:595
204. Drury JL, Dennis RG, Mooney DJ (2004) Biomaterials 25:3187
205. Hill E, Boontheekul T, Mooney DJ (2006) Tissue Eng 12:1295
206. Kim BS, Putnam AJ, Kulik TJ, Mooney DJ (1998) Biotechnol Bioeng 57:46
207. Lee SH, Kim BS, Kim SH, Choi SW, Jeong SI, Kwon IK, Kang SW, Nikolovski J, Mooney DJ, Han YK, Kim YH (2003) J Biomed Mater Res A 66:29
208. Gao FG, Fay JM, Mathew G, Jeevarajan AS, Anderson MM (2005) J Biomed Opt 10:054005
209. Ingram M, Techy GB, Saroufeem R, Yazan O, Narayan KS, Goodwin TJ, Spaulding GF (1997) In Vitro Cell Dev Biol Anim 33:459
210. Tsao YD, Goodwin TJ, Wolf DA, Spaulding GF (1992) Physiologist 35:S49
211. Freed LE, Vunjak-Novakovic G (2002) Adv Space Biol Med 8:177
212. Papadaki M, Bursac N, Langer R, Merok J, Vunjak-Novakovic G, Freed LE (2001) Am J Physiol Heart Circ Physiol 280:168
213. Vunjak-Novakovic G, Freed LE (1998) Adv Drug Deliv Rev 33:15
214. Gonen-Wadmany M, Gepstein L, Seliktar D (2004) Ann N Y Acad Sci 1015:299
215. Calve S, Dennis RG, Kosnik PE, Baar K, Grosh K, Arruda EM (2004) Tissue Eng 10:755

216. Dennis RG, Kosnik PE (2002) Mesenchymal cell culture: instrumentation and methods for evaluating engineered muscle. In: Atala A, Lanza R (eds.) Methods in tissue engineering, chap. 24. Harcourt, Academic Press, San Diego, p. 307
217. Kosnik PE, Dennis RG (2002) Mesenchymal cell culture: functional mammalian skeletal muscle constructs. In: Atala A, Lanza R (eds.) Methods in tissue engineering, chap. 23. Harcourt, Academic Press, San Diego, p. 299
218. Swasdison S, Mayne R (1992) J Cell Sci 102 (Pt 3):643
219. Swasdison S, Mayne R (1991) Exp Cell Res 193:227
220. Swasdison S, Mayne R (1989) Cell Tissue Res 257:537
221. Larkin LM, Calve S, Kostrominova TY, Arruda EM (2006) Tissue Eng 12(11):3149–3158
222. Mittmann C, Eschenhagen T, Munstermann U, Ptak M, Scholz H, Weil J, Zimmermann WH (1997) Naunyn Schmiedebergs Arch Pharmacol 355:356
223. Zimmermann WH, Eschenhagen T, Remmers U, Scholtz H, Wattchow J, Weil J, Wakatsuki T, Elson EL (1996) Circulation 94:1801
224. Zimmermann WH, Schubert P, Schneidberbanger K, Endress B, Hilpert A, Eschenhagen T (2001) Circulation 104:214
225. Dusterhoft S, Pette D (1990) Differentiation 44:178
226. Wehrle U, Dusterhoft S, Pette D (1994) Differentiation 58:37
227. Bayol S, Brownson C, Loughna PT (2005) Cell Biochem Funct 23:361
228. Cooper ST, Maxwell AL, Kizana E, Ghoddusi M, Hardeman EC, Alexander IE, Allen DG, North KN (2004) Cell Motil Cytoskeleton 58:200
229. Kubis HP, Scheibe RJ, Meissner JD, Hornung G, Gros G (2002) J Physiol 541:835
230. Su CT, Huang CF, Schmidt J (1995) FEBS Lett 366:131
231. Connor MK, Irrcher I, Hood DA (2001) J Biol Chem 276:15898
232. Hood DA, Simoneau JA, Kelly AM, Pette D (1992) Am J Physiol 263:C788
233. Irrcher I, Adhihetty PJ, Sheehan T, Joseph AM, Hood DA (2003) Am J Physiol Cell Physiol 284:C1669
234. Irrcher I, Hood DA (2004) J Appl Physiol 97:2207
235. Freyssenet D, Connor MK, Takahashi M, Hood DA (1999) Am J Physiol 277:E26
236. Macpherson PC, Suhr ST, Goldman D (2004) J Cell Biochem 91:821
237. Thelen MH, Simonides WS, van HC (1997) Biochem J 321 (Pt 3):845
238. Valdes JA, Gaggero E, Hidalgo J, Leal N, Jaimonich E, (arrasco MA (2008) Am J Physiol Cell physiol 294(3):C715–C725
239. Brevet A, Pinto E, Peacock J, Stockdale FE (1976) Science 193:1152
240. Coleman AW, Siegel R, Coleman JR (1978) Tissue Cell 10:201
241. Green HJ, Dusterhoft S, Dux L, Pette D (1992) Pflugers Archiv-Eur J Physiol 420:359
242. Green HJ, Pette D (1997) Eur J Appl Physiol Occup Physiol 75:418
243. Hofmann S, Pette D (1994) Eur J Biochem 219:307
244. Klug GA, Reichmann H, Pette D (1983) Med Sci Sports Exerc 15:177
245. Kraus B, Pette D (1997) Eur J Biochem 247:98
246. Lanmuller H, Ashley Z, Unger E, Sutherland H, Reichel M, Russold M, Jarvis J, Mayr W, Salmons S (2005) Med Biol Eng Comput 43:535
247. Leeeuw T, Pette D (1993) Eur J Biochem 213:1039
248. Parra J, Pette D (1995) Biochim Biophys Acta 1251:154
249. Pette D (2002) Can J Appl Physiol 27:423
250. Pette D (2001) J Appl Physiol 90:1119
251. Pette D, Dusterhoft S (1992) Am J Physiol 262:R333
252. Pette D, Heilmann C (1977) Basic Res Cardiol 72:247
253. Pette D, Smith ME, Staudte HW, Vrbova G (1973) Pflugers Arch 338:257
254. Pette D, Staron RS (2001) Histochem Cell Biol 115:359
255. Pette D, Vrbova G (1992) Rev Physiol Biochem Pharmacol 120:115
256. Pette D, Vrbova G (1999) Muscle Nerve 22:666
257. Ramirez BU, Pette D (1974) FEBS Lett 49:188

258. Reichmann H, Hoppeler H, Mathieucostello O, Vonbergen F, Pette D (1985) Pflugers Arch 404:1
259. Reichmann H, Pette D, Vrbova G (1981) FEBS Lett 128:55
260. Reichmann H, Wasl R, Simoneau JA, Pette D (1991) Pflugers Arch 418:572
261. Salmons S, Ashley Z, Sutherland H, Russold MF, Li F, Jarvis JC (2005) Artif Organs 29:199
262. Schuler M, Pette D (1996) Cell Tissue Res 285:297
263. Simoneau JA, Kaufmann M, Pette D (1993) J Physiol Lond 460:573
264. Skorjanc D, Dunstl G, Pette D (2001) J Gerontol A Biol Sci Med Sci 56:B503
265. Skorjanc D, Jaschinski F, Heine G, Pette D (1998) Am J Physiol Cell Physiol 43:C810
266. Dow DE, Cederna PS, Hassett CA, Kostrominova TY, Faulkner JA, Dennis RG (2004) Muscle Nerve 30:77
267. Carrier RL, Papadaki M, Rupnick M, Schoen FJ, Bursac N, Langer R, Freed LE, Vunjak-Novakovic G (1999) Biotechnol Bioeng 64:580
268. Calve S. Ph.D. (2006) Morphological and mechanical characterization of self-assembling tendon constructs and myotendinous junctions. University of Michigan, Ann Arbor, MI
269. Shimko VF, Claycomb WC (2008) Tissue Eng 14(1):49–58
270. Dennis RG (1996) Ph.D. Measurement of Pulse Propagation in Single Permeabilized Muscle Fibers by Optical Diffraction. University of Michigan, Ann Arbor, MI
271. Dennis RG, Cole NM (1993) Biophys J 64:A256
272. Minns HG (1971) J Appl Physiol 30:895
273. Minns HG, Franz GN (1972) J Appl Physiol 33:529
274. Carrier RL, Papadaki M, Rupnick M, Schoen FJ, Bursac N, Langer R, Freed LE, Vunjak Novakovic G (1999) Biotechnol Bioeng 64:580
275. Carrier RL, Rupnick M, Langer R, Schoen FJ, Freed LE, Vunjak-Novakovic G (2002) Tissue Eng 8:175
276. Carrier RL, Rupnick M, Langer R, Schoen FJ, Freed LE, Vunjak-Novakovic G (2002) Biotechnol Bioeng 78:617
277. Kameneva MV, Burgreen GW, Kono K, Repko B, Antaki JF, Umezu M (2004) ASAIO J 50:418
278. Lemoff AV., Lee A.P. (2000) Sens Actuators B Chem 63:178
279. Walker GM, Beebe DJ (2002) Lab Chip 2:131
280. Akins RE, Schroedl NA, Gonda SR, Hartzell CR (1997) In Vitro Cell Devl Biol Anim 33:337
281. Molnar G, Schroedl NA, Gonda SR, Hartzell CR (1997) In Vitro Cell Dev Biol Anim 33:386
282. Villeneuve PE, Dunlop EH (1992) Adv Space Res 12:237
283. Lee SJ, McPherron AC (1999) Curr Opin Genet Dev 9:604
284. Lee SJ, McPherron AC (2001) Proc Natl Acad Sci USA 98:9306
285. Li L, Zhou J, James G, Heller Harrison R, Czech MP, Olson EN (1992) Cell 71:1181
286. Li M, Bernard O (1992) Proc Natl Acad Sci USA 89:3315
287. Deldicque L, Theisen D, Bertrand L, Hespel P, Hue L, Francaux M (2007) Am J Physiol Cell Physiol 293:C1263
288. Florini JR, Ewton DZ (1995) Growth Regul 5:28
289. Florini JR, Ewton DZ (1992) Growth Regul 2:23
290. Florini JR, Ewton DZ (1990) J Biol Chem 265:13435
291. Florini JR, Ewton DZ, Magri KA (1991) Annu Rev Physiol 53:201
292. Florini JR, Ewton DZ, Magri KA, Mangiacapra FJ (1993) Adv Exp Med Biol 343:319
293. Florini JR, Ewton DZ, Roof SL (1991) Mol Endocrinol 5:718
294. Florini JR, Magri KA (1989) Am J Physiol 256:C701
295. Florini JR, Magri KA, Ewton DZ, James PL, Grindstaff K, Rotwein PS (1991) J Biol Chem 266:15917
296. Florini JR, Samuel DS, Ewton DZ, Kirk C, Sklar RM (1996) J Biol Chem 271:12699
297. Downes M, Griggs R, Atkins A, Olson EN, Muscat GE (1993) Cell Growth Differ 4:901

298. Hughes SM, Taylor JM, Tapscott SJ, Gurley CM, Carter WJ, Peterson CA (1993) Development 118:1137
299. Muscat GE, Mynett-Johnson L, Dowhan D, Downes M, Griggs R (1994) Nucleic Acids Res 22:583
300. Everts ME, Clausen T (1988) Am J Physiol 255:E604
301. Simonides WS, Brent GA, Thelen MH, van der Linden CG, Larsen PR, van HC (1996) J Biol Chem 271:32048
302. Simonides WS, Thelen MH, van der Linden CG, Muller A, van HC (2001) Biosci Rep 21:139
303. Edwards JG, Bahl JJ, Flink IL, Cheng SY, Morkin E (1994) Biochem Biophys Res Commun 199:1482
304. Pruliere G, Butler-Browne GS, Cambon N, Toutant M, Whalen RG (1989) Eur J Biochem 185:555
305. Sachs L, de LA, Lebrun JJ, Kelly PA, Demeneix BA (1996) Endocrinology 137:2191
306. Caiozzo VJ, Haddad F (1996) Exerc Sport Sci Rev 24:321
307. Perrone CE, Fenwick-Smith D, Vandenburgh HH (1995) J Biol Chem 270:2099
308. Spangenburg EE, McBride TA (2006) J Appl Physiol 100:129
309. Brooks SV, Opiteck JA, Faulkner JA (2001) J Geront A Biol Sci Med Sci 56:B163
310. Lieber RL, Shah S, Friden J (2002) Clin Orthop Relat Res 403S:S90
311. Lynch GS, Hinkle RT, Chamberlain JS, Brooks SV, Faulkner JA (2001) J Physiol Lond 535:591
312. Patel TJ, Lieber RL (1997) Exerc Sport Sci Rev 25:321
313. Rybakova IN, Patel JR, Davies KE, Yurchenco PD, Ervasti JM (2001) Mol Biol Cell 12:41A
314. Rybakova IN, Patel JR, Davies KE, Yurchenco PD, Ervasti JM (2002) Mol Biol Cell 13:1512
315. Block TA, Aarsvold JN, Matthews KL, Mintzer RA, River LP, Capelli-Schellpfeffer M, Wollmann RL, Tripathi S, Chen CT, Lee RC (1995) J Burn Care Rehabil 16:581
316. Friden J, Lieber RL (1998) Cell Tissue Res 293:165
317. Matsuura N, Kawamata S, Ozawa J, Kai S, Sakakima H, Abiko S (2001) Arch Histol Cytol 64:393
318. Patel TJ, Das R, Friden J, Lutz GJ, Lieber RL (2004) J Appl Physiol 97:1803
319. LaIuppa JA, McAdams TA, Papoutsakis ET, Miller WM (1997) J Biomed Mater Res 36:347
320. Penick KJ, Solchaga LA, Berilla JA, Welter JF (2005) J Biomed Mater Res A 75:168

# Bioreactors for Connective Tissue Engineering: Design and Monitoring Innovations

**A.J. El Haj, K. Hampson, and G. Gogniat**

**Abstract** The challenges for the tissue engineering of connective tissue lie in creating off-the-shelf tissue constructs which are capable of providing organs for transplantation. These strategies aim to grow a complex tissue with the appropriate mechanical integrity necessary for functional load bearing. Monolayer culture systems lack correlation with the in vivo environment and the naturally occurring cell phenotypes. Part of the development of more recent models is to create growth environments or bioreactors which enable three-dimensional culture. Evidence suggests that in order to grow functional load-bearing tissues in a bioreactor, the cells must experience mechanical loading stimuli similar to that experienced in vivo which sets out the requirements for mechanical loading bioreactors. An essential part of developing new bioreactors for tissue growth is identifying ways of routinely and continuously measuring neo-tissue formation and in order to fully identify the successful generation of a tissue implant, the appropriate on-line monitoring must be developed. New technologies are being developed to advance our efforts to grow tissue ex vivo. The bioreactor is a critical part of these developments in supporting growth of biological implants and combining this with new advances in the detection of tissue formation allows us to refine our protocols and move nearer to off-the-shelf implants for clinical applications.

**Keywords** Bioreactors, Connective tissues, On-line monitoring, Mechanics, Tissue engineering

---

A.J. E. Haj (✉), K. Hampson, and G. Gogniath
Institute of Science and Technology in Medicine, Keele University Medical School, Guy Hilton Research Unit, Hartshill, Stoke on Trent ST4 7QB, UK

**Contents**

| | | |
|---|---|---|
| 1 | Introduction | 82 |
| 2 | Development of in Vitro Culture Models | 82 |
| 3 | In Vivo Versus In Vitro Bioreactors | 83 |
| 4 | Application of Physiological Mechanical Environments | 84 |
| 5 | Magnetic Force Bioreactor | 86 |
| 6 | Monitoring On Line | 88 |
| 7 | Conclusions | 91 |
| | References | 91 |

# 1 Introduction

The challenges for tissue engineering lie in creating off-the-shelf tissue constructs which are capable of providing organs for transplantation. In the case of connective tissues, this presents not only the challenge of growing a complex tissue with multiple cell types and matrix orientation outside the body, but also mechanical integrity for functional load bearing. Recent research has been heading in the right direction, however, to date this challenge has not been met. Although a long way off from functional bio-implants, developments are being made in biology and engineering of cells and their growth environments which are moving this field forward. A key part of the process is developing new bioreactors to support tissue growth which presents a requirement for identifying ways of routinely and continuously measuring the growth of ex vivo tissues in a bioreactor environment. In this chapter, we explore the development and constraints of bioreactors for the tissue engineering of connective tissues. The new developments in non-invasive monitoring of tissues in terms of imaging and microscopy as well as on-line measurements of metabolic parameters are also outlined.

# 2 Development of in Vitro Culture Models

The cell culture of connective tissues has historically relied on monolayer culture models. These monolayer cultures have been well characterised and include precursors for the mature bone phenotype such as bone marrow stromal cells or periosteal-derived cells from calvaria [1]. In addition, cell-culture systems for chondrocytes, osteoclasts [2] and osteocytes, which do not grow in confluent layers and present challenges for long-term maintenance in culture conditions, have been well described. These cell culture protocols have been well used in improving our understanding of the underlying biology of the cell populations present in bone and cartilage. They have also been utilised in our attempts to understand the role of physical forces and mechanical loading on bone cells in culture using a variety of 2D mechanical culture systems such as four-point bending [3–5]. The importance of these monolayer systems is that they are sterilisable, scaleable, reproducible, disposable, and visually suited to microscopy systems etc. The problems with these monolayer systems include the lack of correlation with the in vivo environment and the naturally occurring cell phenotypes, and the lack of adaptability to a 3D environment, because of limitations in mass transfer.

The tissue-engineering community is attempting to move away from these monolayer systems and develop 3D models which can be used for biological investigations as well as drug testing. The ultimate aim is to move towards generating tissue for transplantation, and the knowledge developed using these new in vitro models will aid in this. Part of the development of these models is to create growth environments or bioreactors that can replace monolayer culture systems and sets out the requirements for these bioreactors very clearly.

One of the first examples of a commercial bioreactor for tissue engineering which met the above requirements was the rotating-wall bioreactor designed by the National Aeronautics & Space Administration (NASA). These bioreactors have been designed to enable the culture of constructs in a state of free fall which applies low shear stresses to the cells. Previous studies have shown that the use of a rotating bioreactor increases the number of cells present in constructs after 28 days of culture compared to static and perfusion cultures [6]. This is thought to be a result of an improvement in mass transport between the cells seeded within the constructs and the culture media [7]. This is advantageous because in vivo most cells benefit from the close proximity of capillaries that provide the mass transfer requirements of oxygen and nutrients between the cells and the blood [8]. These systems are now being marketed under the label Synthecon or Cellon products. Osteoblast-like cells cultured in these bioreactors display characteristics more representative of in vivo osteoblasts [9, 10] and show a more mature phenotype compared to static 3D cultures [11]. Mesenchymal stem cells (MSCs) cultured in these bioreactors on silk scaffolds have also shown enhanced calcium accumulation compared to static cultures and resulted in constructs that resembled trabecular bone with respect to structure and mineralised tissue [12].

## 3 In Vivo Versus In Vitro Bioreactors

What is the gold standard for bioreactors? There is a growing body of evidence to show that the in vivo environment presents the most optimal environment for engineered tissue implants to grow. Stevens et al. [13] recently published a strategy for in vivo engineering of organs using the periosteal environment as a bone bioreactor. By implanting biomaterials under the periosteal flap in situ on the bone surface in vivo, the authors demonstrate how they can generate bone that is removable and useable for transplantation to other sites in the body. Even more remarkable evidence for an in vivo bioreactor in humans is the work published by Wanke et al.

[14] where titanium meshes were implanted with bone morphogenetic protin 7 in the muscle of the dorsum of a patient. The mineralised bone scaffold was cellularised and generated over a period of months prior to use in transplantation in the jaw of a patient. For the tissue engineer, these studies identify core principles for bone tissue growth which can be taken into account when considering the best bioreactor design for an in vitro environment. The core principles for a bioreactor are: (a) Maintaining sterility, (b) Good mass transfer, (c) Suitable for scale up, (d) Reproducibility of samples, (e) Controlled metabolic environment, (f) Ability to impose physiologically relevant mechanical stimulation to tissues. Added to this is the need to get the growth media and spatial arrangements of the scaffolds in the right configuration.

## 4 Application of Physiological Mechanical Environments

Evidence suggests that in order to grow functional load-bearing tissues in a bioreactor, the cells must experience mechanical loading stimuli similar to those experienced in vivo—in other words, the in vivo stress environment must be mimicked inside the bioreactor. The in vivo environment of bone and cartilage is such that they receive a combination of different types of mechanical loading including tension, compression, bending, shear and torsion and in order to replicate this more closely, recent approaches to bioreactors enable the application of several different types of physical stimulation to the constructs [15]. Application of relevant in vivo physiological loads presents a challenge for the design of bioreactors relevant to a wide range of tissues. These applied stresses are necessary in order to initiate biochemical reaction pathways—for example through the activation of mechanosensitive (MS) $Ca^{++}$ ion channels or the re-orientation of the cytoskeleton—which allow human bone and cartilage to develop their characteristic mechanical properties [5]. It is also evident that many stems cells such as mesenchymal stem cells require mechanical cues in order to direct their differentiation [16]. These cues may even be more important than some biochemical cues [19, 37].

Bioreactors enable the culture of these cells in a 3-dimensional environment and can be used to apply reproducible and accurate regimes of mechanical forces to cell-seeded constructs [17]. Bioreactors can therefore be used to investigate the effects of mechanical stimulation on cells in a 3D environment which has previously been shown to enhance the differentiation of mesenchymal stem cells along numerous lineages ([18]).

Recently, bioreactors have been developed which apply mechanical forces via piston/compression systems, substrate bending, hydrodynamic compression and fluid shear (for review see [19]). Many of these different types of bioreactor are discussed in other chapters in this volume. Figure 1 shows a perfusion compression bioreactor first designed in 1991 by El Haj et al. [20] to maintain viable bone explants ex vivo. The bioreactor has subsequently been adapted to enable the growth of cell-seeded constructs [21]. Yang and El Haj (2005) cultured poly-l-lactic acid (PLLA)

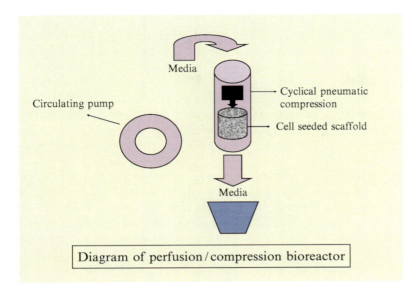

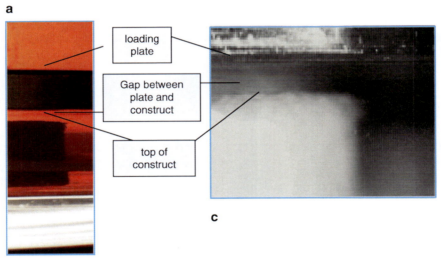

**Fig. 1** (**a**) Diagram of the perfusion compression bioreactor as described by El Haj et al. (1990, 1992) for culture of explants and cell-seeded constructs for tissue engineering; (**b**) and (**c**) show photographs taken of the constructs within the chamber demonstrating the methodology for measuring the amount of compression on the top of the scaffold. Problems arising from the consistency across the top of the construct can be seen

scaffolds seeded with MG63 (osteosarcoma cell line) cells in this adapted system. After 3 weeks of culture under perfusion, the cell-seeded constructs were submitted to a loading regime of 0.1% strain for 1 h per day at a frequency of 1 Hz for 7 days with media being perfused through the system at a rate of 0.1 ml min$^{-1}$. A combination of compression and perfusion resulted in a significant increase in the expression of osteogenic markers compared to static and perfusion only samples. This bioreactor has also been used for the culture of tissue-engineered cartilage and demonstrated that in vivo physiological loads are sometimes too excessive for engineered constructs. It was also demonstrated that low levels of strain can have greater growth-promoting effects [22].

There are, however, problems associated with these types of bioreactors. Any force-producing mechanism that invades the bioreactor (such as in piston/compression systems) may cause infection. The scaffold materials must transmit the force to the cells and in order to withstand the loads in a compression perfusion bioreactor, the scaffolds must be strong. This often results in long degradation times. In order to generate new biological matrices, the support scaffolds are often rapidly degrading which compromises their mechanical properties and makes them weak. These mechanically weak scaffolds therefore may not be capable of transmitting the required forces and may be unsuitable for a large range of available bioreactors. Although the forces required to activate MS ion channels via cell membrane deformation are small (5–100 pN; [23]) this requirement may present technical problems for bioreactor and scaffold design.

## 5 Magnetic Force Bioreactor

Other challenges for bioreactors include continuous profusion of cell nutrients, long-term sterility and, in the case of compression systems, the matrix transmits the applied forces to the cells within the pores. In addition, it is not possible to apply spatially varying stresses in three dimensions in order to form complex tissue structures such as a complete joint with a cartilage/bone interface. There are also serious problems with scale-up when multiple compression systems are applied to large numbers of samples, for example in the case of high-throughput screening applications.

A new development in bioreactor design is based on the theoretical principles and prototype design of a novel mechanical conditioning system—a magnetic force bioreactor (MFB) which is designed to apply stress directly to the cell membrane using forces acting on magnetic nanoparticles—comprising a magnetic core with a polymer outer coating [24] (Fig. 2). In this system, biocompatible magnetic particles are attached to the cell membrane (e.g. via RGD, collagen or integrin receptors) or directly to an ion channel membrane protein. The cells may be in 2D culture or seeded into porous, bioresorbable scaffolds and introduced into a bioreactor.

The applied force simulates mechanical loading of the cell membrane without requiring direct access to the cells inside the bioreactor and without requiring the stress to be transmitted from the scaffold to the cells. Loads can be varied easily by

# Bioreactors for Connective Tissue Engineering

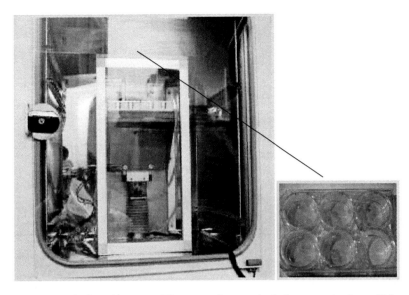

**Fig. 2** The magnetic force bioreactor. The system is comprised of a magnetic plate which can be cyclically adjusted under the base of a table. The table supports a number of configurations which include a standard 6-well plate containing cell-seeded constructs. Magnetic particles are attached to the cells mechanosensitive receptors in a variety of configurations [24]

changing the magnetic field strength and gradient or the magnetic properties of the nanoparticles. Cells carrying particles with different magnetic properties can be seeded into different regions within the 3D scaffold, producing a spatial variation in force using the same magnetic field geometry.

The application of cyclical external magnetic fields (at physiological frequencies such as 1–3 Hz) applies either a translational force (due to the attraction of magnetic nanoparticles along the magnetic field gradient) or a combination of translational force and torque (for larger, magnetically blocked nanoparticles and microparticles) which is transmitted directly to the cell membrane or the cytoskeleton and can be varied in three dimensions within a scaffold. Hughes et al. [25, 26] have demonstrated how magnetic microparticles attached to ion channels through His-tagged clones can be remotely activated by time-varying magnetic fields.

Initial results from static 3D and 2D cultures using these systems demonstrated significant up-regulation of bone-matrix proteins [24, 27]. After 1 week in culture, both osteocalcin, osteopontin and alkaline phosphatase showed up-regulation when compared to controls. Work is now progressing toward building complexity in the biological models with co-cultures and multiple tissues. The system is also allignable to a perfusion bioreactor allowing mass transfer to be addressed. The MFB systems should provide several advantages over current mechanical stimulation systems such as:

- Unprecedented, high precision control of the physical stress parameters through variation of the magnetic force.

- As the stress is applied directly to the cell membrane, the need for mechanically strong scaffold materials is eliminated.
- Particles are remotely coupled to the magnet array with no components connected into the bioreactor. This eliminates a potential infection route.
- The system is scalable and presents the potential to apply a spatially varying load profile via seeding with particles of differing magnetic properties.

In addition to cell culture and 3D scaffold-based culture work, the MFB has been used to investigate the magnetic activation of calcium pathways in human bone marrow stem cells. Observation of $Ca^{2+}$ fluorescence activity in these cells showed significant levels of baseline calcium activity in 30–50% of cells, with many of these demonstrating oscillating intracellular calcium levels. The application of 600 G static magnetic fields did not greatly influence the behaviour of cells already undergoing $Ca^{2+}$ oscillations, however cells with steady baseline $Ca^{2+}$ levels showed clear characteristic transients in response to magnetic stimulation—indicating activation of mechanosensitive calcium pathways [25, 26].

This system should also be adaptable to other biomedical applications in which it is important to mimic the in vivo stress environment in vitro. An example of this is high-throughput drug/toxicity screening in a dynamic culture environment. It is critically important to test and screen drugs on cells and tissue in a dynamic culture environment in order to more clearly understand how the compound will behave in vivo where mechanical forces are constantly being applied to the cells. This cannot be achieved with the static monolayer, or even 3D, culture systems presently in use. The MFB can be easily adapted for this type of screening in both multi-well monolayer cultures and multi-well 3D cultures.

## 6 Monitoring On Line

In order to fully identify the successful generation of a tissue implant, the appropriate on-line monitoring must be developed. In the bio processing industry, the protocols for on-line monitoring are well described. Measurements of pH, $pO_2$ and other metabolites are taken routinely on line in large-scale cell-culture environments and the key markers in these scale-up environments are for levels of infection and cell death. In tissue engineering, on-line monitoring has only recently been considered fully. Ideally, the aim is to monitor tissue growth and formation. This builds a complexity into the analysis which currently relies upon terminal experiments where newly generated tissues can be fully characterised. Ultimately, if engineered tissues are to be delivered to patients for transplantation then routine on-line assessments of the state of the implant are necessary. In this way, identification of when tissues are ready for delivery will be determined routinely.

New ways for non-invasive monitoring by imaging tissue constructs are being developed. Optical technologies such as Optical coherence tomography (OCT) [28] or variations such as Doppler OCT [29] and polarisation OCT [30] are being put forward as cheap and simple methods for monitoring structural tissues. Figure 3 shows an

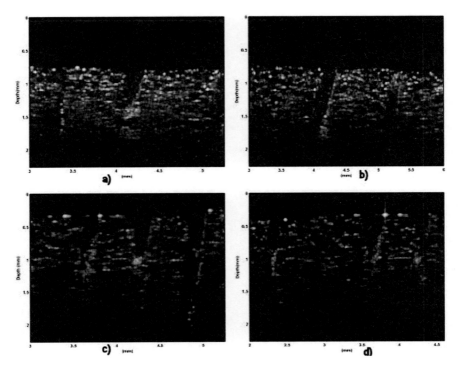

**Fig. 3** OCT images showing the microstructural variations induced by different culture conditions. In the non-stimulated group (static group), cells and extracellular matrix have developed in the middle or at the surface of the channel as seen respectively in (**a**) and (**b**). Morphologically similar microstructural variations are found for the perfusion group (**c**) and (**d**) with a greater degree of matrix laid down in the perfused channels (Image adapted from Bagnaninchi et al. [30])

image of a chitosan scaffold with microchannels seeded with cells and cultured in a flow bioreactor. This technique has been used to quantitatively assess the effects of flow on cell growth in the channels [30]. Combining the OCT with other modalities, such as Doppler, to measure flow through the bioreactor and microchannels within the tissue, or polarisation to determine matrix orientation, allows us to image tissues with time. Polarisation OCT is also being developed to enable the orientation of the matrix during tissue growth to be monitored (Fig. 4). Bioreactors can be designed to enable optical interrogation of the growing tissues, however a key issue in optical modalities is the penetration depth of light. In the case of OCT, this is limited to 2–3 mm depth of imaging which is suitable for some tissue constructs, such as cartilage, but not appropriate for larger scale implants.

Other methods are being explored such as permittivity measurements [32] or NMR spectroscopy systems for the non-invasive measurement of cell numbers within the 3D culture [33]. Alternative strategies for the monitoring of the behaviour of tissue-engineered constructs involves the development of fluorescent scaffolds which can be measured with time to assess decay of fluorescent intensity in relation to

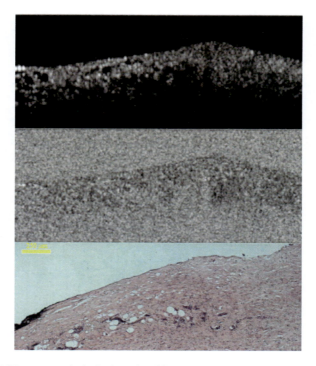

**Fig. 4** (**a**) and (**b**) are respectively the intensity (S0) and phase retardation map of a human tendon. Low area of scattering correlates with fat infiltration as seen on the histology (**c**), while the high scattering sphere correlates with high concentration of cell clusters and matrix formation [31]

degradation of the material. Fluorescent markers can be used in a number of ways but labelling cells with fluorophores may not be suitable for follow on implantation.

Metabolic monitoring relies on existing sensing methods for in vivo and bulk cell bioprocessing systems. In terms of bioreactor process monitoring, there are two possible strategies available using biosensors. One strategy involves the use of sensors being placed inside of the bioreactor. This invasive method, where the biosensor can be located either in the culture fluid or in direct contact with the cells or tissue/scaffold construct can be advantageous in situations when sample extraction and transport could cause difficulties. Boubriak et al. [34] have monitored metabolic gradients within tissue-engineered constructs using embedded sensors within scaffolds. Another example is the use of microelectrodes built into the bioreactor. The microelectrodes are used to measure the concentration of $O_2$ when a change in oxidation potential occurs. There are several disadvantages with this strategy however. The electrode must be introduced into the sample and at different locations, calibration is lengthy and difficult, and problems can arise with sterilisation [35]. Invasive fibre-optic sensors have also been used for $PO_2$ and pH measurements. The fluorescent intensity is measured directly for the pH measurement and the quenching effect of $O_2$ is used for the $PO_2$ measurement [36].

Non-invasive sensing can be carried out using optical methods such as spectrophotometry or fluorimetry. This approach obviously avoids the difficulties of needing to sterilise the sensors, but there is more of a challenge in achieving high specificity and high sensitivity for target molecules such as glucose. Reporter patches fixed to the inside of an optical window in the bioreactor wall are being investigated. Such patches contain ion-sensitive dyes or $O_2$-quenchable fluorophores [36]. The use of biosensors outside the bioreactor offers many advantages. The problems with sensor sterilisation is completely avoided and any sample preparation required, such as dilution to within the linear range of the sensor, pH adjustment by the addition of an appropriate buffer, and temperature measurements, can be accomplished easily.

# 7 Conclusions

New technologies are being developed to advance our efforts to grow tissue ex vivo. Using core elements such as stem cells and biomaterials combined with a range of growth promoting agents—both biochemical and mechanical—we can influence tissue growth and formation. The growth environment or bioreactor is a critical part of these developments in supporting growth of biological implants. Combining this with new advances in detection of tissue formation is allowing us to refine our protocols and move nearer to off-the-shelf implants for clinical applications.

## References

1. Robey PG, Kuznetsov SA, Riminucci M, et al. (2007) Skeletal ("dmesenchymal") stem cells for tissue engineering. Methods Mol Biol 140:83–99
2. Roodman GD (1999) Cell biology of the osteoclast. Exp Hematol 27(8):12229–12241
3. El Haj AJ, Walker LM, Preston MR, et al. (1999) Mechanotransduction pathways in bone: calcium fluxes and the role of voltage operated calcium channels. Med Biol Eng Comput 37(3):403–409
4. Mullender M, El Haj AJ, Yang Y, et al. (2004) Mechanotransduction of bone cells in vitro: mechanobiology of bone tissue. Med Biol Eng Comput 42(1):14–21
5. Walker LM, Publicover SJ, Preston MR, et al. (2000) Calcium channel activation and matrix protein up-regulation in bone cells in response to mechanical strain J Cell Biochem 79(4):648–661
6. Pound JC, Green DW, Chaudhuri JB, et al. (2006) Bioreactor culture of cartilage from mesenchymal populations. J Bone Joint Surg Br 88-B (Supplement III): 405
7. Yu X, Bothway EA, Levine EM,et al. (2004) Bioreactor based bone tissue engineering – The influence of dynamic flow on osteoblast phenotypic expression and matrix mineralization. Proc Natl Acad Sci U S A 101 (31):11203–11208
8. Martin Y, Vermette P (2005) Bioreactors for tissue mass culture: design, characterization, and recent advances. Biomaterials 26(35):7481–7503
9. Botchwey EA, Pollack S, Levine EL, et al. (2001) Bone tissue engineering in a rotating bioreactor using a microcarrier matrix system. J Biomed Mater Res 55(2):242–253

10. Rucci N, Migliaccio S, et al. (2002) Characterization of the osteoblast-like cell phenotype under microgravity conditions in the NASA-approved rotating wall vessel bioreactor (RWV). J Cell Biochem 85(1):167–179
11. Granet C, Laroche N, Vico L, et al. (1998) Rotating-wall vessels, promising bioreactors for osteoblastic cell culture: comparison with other 3D conditions. Med Biol Eng Comput 36(4): 513–519
12. Marolt D, Augst A, Freed LE, et al. (2006) Bone and cartilage tissue constructs grown using human bone marrow stromal cells, silk scaffolds and rotating bioreactors. Biomaterials 27(36):6138–6149
13. Stevens M, Marini RP, Schaefer D, et al. (2005) In vivo engineering of organs. The bone bioreactor 102(32):11450–11455
14. Wanke PH, Springer IN, Wiltgang J, et al. (2004) Growth and transplantation of a custom vascularised bone graft in a man. Lancet 364(9436) 766–770
15. Cartmell, S., El Haj, A.J. (2006) Mechanical Bioreactors for Tissue Engineering in Bioreactors for Tissue Engineering Eds. Dr Julian B Chaudhuri and Dr Mohamed Al Rubeai (Kluwer Academic Publishers)
16. Thomas G, El Haj AJ (1996) Bone marrow stromal cells are load responsive. Calcif Tissue Int 58(2):01–108
17. Démarteau O, Jakob M, Schaefer D, et al. (2003) Development and validation of a bioreactor for physical stimulation of engineered cartilage. Biorheology 40(1–3):331–336
18. Mauney J, Sjostorm S, Horan R, et al. (2004) Mechanical stimulation promotes osteogenic differentiation of human bone marrow stromal cells on 3-D partially demineralized bone scaffolds in vitro. Calcif Tissue Int 74(5):458–468
19. Cartmell S, El Haj AJ (2006) Mechanical bioreactors for tissue engineering in bioreactors for tissue engineering Eds. Dr Julian B Chaudhuri and Dr Mohamed Al Rubeai (Kluwer Academic Publishers)
20. El Haj AJ, Minter SL, Rawlinson S, et al. (1991) Cellular responses to mechanical loading *in vitro*. J Bone Miner Res 5(9):923–932
21. Shelton RM, El Haj AJ (1992) A novel microcarrier bead model to investigate bone cell responses to mechanical compression. J Bone Miner Res 7(2):403–405
22. Freyria AM, Yang Y, Chajra H, et al. (2005) Optimisation of dynamic culture conditions: effects on biosynthetic activities of chondrocytes grown in collagen sponges. Tissue Eng 11:5–6
23. Walker LM, Holm A, Cooling L, et al. (1999) Mechanical manipulation of bone and cartilage cells with optical tweezers. FEBS Lett 459(1): 39–42
24. Dobson J, Cartmell SH, Keramane A, et al. (2007) Principles and design of a novel magnetic force mechanical conditioning bioreactor for tissue engineering, stem cell conditioning, and dynamic in vitro screening. IEEE Trans Nanobiosci 5(3):173–177
25. Hughes S, McBain S, Dobson J, El Haj AJ. (2008) Selective activation of mechanosensitive ion channels using magnetic particles. J R Soc Interface. 2008 Aug 6;5(25):855–863
26. Hughes S, Dobson J, El Haj AJ. (2007) Magnetic targeting of mechanosensors in bone cells for tissue engineering applications. J Biomechanics 40:S96–S104
27. Cartmell SH, Dobson J, Verschueren S, et al. (2002) Development of magnetic particle techniques for long-term culture of bone cells with intermittent mechanical activation. IEEE Trans Nano Biosci 1:92–97
28. Yang Y, Dubois A, Qui XP, et al. (2006) Investigation of OCT as an imaging modality in tissue engineering. Phys Med Biol 51(7):1649–1659
29. Mason C, Markusen JF, Town MA, et al. (2004) Doppler optical coherence tomography for measuring flow in engineered tissues. Biosens Bioelectron 20(3):414–423
30. Bagnaninchi PO, Yang Y, Zghoul N, et al. (2007) Chitosan microchannels scaffolds for tendon tissue engineering characterized by optical coherence tomography. Tissue Eng 13(2):323–331
31. Yang Y, Ahearne M, BAganinchi PO, Hu B, Hampson K, El Haj AJ (2008) Application of polarization OCT in tisuue engineering SPIE Vol 6858,68580k, 1-7

32. Bagnaninchi PO, Dikleakos M, Veres T, et al. (2003) Towards on line monitoring of cell growth in microporous scaffolds. Utilisation and interpretation of complex permitivity measurements. Biotechnol Bioeng 84(3):343–350
33. Stabler CL, Long RC, Sambanis A, Constaninidis I (2005) (H) NMR Spectroscopy to noninvasively quantify viable cell number in tissue engineered substitutes in vitro. Tissue Eng (1193):404–414
34. Boubriak OA, Urban JP, Cui Z (2006) Monitoring of metabolic gradients in tissue engineered constructs. J R Soc Interface 3(10):637–648
35. Malda J, Woodfield TBF, van der Vloodt F, et al. (2004) The effect of PEGT/PBT scaffold architecture on oxygen gradients in tissue engineered cartilaginous constructs. Biomaterials 25(26):5773–5780
36. Rolfe P (2006) Sensing in tissue bioreactors. Meas Sci Technol 17(3):578–583
37. El Haj AJ, Wood M, Thomas P, et al. (2005) Controlling cell biomechanics in orthopaedic tissue engineering and repair. Pathology 53:581–589

# Mechanical Strain Using 2D and 3D Bioreactors Induces Osteogenesis: Implications for Bone Tissue Engineering

**M. van Griensven, S. Diederichs, S. Roeker, S. Boehm, A. Peterbauer, S. Wolbank, D. Riechers, F. Stahl, and C. Kasper**

**Abstract** Fracture healing is a complicated process involving many growth factors, cells, and physical forces. In cases, where natural healing is not able, efforts have to be undertaken to improve healing. For this purpose, tissue engineering may be an option. In order to stimulate cells to form a bone tissue several factors are needed: cells, scaffold, and growth factors. Stem cells derived from bone marrow or adipose tissues are the most useful in this regard. The differentiation of the cells can be accelerated using mechanical stimulation. The first part of this chapter describes the influence of longitudinal strain application. The second part uses a sophisticated approach with stem cells on a newly developed biomaterial (Sponceram) in a rotating bed bioreactor with the administration of bone morphogenetic protein-2. It is shown that such an approach is able to produce bone tissue constructs. This may lead to production of larger constructs that can be used in clinical applications.

**Keywords** Biomaterials, Bone, Mechanical strain, Rotating bed bioreactor, Tissue engineering

---

M. van Griensven (✉), A. Peterbauer, and S. Wolbank
Ludwig Boltzmann Institute for Experimental and Clinical Traumatology, Donaueschingenstraße 13, A-1200 Vienna, Austria
e-mail: Martijn.van.Griensven@lbitrauma.org, Anja.Peterbauer@lbitrauma.org, Susanne.Wolbank@lbitrauma.org

S. Roeker, S. Boehm, D. Riechers, F. Stahl, and C. Kasper
Institut für Technische Chemie, Leibniz University Hannover, Callinstraße 3, 30167 Hannover, Germany
e-mail: Roeker@iftc.uni-hannover.de, Boehm@iftc.uni-hannover.de, Riechers@iftc.uni-hannover.de, Stahl@iftc.uni-hannover.de, Kasper@iftc.uni-hannover.de

S. Diederichs
Ludwig Boltzmann Institute for Experimental and Clinical Traumatology, Donaueschingenstraße 13, A-1200 Vienna, Austria and
Institut für Technische Chemie, Leibniz University Hannover, Callinstraße 3, 30167 Hannover, Germany
e-mail: Diederichs@iftc.uni-hannover.de

## Contents

1   Introduction ........................................................................................................... 97
    1.1   Clinical Problem ........................................................................................... 97
    1.2   Mechanical Forces and Bone Formation ...................................................... 97
    1.3   Effects of Strain ........................................................................................... 98
    1.4   Cell Source .................................................................................................. 98
    1.5   Strain Elicits Cellular Signals ...................................................................... 99
    1.6   Biomaterials ................................................................................................. 100
    1.7   Bone Engineering ........................................................................................ 100
2   2D Culture (Monolayer) ........................................................................................ 101
    2.1   BMSC .......................................................................................................... 101
    2.2   Adipose-Derived Mesenchymal Stem Cells (AdMSC) ............................... 106
3   3D Culture in a Rotating Bed Bioreactor .............................................................. 113
    3.1   Sponceram® .................................................................................................. 113
    3.2   Cell Cultivation in the ZRP-Bioreactor ....................................................... 113
    3.3   Glucose Assay ............................................................................................. 114
    3.4   Matrix Mineralization .................................................................................. 115
    3.5   Scanning Electron Microscopy .................................................................... 115
    3.6   Reverse Transcriptase-Polymerase Chain Reaction (RT-PCR) ................... 117
4   Discussion ............................................................................................................. 118
    4.1   2D Culture ................................................................................................... 118
    4.2   Rotating Bed Bioreactor Culture ................................................................. 120
References ................................................................................................................... 121

## Abbreviations

| | |
|---|---|
| AdMSC | Adipose-derived mesenchymal stem cells |
| AP | Alkaline phosphatase |
| BMP | Bone morphogenetic proteins |
| BMSCs | Bone marrow stromal cells |
| BSP | Bone sialoprotein |
| cAMP | Cyclic adenosine-mono-phosphate |
| cbfa1 | Core binding factor alpha1 |
| cGMP | Cyclic guanosine-mono-phosphate |
| CICP | Procollagen I propeptide |
| COL I | Collagen I |
| COX-2 | Cyclooxygenase-2 |
| DM | Differentiation medium |
| ECIS | Electric cell–substrate impedance sensing |
| ECL | Enhanced chemiluminescence |
| ECM | Extracellular matrix |
| EDTA | Ethylene diamine tetraacetic acid |
| FACS | Fluorescent activated cell scanner |
| HA | Hydroxyapatite |
| iNOS | Inducible nitric oxide synthase |
| IP3 | Inositol tri-phosphate |
| JNK/SAPK | Jun-N-terminal-kinase/stress-activated protein kinase |
| MAPK | Mitogen-activated protein kinase |

| | |
|---|---|
| MTT | 3-(4,5-Dimethylthiazol-2-yl)-2,5-diphenyl-tetrazolium-bromide |
| NFκB | Nuclear factor-kappa B |
| NM | Normal medium |
| NO | Nitric oxide |
| OC | Osteocalcin |
| PDMS | Polydimethylsiloxan |
| OPN | Osteopontin |
| PPS | Photopatternable silicone |
| PVD | Physical vapour deposition |
| RT-PCR | Reverse transcriptase-polymerase chain reaction |
| RWVR | Rotating-wall vessel reactor |
| SDS | Sodium dodecyl sulfate |

# 1  Introduction

## 1.1  Clinical Problem

Large bone defects caused by tumors, infectious diseases, or trauma clearly result in the medical need for bone regeneration. If possible an autologous bone graft (mostly derived from the iliac crest) is carried out. However, the quantity of the obtained material is often not sufficient in case of a large bone defect. Alternatively, allogeneic bone from donor (corpse) patients (e.g., cryopreserved) is transplanted, which is critical with regards to infection risks (AIDS, hepatitis). Therefore, alternative methods have been developed.

## 1.2  Mechanical Forces and Bone Formation

The effects of biophysical force on bone remodeling have become increasingly evident in recent years. It is well known that extended periods of immobilization lead to bone loss. This is especially apparent in situations of weightlessness. Subjects exposed to weightlessness have shown diminished or arrested bone formation [1], reduced collagen production [2], increased osteoclast numbers [3], and consequently a decrease in mechanical properties of bone. Mechanical loading is one of the few positive stimuli for bone formation, and the use of suitable exercise regimes has been proposed as being potentially of significant benefit in maintaining bone mass in postmenopausal women and accelerating bone mass recovery after bed rest [4, 5].

Various studies demonstrate that mechanical loading is an essential factor for bone metabolism. Mechanical passive states of the skeletal system due to zero gravity, functional immobilization, or postoperative bedfast have been shown to result in decreased bone formation and mineralization as well as reduced protein synthesis [1, 6]. On the other hand, bone mass increases upon application of elevated skeletal load [7]. In vitro and in vivo studies have shown that physical forces like load, fluid flow, and electromagnetic fields can regulate the function of mesenchymal progenitor cells. For bone tissue engineering, mechanical force types like linear straining or pressure load correlate most closely to the physiologic conditions and therefore are most widely used in connection with cultivating bone cells and generating bone-like tissue. However, the methods of strain application vary widely concerning substrate materials and geometry as well as physical parameters like strain duration, elongation, and frequency.

## 1.3 Effects of Strain

Strain application described in the literature differs widely concerning not only substrate geometry and material but also strain parameters like frequency, elongation, and strain duration. An optimal frequency of 1 Hz has been reported by Kaspar et al. [8] and has been used by many other groups [9–11]. Strain amplitudes are chosen depending on the utilized cell types. Cyclic elongation of bone marrow stromal cells (BMSCs) in physiological ranges (0.035–0.25%) did not show any effect on proliferation rates and type I collagen synthesis [9]. Another study finds significant increases ($p < 0.05$) of alkaline phosphatase activity and type I collagen synthesis with 8% elongation [12]. However, this high elongation may have damaged the cells as they produced more type III collagen. Therefore, we used a moderate elongation of 5% in order to induce notable cellular reactions without causing any cell damage.

Mechanical load aligns collagen fibers and this tissue reorganization increases functionality [13]. Thus, mechanical loading is important for maintaining the physiological properties of mature bone. There have been several investigations dedicated to examining the influence of cyclic mechanical stretching upon osteoblasts obtained from cancellous bone chips [14, 15]. Mechanical stress has been demonstrated to stimulate the secretion of osteogenic proteins [14]. It could be demonstrated that cyclic stretching stimulates osteoblast proliferation and CICP (procollagen I propeptide) production but decreases the synthesis of alkaline phosphatase and osteocalcin [14].

## 1.4 Cell Source

This raised the question whether mechanical loading may also be able to enhance precursor cell differentiation to osteoblasts. This would be important in light of tissue engineering of bone constructs for regenerating bone tissue after trauma or osteoporosis linked complications. As an alternative to obtaining autologous bone from the iliac crest, which is associated with considerable donor site morbidity, the field of tissue engineering promises the opportunity to develop a synthetic construct based on cells seeded onto an appropriate matrix. Bone marrow stromal cells have been identified as an attractive cell source for a wide variety of tissue-engineering strategies. Since even in older individuals, bone marrow stroma harvesting is a relatively easy procedure, bone marrow contains a pluripotent population of cells capable of differentiating along multiple mesenchymal lineages (e.g., bone, ligament, adipose tissue, cartilage, muscle tissue), and can easily be expanded ex vivo utilizing routine cell-culture techniques [16–22]. Thus, BMSCs can serve as a basis for tissue engineering of autologous implants without concerns over transplant rejection.

Cyclic compressive loading of rabbit BMSC in agarose cultures stimulated chondrogenesis [23]. Furthermore, mechanical strain has been shown to promote osteogenesis of BMSCs in vitro, verified by the upregulation of osteogenic marker proteins like alkaline phosphatase [9, 12, 24], osteocalcin [12, 24], osteopontin [25], and type I collagen [12, 26]. These results, however, are dependent on the type and intensity of strain. Mechanical strain may act with different frequencies and strength, which appears to have relevance under normal physiological circumstances [27, 28]. In vitro, there are several devices for the application of strain. The three main systems are (1) circular membranes (Flexercell), (2) longitudinal strain [12, 29], and (3) 4-point bending [30]. The disadvantage of the circular membranes is that the strain distribution across the membrane is heterogenous. Therefore, in our studies we have used a longitudinal strain device.

## 1.5 Strain Elicits Cellular Signals

As shown above, mechanical strain can influence the differentiation of BMSC into osteoblasts. However, what is not clear, are the pathways by which mechanical strain is transmitted into biological signals. In previous studies, in human patellar tendon fibroblasts, we have shown that cyclic longitudinal mechanical strain induces the secretion of nitric oxide (NO) [31] and the activation of Jun-N-terminal-kinase/stress-activated protein kinase (JNK/SAPK) in a time-dependent manner [32]. Montaner et al. observed a link between JNK/SAPK and Nuclear Factor-kappa B (NFkB) transduction pathways [33]. Both JNK/SAPK and NFkB are, among others, involved in proliferation and apoptosis of different cell types [34]. Indeed, cyclic longitudinal mechanical strain can modulate proliferation [35] and apoptosis [36] of patellar tendon fibroblasts. Also in bone cells, similar strain-induced mechanotransduction pathways have been recognized. These include cyclic adenosine-mono-phosphate (cAMP), cyclic guanosine-mono-phosphate (cGMP) [37, 38], cfos [39], inositol tri-phosphate (IP3) [40], intracellular calcium [41], cyclooxygenase-2 (COX-2) and prostaglandins [39] and inducible nitric oxide synthase (iNOS) and NO [39]. Furthermore, cyclic longitudinal strain induced core binding factor alpha1 (cbfa1) [12]. Parts of these pathways are mediated via specific mechanosensitive calcium-channels [42] or integrins [43]. The latter ones induce enhanced phosphorylation of cytoskeletally anchored proteins such as mitogen-activated protein kinase (MAPK) [43].

## 1.6 Biomaterials

Depending on the tissue that is supposed to be replaced, substrate materials have to be provided for cell growth, which need to fulfill different requirements with regard to their mechanical stability, biodegradability, and porosity. These matrices have to be biocompatible, should support cell attachment, growth and differentiation towards the desired phenotype. Highly porous materials provide space for the bone tissue and allow an optimal cell growth inside the scaffold. Many materials have been reported to fulfill these requirements including natural/synthetic polymers, metals, and ceramics [44–48]. Ceramics like hydroxyapatite, calciumphosphates, alumina, zirconia and composite materials, etc. are widely used as scaffold material [49–54]. Moreover, hydroxyapatites, calcium phosphates, or composite materials are known to enhance the osteogenic differentiation when seeded with progenitor cells or preosteoblastic MC3T3-E1 cells [55, 56]. Furthermore, a controllable degradation of scaffolds is often desired, which should not cause any inflammatory reactions in vivo. After transplantation, cells of the surrounding tissue invade the scaffold.

## 1.7 Bone Engineering

One innovative approach to bone tissue engineering is to seed osteoblasts or their precursor cells onto an appropriate 3D matrix and to culture this scaffold in vitro before implantation into the defect of a patient. During the generation of bone tissue different osteoblastic markers such as alkaline phosphatase (AP), collagen I (COL I), osteocalcin (OC), and bone sialoprotein (BSP) are chronologically expressed [57]. Finally, the cells are embedded in their extracellular matrix (ECM) and begin to mineralize by depositing mineral along and within the grooves of the collagen fibrils [58].

### 1.7.1 Growth Factors

The differentiation of cells into bone cells is mediated by growth factors. Specific growth factors for bone differentiation are bone morphogenetic proteins (BMPs), members of the TGF-$\beta$ superfamily [59–61]. BMPs stimulate the differentiation of different cell types to osteoblasts including undifferentiated mesenchymal cells, BMSCs, and preosteoblasts [62]. For the in vitro promotion of the differentiation process into bone tissue mainly BMP-7 and BMP-2 are used [62, 63], whereas BMP-2 increases the AP and OC level in MC3T3-E1 cells [64]. This certifies the eminent influence of BMP-2 and of BMPs in general during the process of bone tissue formation and is thus considered as vital for bone tissue engineering.

### 1.7.2 Bioreactor Cultivation

For an optimal supply of the cells with oxygen and nutrients within the scaffold the cultivation is ideally performed in a bioreactor system, since static cultures are insufficient to mimic the in vivo conditions. In a bioreactor, oxygen, pH and the transport of nutrients and metabolic waste in the tissue microenvironment can easily be controlled. The most commonly used bioreactors for bone tissue engineering are the spinner flask, perfusion culture systems, and the Rotating-Wall Vessel Reactor (RWVR). In spinner flasks, the cells can be cultured either on scaffolds fixed on needles or seeded onto microcarriers [65, 66]. Perfusion culture systems provide a continuous exchange of medium, which ensures the removal of metabolic waste and supplementation of essential nutrients throughout the cultivation. Perfusion systems are frequently applied for the cultivation of bone tissue [67–71]. The RWVR was originally developed by NASA to mimic gravity [65]. Gravity is an important factor for bone stability and integrity (*astronaut's disease*). Using the RWVR, the scaffolds are cultured in a free-fall manner. The RWVR has already been used successfully for the cultivation of bone and cartilage cells [72].

This chapter deals with the 2D and 3D cultivation of stem cells for bone tissue engineering. First the 2D aspects are described. Thereafter the 3D part is presented.

## 2  2D Culture (Monolayer)

### 2.1  BMSC

#### 2.1.1  BMSC Isolation and Cultivation

Human bone marrow aspirates were obtained during routine orthopedic surgical procedures involving exposure of the iliac crest. The Hannover Medical School's ethics committee approved all procedures, and informed consent was obtained from all donors.

For cell isolation, bone marrow aspirates were washed with cell culture medium [DMEM/Ham's F12 (1:1) (Biochrom, Berlin, Germany)] supplemented with 10% fetal calf serum (Biochrom, Berlin, Germany), 100 µg ml$^{-1}$ penicillin/streptomycin (Biochrom, Berlin, Germany), 2.5 µg ml$^{-1}$ amphotericin B (Biochrom, Berlin, Germany), 7.5% sodium hydrogen carbonate (Merck, Darmstadt, Germany), buffered with 4-(2-hydroxyethyl)-piperazin-1-ethane sulfonic acid (HEPES) buffer (Roth, Karlsruhe, Germany; pH 7.0). The cell pellet was centrifuged over a Percoll gradient (Amersham Biosciences, Buckinghamshire, UK; $d$ = 1.131 g ml$^{-1}$) for 15 min at 1,750×$g$. The supernatant was washed again with cell culture medium and was then cultivated in standard cell culture flasks at 37°C and 5% $CO_2$ in humidified atmosphere for at least 5 days. The culture medium with non-adherent cells was removed. In order to obtain enough BMSC, cells of the third passage were used.

BMSCs were incubated with differentiation medium (culture medium as above supplemented with 50 μg ml$^{-1}$ β-ascorbic acid, 10-mM β-glycerol phosphate, and 10-nM dexamethasone 21-dihydrogen phosphate) for at least 1 week before the start of the experiments.

### 2.1.2 Cyclic Longitudinal Strain

The cell-stretching system consisted of rectangular, elastic silicone dishes in which the whole dish, not only the cell culture surface, was deformable. The dishes were designed for use in a stimulation apparatus driven by an eccentric motor that allowed variation in amplitude (0.5–10%) and frequency (0.5–2 Hz) of applied strain (Fig. 1). The dishes consisted of a two-component silicone elastomer (Silbione® RTV 71556 A + B, Rhône-Poulenc Silicon GmbH, Lübeck, Germany) at a ratio of 10:1 of silicone oil:crosslinker. The rectangular dishes were 8 cm long × 3 cm wide × 1 cm high, and the wells had a 5 cm × 2.3 cm cell culture surface. New dishes were autoclaved at 121°C and preconditioned for 3 days in culture medium before the cells were seeded.

BMSCs of the second passage were harvested, counted and an overall viability of more than 90% was observed using the trypan blue exclusion test. 1.5 × 10$^5$ cells were seeded on each silicone dish. After 24 h of culture, the concentration of fetal calf serum was reduced to 1% for 24 h in order to align most cells into the $G_0$ phase of the cell cycle.

The cells in the silicone dishes were cyclic-longitudinally strained at a frequency of 1 Hz and amplitude of 5%. Short time strain was applied for either 15 or 60 min. The observation periods after cessation of strain were 6, 12, and 24 h. Long time strain was applied three times for 8 h with pauses of 15 h between the single straining periods. As a control, cells were grown on silicone dishes, but did not receive any strain.

### 2.1.3 Proliferation

Cell proliferation was monitored using BrdU incorporation (Roche, Mannheim, Germany). BrdU was added to the cells on the silicone dishes directly before every

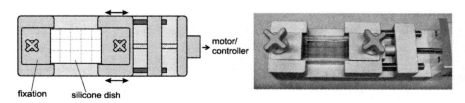

**Fig. 1** Stimulation apparatus driven by an eccentric motor for cyclic stretching of silicone dishes

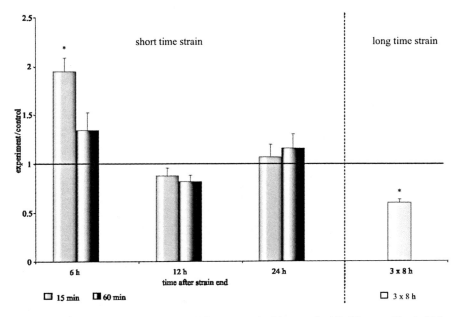

**Fig. 2** Proliferation rates of strained BMSCs measured with a standard BrdU assay. The *bold line* represents values of static controls. Values are given as mean of six samples ± SEM. Statistically significant values are designated by an *asterisk* (*)

stretching period of 8 h. BrdU detection was performed according to the manufacturer's instructions. Relative proliferation rates were determined by comparing strained cells with static control cells.

Six hours after 15 min of cyclic longitudinal strain, proliferation rates of BMSCs were significantly increased to 1.95 ± 0.14 compared to static control cells ($p < 0.05$) (Fig. 2). Twelve and 24 h after cessation of this 15 minutes of strain, no differences in proliferation from control cells could be detected. After 60 min of cyclic longitudinal strain a similar pattern to the 15 min of strain was observed. However, no significant differences could be detected in comparison to static controls. Highest proliferation rates were observed 6 h after cessation of the 60-minutes' strain (1.34 ± 0.19). Again, proliferation rates returned to levels as seen in static controls at 12 and 24 h. Application of repetitive long time cyclic longitudinal strain resulted in lower proliferation rates compared to static control cells.

### 2.1.4 Apoptosis

After the application of mechanical strain, cells on the silicone dishes were washed with PBS and then 1-ml fluorescent activated cell scanner (FACS) binding buffer was added. The cells were detached from the dishes using a cell scraper and spun down for 5 min at $1,750 \times g$ and 4°C. Cell pellets were resuspended in 100-ml FACS binding

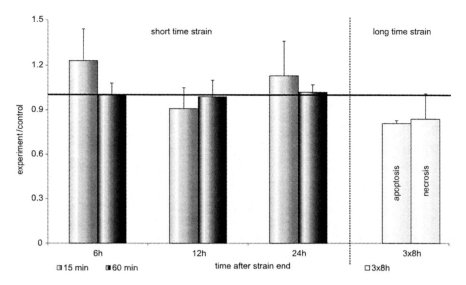

**Fig. 3** Apoptosis and necrosis rates of long time strained (3 × 8 h) BMSCs measured with flow cytometry after dyeing with FITC coupled annexin and propidium iodide. The *bold line* represents values of static controls. Values are given as mean of six samples ± SEM

buffer and incubated for 20 min in the dark with 5-ml fluorescein isothiocyanate (FITC)-labeled Annexin V (Bender MedSystems Diagnostics GmbH, Vienna, Austria) to detect early apoptotic cells and 5-ml propidium iodide (Bender MedSystems Diagnostics GmbH, Vienna, Austria) to detect late apoptotic and dead cells. Cells were washed and analyzed by flow-cytometry. Relative rates of apoptosis and cell death were calculated by comparing strained versus static control cells (Fig. 3).

Apoptosis rates were slightly increased 6 h after 15 min of strain application. 60 min of cyclic longitudinal strain induced apoptosis after 12 and 24 h. Both apoptosis and cell death rates after long time strain decreased to 0.81 ± 0.02 and 0.84 ± 0.17, respectively, compared to static control cells (the difference was statistically insignificant).

### 2.1.5 Mineralization

Mineralization was detected by using von Kossa staining. Cells on the silicon dishes were washed twice with phosphate buffered saline. Subsequently, cells were fixed for 10 min using 3% phosphate buffered formaldehyde. A 3% silver nitrate solution was added to the cells for 30 min in the dark. Silver-calcium precipitation in the matrix was developed using 1% pyrogallol for 3 min. Finally, surplus silver was removed and the precipitates were fixated using 5% sodium thiosulfate for 3 min. Lime precipitates were dark colored. Nuclei were counterstained using nuclear fast red.

After 60 min of strain, cells showed significant mineralization. Static control cells displayed no mineralization at all. After 15 min of strain, no mineralization was observed.

### 2.1.6 Western Blotting

The influence of cyclic longitudinal mechanical load on expression and activation of signal transduction proteins (JNK, ERK, p38) was studied by western blot. Cells were washed with PBS and lysed using 100-μl Laemmli buffer (2.5% sodium dodecyl sulfate (SDS), 12.5% glycerol, 0.025 M TRIS, 0.5 mM ethylene diamine tetraacetic acid (EDTA), 2.5% mercapto ethanol, 0.01% bromphenol blue) and vigorously detached from the dishes.

Gel electrophoresis and blotting was performed onto a nitrocellulose membrane. After incubation with the specific antibodies, bands were visualized using the enhanced chemiluminescence (ECL) system (Amersham, Biosciences, Buckinghamshire, UK). Band intensity was analyzed densitometrically and semi quantified relating to the band intensity of β-actin. The amount of activated MAP kinase was related to the amount of total corresponding protein. Strained cells were compared with the respective static controls (Fig. 4).

Activation of p38, ERK, and JNK was determined only in long time strained cells (Fig. 4). Phosphorylated p38 could not be detected. p38 was less expressed in strained cells compared to the static controls. Phosphorylation rates of all detected proteins in strained cells were not significantly different from static control cells.

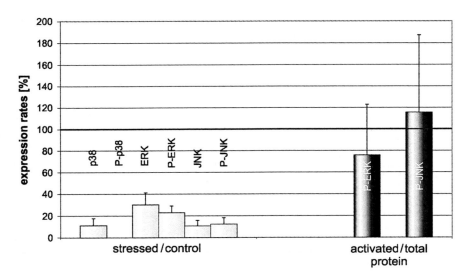

**Fig. 4** Expression rates of MAP Kinases (p38, ERK, JNK) in long time strained cells. *Left columns*: protein levels of strained cells related to static controls. *Right columns*: activated proteins in strained cells related to the respective protein level of activated and non-activated protein

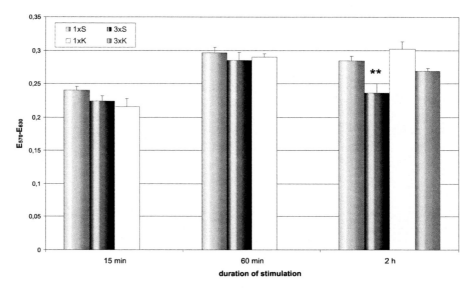

**Fig. 5** Viability of adMSCs after mechanical strain of once (1×S) or thrice (3×S) 15 min, 60 min, or 2 h of cyclic elongation (5%, 1 Hz). Values are given as mean of six samples SEM. Statistically significant values ($p < 0.001$) are designated by an *asterisk* (\*\*)

## 2.2 Adipose-Derived Mesenchymal Stem Cells (AdMSC)

### 2.2.1 AdMSC Isolation and Cultivation

AdMSCs were isolated from adipose tissue according to established methodologies [73]. AdMSCs were cultured with standard proliferation medium (normal medium = NM, Dulbecco's Modified Eagle's Medium, 10% fetal calf serum, 2-mM glutamine, antibiotics). For experiments, only third and fourth passage cells were used. For osteogenic differentiation, adMSCs were cultured with differentiation medium (DM = NM supplemented with 10-nM dexamethasone, 50 mg ml$^{-1}$ ascorbic acid and 10-mM beta-glycerol phosphate) or BMP-2-containing differentiation medium (10 ng ml$^{-1}$).

### 2.2.2 Strain Experiments

The cell straining was the same as for the experiments using BMSC. Preceding strain application, adMSCs were cultivated with DM for 7 days. Serum was reduced to 1% for 24 h directly before experiments started. Cyclic mechanical strain was applied with the frequency of 1 Hz and 5% elongation. Strain duration was 15, 60 min, and 2 h, respectively, and a repetitive strain of three-times 15, 60 min, or 2 h with each having an intermission of double the strain time (i.e., 30, 60 min, and 4 h, respectively). As a control, cells were grown on silicone dishes, but did not receive any strain. Each experiment was carried out with $n = 6$. After strain experiments, an MTT (3-(4,5-dimethylthiazol-2-yl)-2,5-diphenyl-tetrazolium-bromide) assay was performed, or cells were harvested for RNA isolation and Reverse Transcriptase Polymerase Chain Reaction (RT-PCR) or for alkaline phosphatase activity test.

### 2.2.3 MTT Assay

After finishing the one or three cycles of mechanical stimulation, the cultivation medium was removed. 3-ml fresh medium and 300-ml MTT solution 5 mg ml$^{-1}$ in PBS) were added. After 4 h incubation at 37°C, 3-ml SDS (10% in 0.01 N HCl) were added to dissolve the formazan crystals. After overnight incubation, formazan absorption of the supernatant was measured at 570 nm in a plate reader with 630 nm as absorption reference. Non-cell seeded silicone was used as negative control.

An MTT assay was performed after one and three 2-h periods of strain as well as 15 and 60 min of strain and revealed no significant differences between strained cells and controls (Fig. 6). Obviously, 2 h of strain did not cause any detrimental effects in so far that cell viability was considerably affected.

### 2.2.4 RT-PCR

For RT-PCR, RNA was isolated with the Trizol® method and cDNA produced from each sample and a PCR for the housekeeping gene glyceraldehyde 3-phosphate dehydrogenase (GAPDH) was performed for each individual sample. For bone marker PCRs the six respective samples were pooled. Agarose gel bands were densitometrically semiquantified and related to GAPDH expression. Investigated bone markers were type I collagen, core binding factor alpha 1 (cbfa1, runx2), osteopontin (OPN), AP, bone morphogenetic protein 4 (BMP-4), and OC.

*BMP-2 Addition*: AdMSCs were cultivated with standard proliferation medium NM, osteogenic differentiation medium, and with BMP-2-containing medium in tissue culture flasks. The expression of osteogenic markers was determined via RT-PCR. Type I collagen, cbfa1/runx2, osteopontin, BMP-4, and osteocalcin were detected (Fig. 7). Interestingly, all these markers were detected even after cultivation with NM without any differentiation supplements. Probably, this was due to confluent growth of the cells which is known to sometimes trigger differentiation processes. Osteopontin, osteocalcin, and alkaline phosphatase expression were lowest in the standard cultivation medium, higher in osteogenic differentiation medium and highest in BMP-2-containing medium. Since the differentiation stimulus

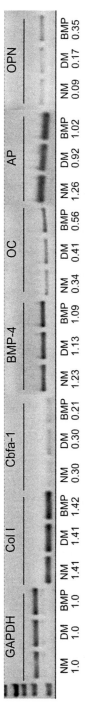

**Fig. 6** Expression of bone marker genes detected with RT-PCR in adMSCs treated with standard proliferation medium NM, osteogenic differentiation medium DM and with BMP-2-containing medium. GAPDH: glyceraldehyde 3-phosphate dehydrogenase, Col I: type I collagen, OC: osteocalcin, AP: alkaline phosphatase, OPN: osteopontin. Relative band intensity related to GAPDH band intensity is given underneath each band

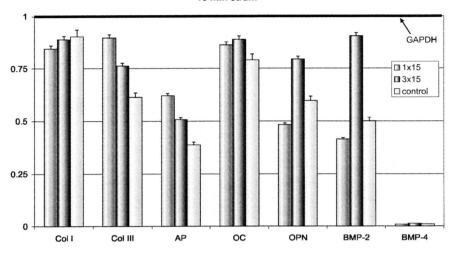

**Fig. 7** Expression rates of bone marker genes detected with RT-PCR in adMSCs strained for 15 min. GAPDH: glyceraldehyde 3-phosphate dehydrogenase, ColI: type I collagen, OC: osteocalcin, AP: alkaline phosphatase, OPN: osteopontin. Relative band intensities are related to GAPDH band intensity. All values are given as experimental mean of six samples SEM of individually performed GAPDH PCRs

of BMP-2 is supposed to be higher than that of dexamethasone, these findings may be related to the BMP-2 supplemented to the medium.

*Strain*: With RT-PCR of strained cells, bone markers like collagen I, alkaline phosphatase, osteocalcin, osteopontin, and BMP-2 and -4 were detected. So the cells exhibited the desired osteogenic phenotype. Mechanical strain affected osteogenic marker expression differently, depending on strain duration and repetition. Additionally, collagen III PCRs were performed. Collagen III is typical for scar tissue, thus, indicating the cells initiated repair mechanisms. Therefore, changes in collagen III expression can be interpreted with regard to detrimental effects of the applied strain.

Short time strain of one period of 15 min did not have distinct effects on osteogenic marker expression compared to static control cells (Fig. 8). Collagen III expression in 15-min strained cells was higher than in controls indicating the cells had initiated repair mechanisms. Three periods of 15 min of strain, however, resulted in increased osteopontin and BMP-2 expression compared to controls and once-strained cells, while collagen III expression was lower than after singular strain, but still above the control level. So obviously, repetition of strain provides an extra stimulus regarding osteogenic markers and moreover, cells seem to become customaced to a physically active environment when the strain is repeated.

One period of 60 min of strain had significant effects on bone marker expression (Fig. 9). Alkaline phosphatase, osteopontin, osteocalcin, BMP-2 and BMP-4 showed considerably higher expression levels in strained cells than in controls.

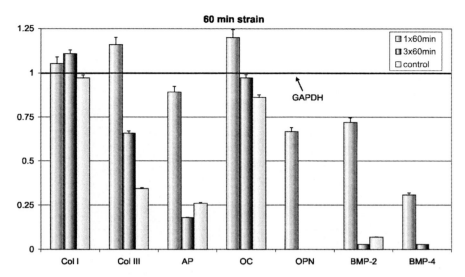

**Fig. 8** Expression rates of bone marker genes detected with RT-PCR in adMSCs strained for 60 min. GAPDH: glyceraldehyde 3-phosphate dehydrogenase, ColI: type I collagen, OC: osteocalcin, AP: alkaline phosphatase, OPN: osteopontin. Relative band intensities are related to GAPDH band intensity. All values are given as experimental mean of six samples SEM of individually performed GAPDH PCRs

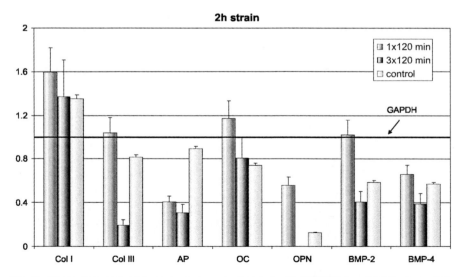

**Fig. 9** Expression rates of bone marker genes detected with RT-PCR in adMSCs strained for 2 h. GAPDH: glyceraldehyde 3-phosphate dehydrogenase, ColI: type I collagen, OC: osteocalcin, AP: alkaline phosphatase, OPN: osteopontin. Relative band intensities are related to GAPDH band intensity. All values are given as experimental mean of six samples SEM of individually performed GAPDH PCRs

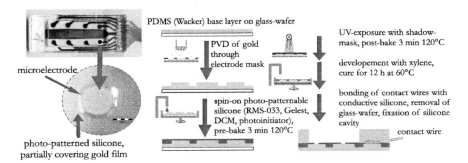

**Fig. 10** Fabrication of flexible microelectrode dishes

The 60-min strained cells obviously possessed a different differentiation status than the control cells. But collagen III levels were very high indicating cellular damage. Repetition of 60 min strain again yielded lower collagen III levels, but still the control level was not achieved. Additionally, repetition of 60 min strain did not induce bone markers like one 60-min period did, and no further differences between thrice-strained cells and controls were discovered.

One period of 2 h of strain yielded slightly elevated expression rates of osteocalcin, osteopontin, and BMP-2 compared to static controls, while three periods of 2 h of strain had no such effects (Fig. 10). However, collagen III levels in these experiments were comparable or even lower than in static controls indicating cell adjustment to mechanical strain.

### 2.2.5 Fabrication of Flexible Microelectrode Dishes

Flexible microelectrode dishes were prepared by Physical Vapour Deposition (PVD) of gold (99.99%, ABCR, Karlsruhe, Germany) through an appropriate mask on a 1-mm Polydimethylsiloxan (PDMS) base layer (Elastosil RT601, Wacker, München, Germany). Photopatternable silicone (PPS) was created by mixing 200 mg 2,2-dimethoxy-2-phenylacetophenone (Sigma, Schnelldorf, Germany) in 400 µl DCM with 10 g RMS-033 (Gelest, Tullytown, PA). The gold film on PDMS was spin-coated with PPS at 3,000 rpm, followed by a prebake of 3 min at 120°C. The PPS-film, covered with a shadow-mask, defining the microelectrode structure, was exposed to UV radiation. Unpolymerized silicone was removed after a 3-min postbake at 120°C by flushing with xylene at 3,000 rpm. Final curing and removal of excess photoinitiator and xylene occurred overnight at 60°C. Contact pads were bonded with conductive silicone (Emerson&Cuming) to a flat cable for connecting the impedance measurement system. In a last step, completing the dish, a silicone cavity was fixed with PDMS around the electrode array (Fig. 11).

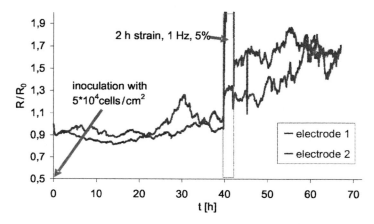

**Fig. 11** Electric cell substrate impedance sensing (ECIS)

**Fig. 12** Sponceram carrier disc and microscopic structure (www.zellwerk.biz)

### 2.2.6 Electric Cell–Substrate Impedance Sensing (ECIS)

A flexible microelectrode dish was equilibrated with DMEM (10% FCS, P/S) overnight and seeded to 50% confluence with MC3T3-E1 cells ($5 \times 10^4$ cells cm$^{-2}$). Impedance measurements were carried out by applying 1 $V_{p-p}$ AC, 4 kHz through a 1 MΩ resistor to the microelectrodes (Fig. 12). In-phase and out of-phase electrode-current were recorded by a Lock-In Amplifier and a homebrew Labview-tool. At 40 h mechanical stimulation was carried out (5% strain, 1 Hz, 2 h). At 208 h cells were killed with triton-X 100. Microscopic pictures of the dry electrodes were shot directly after fabrication of the dish and at the end of the cultivation.

ECIS measurements combined with mechanical stress were carried out on MC3T3-E1 cells (Fig. 12). After the attachment of cells and complete spreading, mechanical stimulation was carried out at 40 h. The cyclic stimulation led to a significant increase in the impedance of electrode No. 3 due to microfractioning, but the electrode remained operational.

Microscopy of electrodes 1, 2, and 4 confirmed that these electrodes were not compromised by the cyclic stress as much as electrode 3 (pictures not shown). Impedance of all four electrodes could be measured until the end of the cultivation at 208 h when cells were killed with triton-X. The next generation of electrodes will be outfitted with a chromium adhesion layer between the gold film and silicone to prevent microfractioning and signal alteration caused by strain. With these dishes ECIS-measurements under strain will be carried out to determine the influence of cyclic mechanical stimulation on cell morphology and proliferation.

## 3 3D Culture in a Rotating Bed Bioreactor

### 3.1 Sponceram®

Sponceram® is a ceramic support material consisting of doped zirconium oxide (Fig. 13). The structure of the material combines a unique mixture of macro- and micropores. The large surface ($2 \text{ m}^2 \text{ g}^{-1}$) enhances cell adherence and stimulates the formation of extracellular matrix. Both Sponceram®(pore size: 900 µm) as well as Sponceram® with hydroxyapatite (HA) coating (pore size 600 mm) were used as carrier discs (65 µm in diameter, 3 mm thickness) for the ZRP-Bioreactor.

### 3.2 Cell Cultivation in the ZRP-Bioreactor

The ZRP-Bioreactor can be equipped with up to 20 thin Sponceram® discs (Fig. 14). In our study, the reactor was equipped with four Sponceram® carrier discs (65 mm in diameter, 3 mm thickness) and respective spacers for the cultivation

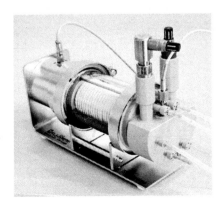 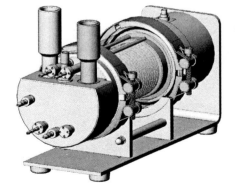

**Fig. 13** ZRP® bioreactor with rotating bed

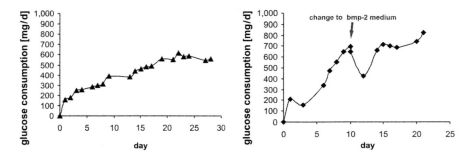

**Fig. 14** Glucose consumption of MC3T3-E1 cells on Sponceram® in the ZRP® bioreactor. *Left*: cultivation in standard proliferation medium. *Right*: cultivation initially with standard proliferation medium followed by change on day 10 to BMP-2-containing medium (10 ng mL$^{-1}$)

of MC3T3-E1 cells. Cell inoculation was carried out with a total volume of 2 ml cell suspension/disc (seeding cell number: 55 × 10$^6$ per disc). For the cultivation of primary osteoblasts the reactor was equipped with two Sponceram® and two Sponceram/HA® carrier discs and respective spacers for the cultivation. Osteoblasts of the third passage were used and the inoculation was carried out with a total volume of 2 ml cell suspension/disc (seeding cell number: 1 × 10$^7$ per disc).

The cell suspensions for the cultivations were injected through a feeding pipe onto the carrier discs (Fig. 14). To distribute the cell suspension homogenously onto the discs a rotation speed of 4 rpm during cell seeding was applied. To allow adhesion onto the Sponceram® surface the reactor was filled with 300 ml of medium 30 min after cell inoculation. The following cultivation was performed at 37°C, 2 rpm, and a pH of 7.3.

One culture experiment of the MC3T3-E1 cells was performed for 21 days using standard medium. The second one was performed for 10 days using standard medium followed by 11 days cultivation using BMP-2 medium (medium composition: see above). The cultivation of the osteoblasts was performed for 26 days in differentiation medium. After the cultivations in the ZRP-Bioreactor the disc-shaped scaffolds were used for the investigation of matrix mineralization, scanning electron microscopy, and RT-PCR analysis.

## 3.3 Glucose Assay

Cell growth during the bioreactor cultivations was determined by the estimation of glucose consumption using the YSI 2700 automated glucose analyzer (Yellow Springs Instruments, USA).

As an indicator for cell proliferation the glucose consumption was measured during the cultivation in the ZRP-Bioreactor (Fig. 15). The results revealed that the glucose consumption was 8% higher in BMP-2 medium (total consumption: 9.142 g) than using standard medium (total consumption: 8.389 g).

**Fig. 15** Matrix mineralization after ZRP® cultivation of MC3T3-E1 with standard medium and BMP-2-containing medium, respectively. *Left*: von Kossa staining, *right*: Alizarin Red staining

## 3.4 Matrix Mineralization

The Sponceram® discs were removed from the bioreactor after cultivation, washed with PBS and fixed in ice-cold 100% ethanol for 20 min at room temperature. The Alizarin Red staining was performed by washing with PBS followed by staining with 1% Alizarin Red in 2% of ethanol for 15 min at room temperature. For von Kossa staining fixed cells were washed with deionized water and incubated for 30 min in 5% $AgNO_3$ in the dark, washed again and exposed to ultraviolet light for 2 min. Cells were fixed with 5% sodium thiosulfate for 2 min and washed three times with deionized water.

The mineralization of the extracellular matrix was qualitatively determined by histochemical staining with Alizarin Red and von Kossa. Cells cultivated in BMP-2 medium showed a highly intense von Kossa and Alizarin Red staining (Fig. 16).

Interestingly, MC3T3-E1 cells cultured in standard medium in the ZRP-Bioreactor showed mineralization of the extracellular matrix.

## 3.5 Scanning Electron Microscopy

Cell grown Sponceram® discs were fixed in Karnovsky buffer at 4°C over night prior to scanning electron microscopy. Samples were then dehydrated in solutions containing increasing percentages of acetone (10, 30, 50, 70, 90, 100%) and subsequently imaged with a JEOL JSM-6700F scanning electron microscope.

Scanning electron micrographs showed the cell morphology of MC3T3-E1 cells cultured on Sponceram®. The first experiments under static conditions in 96-well dishes showed that the cells grew well inside the macroporous structure of Sponceram® showing a cuboid morphology of osteoblast-like cells with and without BMP-2 (Fig. 17a–d). Cells were cultured for 10 days. They grew as an

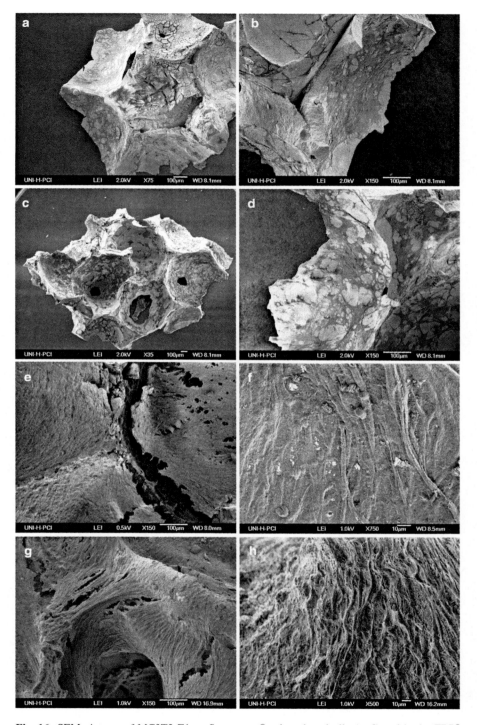

**Fig. 16** SEM pictures of MC3T3-E1 on Sponceram® cultured statically (**a–d**) and in the ZRP® bioreactor (**e–h**). (**a, b, e, f**) Standard medium. (**c, d, g, h**) BMP-2-containing medium

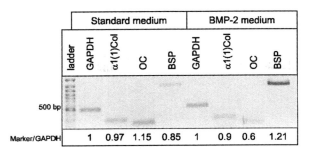

**Fig. 17** RT-PCR for the bone markers type I collagen (a1(1)Col), osteocalcin (OC) and bone sialoprotein-2 (BSP) of MC3T3-E1 cultured in the ZRP® bioreactor. *Numbers* underneath the gel represent band intensity related to GAPDH band intensity

interconnecting network having intercellular contacts to the surrounding cells. In comparison to the scaffold cultured in the bioreactor the cell density under static conditions is low without a visible ECM formation. Cultivating the cells for 21 and 28 days respectively in the bioreactor resulted in a dense cell layer thus the ceramic scaffold was no longer visible (Fig. 17e–h). A tight layer of ECM covered the scaffold as no scaffold structure could be observed. The examination of the cell morphology after cultivating the scaffolds with or without BMP-2 did not show significant differences. Fibrils in the ECM can be observed whereas few are arranged in a parallel structure.

## 3.6 Reverse Transcriptase-Polymerase Chain Reaction (RT-PCR)

MC3T3-E1 cells were cultivated on scaffolds in the ZRP-Bioreactor as described above. Cells were removed from the disc by incubation in the enzyme mix ZW-DT-04 (Zellwerk GmbH, Oberkraemer, Germany) at 37°C for 2 h and centrifuged at 400×$g$ for 5 min. Cells were disrupted with RiboLyse tubes green (Hybaid, Heidelberg, Germany) for 40 s at 6.0 Fast Prep FP 120 (Bio 101® Systems, Qbiogene, Heidelberg, Germany). The RNA was isolated by the SV total RNA Isolation System (Promega, Mannheim, Germany).

RT-PCR was carried out with 2 μg RNA using the Superscript II system (Promega, Mannheim, Germany) with oligo dT primer in a total volume of 40 μl. PCR was performed in a PCR-Thermocycler (MWG Biotech, Ebersberg, Germany) using specific primers. The reaction volume was 50 μl with an equivalent RNA concentration of 0.1 μg.

Amplification reactions were performed using the following primers (each 10 pmol) and protocols (30 cycles):

α1 (I) collagen:          Forward: 5'-TTC TCC TGG TAA AGA TGG TGC-3'
                          Reverse: 5'-GGA CCA GCA TCA CCT TTA ACA-3'
                          Annealing 57°C, 255 bp product
Osteocalcin (OC):         Forward: 5'-ACA AGT CCC ACA CAG CAG CTT-3'
                          Reverse: 5'-GCC GGA GTC TGT TCA CTA CCT-3'
                          Annealing 62°C, 187 bp product
Bone sialoprotein (BSP):  Forward: 5'-CTG TAG CAC CAT TCC ACA CT-3'
                          Reverse: 5'-ATG GCC TGT GCT TTC TCG AT-3'
                          Annealing 56°C, 1,055 bp product
GAPDH:                    Forward: 5'-GCC ACC CAG AAG ACT GTG GAT-3'
                          Reverse: 5'-TGG TCC AGG GTT TCT TAC TCC-3'
                          Annealing 60°C, 455 bp product

The differentiation process into bone cells in both cultures including the ECM mineralization was confirmed by analyzing osteoblastic marker proteins. Results of mRNA levels showed that collagen I, osteocalcin, and bone sialoprotein were expressed independently of BMP-2 medium (Fig. 18). The marker protein/GAPDH ratio revealed that the osteocalcin expression was higher using standard medium conditions cultivating in the ZRP-Bioreactor compared to BMP-2 medium. Collagen I expression level was similar in both cultures; bone sialoprotein level was higher using the BMP-2 medium.

## 4 Discussion

The significance of mechanical loading for bone metabolism has been demonstrated extensively in various studies. Mechanical passive states of the skeletal system due to weightlessness, functional immobilization, or prolonged postoperative bed rest have been shown to result in decreased bone formation and mineralization as well as reduced protein synthesis [1, 2]. On the other hand, bone mass increases with increased skeletal loading [4, 5]. In the mean time, the application of physical forces has entered the field of tissue engineering using electromagnetic fields, ultrasound or mechanical loading including pressure, fluid flow, torsion, and tension. For bone tissue engineering, mechanical forces like linear stretching or pressure correlate most closely with the physiological conditions and, therefore, are most widely used in connection with bone cells or bone-like tissue. However, the methods of strain application vary widely according to substrate material and geometry. Physical parameters like strain duration, elongation and frequency can also be varied to suit differentiation potential of stem cells.

### 4.1 2D Culture

In order to differentiate BMSCs into osteoblasts and additionally precondition them, BMSCs were strained on flexible silicone dishes in a cyclic longitudinal way. The applied frequency (1 Hz) and elongation (5%) were chosen in accordance with other studies that implied that these values are optimal [8, 9, 74]. Since it has not yet been described what the optimal strain duration is, at first short time strain (15 or 60 min) was applied and the strained cells were tested for proliferation rates and mineralization.

Fifteen minutes of strain gave rise to a significant twofold increase of the proliferation rate 6 h after strain ended but 12 h after strain end the proliferation rate was back to normal. Sixty minutes of strain, however, did not yield an increased proliferation rate. Overall, 15 min short time strain appears to be the better strain duration yielding increased proliferation. However, all observed effects were only short-lasting and cell metabolism seemed to be back to prestrain levels 12–24 h after the cessation of strain application. Application of repetitive strain regimens to fibroblasts are known to induce sustained increased proliferation rates with hardly any apoptosis [75]. This protective effect was accounted for by the induction of HSP72 [75]. Therefore, in the present study, repeated long time strain (3 × 8 h) was also applied.

Long time strained BMSCs showed a 40% reduction of proliferation rate which could be a sign for an advanced differentiation status. Proliferation rates are decreased during differentiation and terminally differentiated cells often do not proliferate at all. Furthermore, FACS analysis after repeated long time strain showed no increase in the percentage of either early apoptotic or late apoptotic cells. As a matter of fact, apoptosis rates were even lower than those observed in static controls. In fact, earlier studies indicate an association between apoptosis rates and differentiation status. Weyts et al. induced apoptosis in osteoblasts by mechanical strain but observed decreasing apoptosis rates with longer cultivation in osteogenic medium [76]. Therefore, the low apoptosis rate after repeated long time strain observed in this study may not only indicate the development of strain tolerance but also an advanced cell differentiation status. The increased differentiation of the BMSCs is strongly supported by the pronounced mineralization as detected by von Kossa staining.

We were not able to relate MAP kinase phosphorylation to any cellular reaction after repeated long time strain. Obviously BMSCs do not activate MAP kinases permanently, but only for a short time to induce the subsequent reactions. Interestingly, low MAP kinase levels after repeated long time strain suggest increased MAP kinase degradation after numerous activations due to the applied strain. Thus, it is very possible that the MAP kinases have been phosphorylated extensively. Further experiments will be carried out in order to investigate early MAP kinase activation. The increased expression of ERK in strained cells without phosphorylation should also be investigated.

In conclusion, this 2D study implies that short time strain of up to 1 h does not lead to persistent induction of human BMSC osteogenic differentiation. Thus, longer and/or repeated strain seems to be necessary in order to maintain a continuous differentiation stimulus. Moreover, this prepares the cells for a mechanically active environment. Mechanical strain may be a useful tool to help differentiate BMSCs to osteoblasts. Such differentiated cells can be seeded onto scaffolds and implanted to treat bone defects.

## 4.2 Rotating Bed Bioreactor Culture

The MC3T3-E1 cell line derived from mouse embryo calvaria at the osteoprogenitor stage was used to study the development of bone cells on Sponceram® scaffolds. It is a well-established cell line to investigate the differentiation process into bone [77].

The scaffold used for tissue engineering plays a critical role as it has to provide a tissue specific environment and architecture. Biocompatibility is essential as well as the supply with nutrients and cytokines. Additionally, the scaffold should promote or enhance cell proliferation and differentiation.

To guarantee the generation of a functional bone tissue substitute, it is necessary to develop an appropriate scaffold culture system that mimics the in vivo environment. Developing the scaffold using static culture conditions is not sufficient since the transport and distribution of nutrients and metabolic waste is inhomogenous without mixing of the medium. Additionally, using static conditions the dynamics of the in vivo environment found in bone-like mechanical stimulation are missing. The results of static cultivations showed that Sponceram® is an appropriate scaffold for fast cell proliferation. The use of static culture conditions is only limited by the insufficient nutrient and oxygen supply. In contrast, the course of glucose consumption showed that the cell proliferation was not limited during the bioreactor cultivation.

The mineralization of the ECM was independent of the addition of BMP-2 as demonstrated by von Kossa and Alizarin Red staining. Both cultivations had a positive von Kossa and Alizarin Red staining with a higher intensity on scaffolds cultured in BMP-2 medium. This fact seems to be obvious due to the differentiation induction of the growth factor. However, by analyzing the glucose concentration it was shown that there was an 8% higher consumption during the cultivation with the growth factor even though this cultivation was 7 days shorter. This leads to the conclusion that the higher mineralization could additionally be induced by an 8% higher cell number on the scaffold or because of a stronger calcification activity. To investigate the differentiation in more detail the expression of several proteins correlating with bone differentiation were determined. Alkaline phosphatase and collagen I are both proteins found in early osteoblastic differentiation [78]. The expression of collagen I as the primary and main matrix protein of bone tissue was identical after both bioreactor cultivations as revealed by RT-PCR results. In contrast the results of the static cultures showed an increase in AP activity only in the presence of BMP-2 or differentiation medium, respectively. These findings suggest that Sponceram® accelerates the early osteoblastic differentiation even in the absence of differentiation inductors when cultured in the ZRP-bioreactor.

Osteocalcin and bone sialoprotein as non-collagenous proteins are known to play an important role in matrix mineralization. OC is known to bind calcium and is likely to be involved in hydroxyapatite regulation [79]. BSP contains an RGD-sequence that interacts directly with integrins [80]. Results of RT-PCR analysis of the two ZRP-Bioreactor RBS cultivations demonstrated that Col I, OC, and BSP were expressed in both dynamic cultures, independent of the addition of BMP-2. These results confirm the above-mentioned findings, that the scaffold itself is able to induce differentiation in bone cells with concomitant mineralization when cultured in the bioreactor system. This promotion can be due to the composition of the Sponceram® scaffold and/or is related to its 3D structure. Scaffold materials like hydroxyapatite, tricalciumphosphates, or composites are known to induce differentiation via an interaction of the inorganic material with cellular components. This fact should also be considered here although Sponceram® does not comprise these components. In addition, the ideal culture conditions in the ZRP-bioreactor ensure a homogenous nutrient distribution throughout the scaffolds. This supported rapid cell proliferation and differentiation.

Considering the marker gene/GAPDH ratios there was a higher expression of osteocalcin using the standard medium and a higher expression of BSP using BMP-2 medium. This difference can be due to the fact that the two proteins underlie time-dependent up and down regulations thus different expression levels correlate with the two different culture durations.

In summary, this study demonstrated that Sponceram® within the ZRP-bioreactor is applicable for the differentiation of MC3T3-E1 cells into the osteoblastic phenotype. The macroporous structure of the scaffold contributed to a fast cell attachment and proliferation. The ultimate shape of the used scaffold and cultivation procedure provided the differentiation. Additionally, the alternate contact of cells to the medium and the oxygen atmosphere supported the proliferation and differentiation process of the MC3T3-E1 cells within the scaffold.

**Acknowledgements** This work was partially carried out under the scope of the European NoE "EXPERTISSUES" (NMP3-CT-2004-500283). Solvig Diederichs was a student in the early research training program Marie Curie "Alea Jacta EST" (MEST-CT-2004-8104). The Sponceram ceramics were kindly provided by Zellwerk GmbH (Oberkrämer, Germany) and the BMP-2 was a gift from Prof. Walter Sebald (University of Würzburg).

# References

1. Morey ER, Baylink DJ (1978) *Science* 201:1138
2. Simmons DJ, Russell JE, Grynpas MD (1986) *Bone Miner* 1:485
3. Vico L, Chappard D, Palle S, Bakulin AV, Novikov VE, Alexandre C (1988) *Am J Physiol* 255:R243
4. Leblanc AD, Schneider VS, Evans HJ, Engelbretson DA, Krebs JM (1990) *J Bone Miner Res* 5:843
5. Simkin A, Ayalon J, Leichter I (1987) *Calcif Tissue Int* 40:59
6. Rambaut PC, Goode AW (1985) *Lancet* 2:1050
7. Prince RL, Smith M, Dick IM, Price RI, Webb PG, Henderson NK, Harris MM (1991) *N Engl J Med* 325:1189
8. Kaspar D, Seidl W, Neidlinger-Wilke C, Beck A, Claes L, Ignatius A (2002) *J Biomech* 35:873
9. Thomas GP, el Haj AJ (1996) *Calcif Tissue Int* 58:101
10. Schaffer JL, Rizen M, L'Italien GJ, Benbrahim A, Megerman J, Gerstenfeld LC, Gray ML (1994) *J Orthop Res* 12:709

11. Wang H, Ip W, Boissy R, Grood ES (1995) *J Biomech* 28:1543
12. Jagodzinski M, Drescher M, Zeichen J, Hankemeier S, Krettek C, Bosch U, van Griensven M (2004) *Eur Cell Mater* 7:35
13. Grenier G, Remy-Zolghadri M, Larouche D, Gauvin R, Baker K, Bergeron F, Dupuis D, Langelier E, Rancourt D, Auger FA, Germain L (2005) *Tissue Eng* 11:90
14. Kaspar D, Seidl W, Neidlinger-Wilke C, Claes L (2000) *J Musculoskelet Neuronal Interact* 1:161
15. Neidlinger-Wilke C, Grood ES, Wang, JHC, Brand RA, Claes L (2001) *J Orthop Res* 19:286
16. Prockop DJ (1997) *Science* 276:71
17. Pittenger MF, Mackay AM, Beck SC, Jaiswal RK, Douglas R, Mosca JD, Moorman MA, Simonetti DW, Craig S, Marshak DR (1999) *Science* 284:143
18. Beresford JN, Bennett JH, Devlin C, Leboy PS, Owen ME (1992) *J Cell Sci* 102 (Pt 2):341
19. Wakitani S, Goto T, Pineda SJ, Young RG, Mansour JM, Caplan AI, Goldberg VM (1994) *J Bone Joint Surg Am* 76:579
20. Seshi B, Kumar S, Sellers D (2000) *Blood Cells Mol Dis* 26:234
21. Altman GH, Horan RL, Martin I, Farhadi J, Stark PR, Volloch V, Richmond JC, Vunjak-Novakovic G, Kaplan DL (2002) *FASEB J* 16:270
22. Bianco, P Gehron RP (2000) *J Clin Invest* 105:1663
23. Huang CY, Reuben PM, Cheung HS (2005) *Stem Cells* 23:1113
24. Yoshikawa T, Peel SA, Gladstone JR, Davies JE (1997) *Biomed Mater Eng* 7:369
25. Wozniak M, Fausto A, Carron CP, Meyer DM, Hruska KA (2000) *J Bone Miner Res* 15:1731
26. Zaman G, Dallas SL, Lanyon LE (1992) *Calcif Tissue Int* 51:132
27. Wang N, Ingber DE (1994) *Biophys J* 66:2181
28. Sandy JR, Meghji S, Scutt AM, Harvey W, Harris M, Meikle MC (1989) *Bone Miner* 5:155
29. Neidlinger-Wilke C, Wilke, H-J, Claes L (1994) *J Orthop Res* 12:70
30. Jones DB, Nolte H, Scholubbers JG, Turner E, Veltel D (1991) *Biomaterials* 12:101
31. van Griensven M, Zeichen J, Skutek M, Barkhausen T, Krettek C, Bosch U (2003) *Exp Toxicol Pathol* 54:335
32. Skutek M, van Griensven M, Zeichen J, Brauer N, Bosch U (2003) *Knee Surg Sports Traumatol Arthrosc* 11:122
33. Montaner S, Perona R, Saniger L, Lacal JC (1998) *J Biol Chem* 273:12779
34. Kuhnel F, Zender L, Paul Y, Tietze MK, Trautwein C, Manns M, Kubicka S (2000) *J Biol Chem* 275:6421
35. Zeichen J, van Griensven M, Bosch U (2000) *Am J Sports Med* 28:888
36. Kuroki Y, Shiozawa S, Sugimoto T, Fujita T (1992) *Biochem Biophys Res Commun* 182:1389
37. Binderman I, Zor U, Kaye AM, Shimshoni Z, Harell A, Somjen D (1988) *Calcif Tissue Int* 42:261
38. el Haj AJ, Minter SL, Rawlinson SC, Suswillo R, Lanyon LE (1990) *J Bone Miner Res* 5:923
39. Kawata A, Mikuni-Takagaki Y (1998) *Biochem Biophys Res Commun* 246:404
40. Brighton CT, Sennett BJ, Farmer JC, Iannotti JP, Hansen CA, Williams JL, Williamson J (1992) *J Orthop Res* 10:385
41. Walker LM, Publicover SJ, Preston MR, Said Ahmed MA, el Haj AJ (2000) *J Cell Biochem* 79:648
42. el Haj AJ, Walker LM, Preston MR, Publicover SJ (1999) *Med Biol Eng Comput* 37:403
43. Schmidt C, Pommerenke H, Durr F, Nebe B, Rychly J (1998) *J Biol Chem* 273:5081
44. Shin H, Jo S, Mikos AG (2003) *Biomaterials* 24:4353
45. Sefton MV, Woodhouse KA (1998) *J Cutan Med Surg* 3:18
46. Langer R (2000) *Acc Chem Res* 33:94
47. Hutmacher DW, Goh JC, Teoh SH (2001) *Ann Acad Med Singapore* 30:183
48. Hubbell JA (1995) *Biotechnology (NY)* 13:565
49. Zhao L, Chang J (2004) *J Mater Sci Mater Med* 15:625
50. Hardouin P, Anselme K, Flautre B, Bianchi F, Bascoulenguet G, Bouxin B (2000) *Joint Bone Spine* 67:419

51. Burg KJ, Porter S, Kellam JF (2000) *Biomaterials* 21:2347
52. Bohner M (2000) *Injury* 31:37
53. Arinzeh TL, Tran T, McAlary J, Daculsi G (2005) *Biomaterials* 26:3631
54. Alam MI, Asahina I, Ohmamiuda K, Takahashi K, Yokota S, Enomoto S (2001) *Biomaterials* 22:1643
55. Shu R, McMullen R, Baumann MJ, McCabe LR (2008) *J Biomed Mater Res A* 67:1196
56. Chou YF, Dunn JC, Wu BM (2005) *J Biomed Mater Res B* 75:81
57. Owen TA, Aronow M, Shalhoub V, Barone LM, Wilming L, Tassinari MS, Kennedy MB, Lian JB, Stein GS (1990) *J Cell Physiol* 143:420
58. Lian JB, Stein GS (1995) *Iowa Orthop J* 15:118
59. Reddi AH (1992) *Curr Opin Cell Biol* 4:850
60. Kawabata M, Imamura T, Miyazono K (1998) *Cytokine Growth Factor Rev* 9:49
61. Sebald W, Nickel J, Zhang JL, Mueller TD (2004) *Biol Chem* 385:697
62. Yamaguchi A, Komori T, Suda T (2000) *Endocr Rev* 21:393
63. Hollinger JO, Schmitt JM, Buck DC, Shannon R, Joh SP, Zegzula HD, Wozney J (1998) *J Biomed Mater Res A* 43:356
64. Takuwa Y, Ohse C, Wang EA, Wozney JM, Yamashita K (1991) *Biochem Biophys Res Commun* 174:96
65. Temenoff J, Mikos A (2000) *Biomaterials* 21:2405
66. Sikavitsas VI, Bancroft GN, Mikos AG (2002) *J Biomed Mater Res A* 62:136
67. Glowacki J, Mizuno S, Greenberger JS (1998) *Cell Transplant* 7:319
68. Mueller SM, Mizuno S, Gerstenfeld LC, Glowacki J (1999) *J Bone Miner Res* 14:2118
69. Sikavitsas VI, Bancroft GN, Holtorf HL, Jansen JA, Mikos AG (2003) *Proc Natl Acad Sci USA* 100:14683
70. Wang Y, Uemura T, Dong J, Kojima H, Tanaka J, Tateishi T (2003) *Tissue Eng* 9:1205
71. Yu X, Botchwey EA, Levine EM, Pollack SR, Laurencin CT (2004) *Proc Natl Acad Sci USA* 101:11203
72. Qiu QQ, Ducheyne P, Ayyaswamy PS (1999) *Biomaterials* 20:989
73. Zuk PA, Zhu M, Mizuno H, Huang J, Futrell JW, Katz AJ, Benhaim P, Lorenz HP, Hedrick MH (2001) *Tissue Eng* 7:211
74. Jagodzinski M, Cebotari S, Tudorache I, Zeichen J, Hankemeier S, Krettek C, van Griensven M, Mertisching H (2004) *Orthopade* 33:1394–1400
75. Barkhausen T, van Griensven M, Zeichen J, Bosch U (2003) *Exp Toxicol Pathol* 55:153
76. Weyts FA, Bosmans B, Niesing R, van Leeuwen JP, Weinans H (2003) *Calcif Tissue Int* 72:505
77. Quarles LD, Yohay DA, Lever LW, Caton R, Wenstrup RJ (1992) *J Bone Miner Res* 7:683
78. Aubin JE (1996) The Osteoblast lineage. In: Bilezikian J, Raisz L, Rodan G (eds) Principles of bone biology. Academia Press, New York p 51
79. Hauschka PV, Lian JB, Cole DE, Gundberg CM (1989) *Physiol Rev* 69:990
80. Oldberg A, Franzen A, Heinegard D, Pierschbacher M, Ruoslahti E (1988) *J Biol Chem* 263:19433

# Bioreactors for Tissue Engineering of Cartilage

S. Concaro, F. Gustavson, and P. Gatenholm

**Abstract** The cartilage regenerative medicine field has evolved during the last decades. The first-generation technology, autologous chondrocyte transplantation (ACT) involved the transplantation of in vitro expanded chondrocytes to cartilage defects. The second generation involves the seeding of chondrocytes in a three-dimensional scaffold. The technique has several potential advantages such as the ability of arthroscopic implantation, in vitro pre-differentiation of cells and implant stability among others (Brittberg M, Lindahl A, Nilsson A, Ohlsson C, Isaksson O, Peterson L, N Engl J Med 331(14):889–895, 1994; Henderson I, Francisco R, Oakes B, Cameron J, Knee 12(3):209–216, 2005; Peterson L, Minas T, Brittberg M, Nilsson A, Sjogren-Jansson E, Lindahl A, Clin Orthop (374):212–234, 2000; Nagel-Heyer S, Goepfert C, Feyerabend F, Petersen JP, Adamietz P, Meenen NM, et al. Bioprocess Biosyst Eng 27(4):273–280, 2005; Portner R, Nagel-Heyer S, Goepfert C, Adamietz P, Meenen NM, J Biosci Bioeng 100(3):235–245, 2005; Nagel-Heyer S, Goepfert C, Adamietz P, Meenen NM, Portner R, J Biotechnol 121(4):486–497, 2006; Heyland J, Wiegandt K, Goepfert C, Nagel-Heyer S, Ilinich E, Schumacher U, et al. Biotechnol Lett 28(20):1641–1648, 2006). The nutritional requirements of cells that are synthesizing extra-cellular matrix increase along the differentiation process. The mass transfer must be increased according to the tissue properties. Bioreactors represent an attractive tool to accelerate the biochemical and mechanical properties of the engineered tissues providing adequate mass transfer and physical stimuli. Different reactor systems have been [5] developed during the last decades based on different physical stimulation concepts. Static and dynamic compression, confined and nonconfined compression-based reactors have been described in this review. Perfusion systems represent an attractive way of culturing constructs under dynamic conditions. Several groups showed increased matrix production using confined and unconfined systems. Development of automatic culture systems and noninvasive monitoring of matrix production will take place during the next few years in order to improve the cost affectivity of tissue-engineered products.

---

S. Concaro
Department of Molecular Biology and Regenerative Medicine, University of Gothenburg, Gothenburg, Sweden

S. Concaro, F. Gustavson, and P. Gatenholm (✉)
Department of Chemical and Biological Engineering, Biopolymer Technology,
Chalmers University of Technology, SE41296 Gothenburg, Sweden

**Keywords** Bioreactors, Cartilage, Chitosan, Human chondrocytes, Regenerative medicine, Tissue engineering

**Contents**

1   Introduction............................................................................................................ 126
2   Articular Cartilage ................................................................................................. 127
    2.1   Composition................................................................................................. 127
    2.2   Collagen....................................................................................................... 127
    2.3   Proteoglycans............................................................................................... 127
    2.4   Tissue Fluid.................................................................................................. 128
    2.5   Interaction Between Components in Cartilage ........................................... 128
    2.6   Biomechanics of Articular Cartilage ........................................................... 129
    2.7   Permeability of Articular Cartilage.............................................................. 129
3   Bioreactor Systems ................................................................................................ 130
    3.1   Spinner Flasks.............................................................................................. 130
    3.2   Rotating-Wall Vessels ................................................................................. 131
    3.3   Hollow-Fiber Bioreactors ............................................................................ 131
    3.4   Dynamic Compression Bioreactors ............................................................. 131
4   Perfusion Systems.................................................................................................. 131
5   Physical Environment ........................................................................................... 132
6   Perfusion Systems for Cartilage Generation......................................................... 134
    6.1   Flow Velocity Calculations ......................................................................... 138
    6.2   Metabolic Parameter Determination............................................................ 138
    6.3   Histology...................................................................................................... 138
    6.4   Results ......................................................................................................... 139
7   Conclusion ............................................................................................................. 141
References.................................................................................................................... 141

# 1 Introduction

The cartilage regenerative medicine field has evolved over the last decades. The first-generation technology involved the transplantation of in vitro expanded chondrocytes to cartilage defects. This technique showed good long-term clinical results as reported by several independent orthopedic scientists [1–3]. Despite the good results of the first-generation technology of autologous chondrocyte transplantation (ACT) some of the concerns regarding the transplantation of immature chondrocytes in suspension led to the development of a second generation of ACT. This technology involves the seeding of chondrocytes in a three-dimensional scaffold. Other techniques include the use of in vitro-generated neocartilage using cartilage-carrier-constructs [4–8]. The technique has several potential advantages such as the ability of arthroscopic implantation, in vitro pre-differentiation of cells, and implant stability among others [9–15]. The nutritional requirements of cells that are synthesizing extra-cellular matrix (ECM) increase along the differentiation process. The mass transfer must be increased according to the tissue properties. Bioreactors represent a great tool to accelerate the biochemical and mechanical properties of the engineered tissues providing adequate mass transfer and physical stimuli [13]. Nondestructive monitoring of the cell–scaffold constructs is of profound importance to evaluate the evolution of three-dimensional dynamic cultures for human applications.

## 2 Articular Cartilage

### 2.1 Composition

Chondrocytes are the cells that make up cartilage. The amount of chondrocyte cells in the cartilage is less than 10% of the tissue volume. The chondrocyte cells produce an extra-cellular matrix that is composed of a dense network of collagen fibers (collagen II) and proteoglycans (PGs). The collagen content in cartilage is about 10–30% while the content of PGs is 3–10% (wet weight). The remaining composition is water. Other compounds are inorganic salts and small amounts of other matrix proteins, glycoproteins and lipids. It is the collagen and PGs that provide structure for the tissue and together with water determine the biomechanical properties and functional behavior of cartilage. The articular cartilage can be divided into different zones: the superficial tangential zone (10–20%), middle zone (40–60%), and deep zone (30%). There is also a calcified zone close to the bone.

### 2.2 Collagen

Collagen is a protein that is very common in the body. Three procollagen polypeptide chains (alfa-chains) that are coiled into left-handed helices, which are then coiled in a right-hand helix around each other, form the basic biological unit of collagen called tropocollagen. The tropocollagen then assembles into larger collagen fibrils. This crosslinking of tropocollagen is responsible for the tensile strength of collagen. The diameter of the collagen varies but the average diameter in articular cartilage is 25–40 nm. The distribution of collagen in the articular cartilage is not homogenous and varies through the different zones of the cartilage. In the superficial tangential zone the collagen fibers are densely packed and randomly orientated in planes parallel to the articular surface.

### 2.3 Proteoglycans

Proteoglycans are polysaccharide–protein compounds that can exist as either monomer or as aggregates. The polysaccharide part is about 95% with the other 5% being protein. The PG monomer is a protein core, about 200 nm long, covalently bound to several glycosaminoglycans (GAGs) and oligosaccharides (both O-linked and N-linked). There are about 150 GAG chains attached to the protein core. Some of the polysaccharides found in the articular cartilage are keratan sulfate, chondroitin sulfate, hyaluronic acid, dermatan sulfate, and heparan sulfate. The length of the keratan sulfate chain is about 13 disaccharide units while the chain length for the chondroitin sulfate is 25–30 disaccharide units. The GAGs are negatively charged, at least one per disaccharide making the GAGs repel each other but it also causes them to attract cations and interact with water. The amounts of these GAGs change as the cartilage ages. Chondroitin sulfate decreases while keratan sulfate increases. Most of the proteoglycan monomers form aggregates with hyaluronate [16].

## 2.4 Tissue Fluid

About 80% of the wet weight of the cartilage is fluid. The fluid contains mostly water but also gases, metabolites, and a large amount of cations which stabilizes the negative charges from the GAGs. The nutrient and oxygen transport and waste removal in the cartilage take place through diffusion exchange between the tissue fluid and the synovial fluid. Only a small percentage of the fluid is intracellular. About 30% of the fluid is believed to have a strong association with the collagen fibers thus being very important for the structural organization of the ECM. This interaction with the ECM provides the ability to resist and recover from compression. The rest of the fluid (about 70%) can move freely during loading [17–22].

## 2.5 Interaction Between Components in Cartilage

The negatively charged GAGs attract mobile cations in the tissue fluid such as sodium and calcium, which creates an osmotic pressure (Donnan Osmotic Effect) of approximately 0.35 MPa. The collagen network inhibits swelling, leading to a pre-stress in the collagen network. When the cartilage is compressed the internal pressure in the matrix exceeds the osmotic pressure and the fluid begins to flow out of the cartilage. This leads to an increase in the charge density from the GAGs, which then increases the osmotic pressure and charge–charge repulsion. This finally leads to equilibrium with the external stress. This property complements the tensile strength of the collagen fibers. The compression strength of the proteoglycans is derived from the osmotic swelling pressure and from the PG aggregates that are entangled in the collagen network. The elastic modulus for the collagen–PG matrix is approximately 0.78 MPa [17, 19, 21–23].

## 2.6 Biomechanics of Articular Cartilage

To understand the mechanical response of cartilage one can view it as biphasic tissue. First is the fluid phase, which comprises interstitial fluid and the inorganic salts. The other phase is the solid phase, which is represented by the extra-cellular matrix. So the cartilage can be seen as a fluid-filled, porous, and permeable material. Since the cartilage has a fluid part and a solid part its mechanical response is viscoelastic, a combination of the viscous response from the fluid and the elastic response of the solid. Because of this viscoelasticity the response to a constant load or constant deformation is time dependent. These responses are called creep and stress relaxation, respectively. If the response from a constant mechanical load is a quick initial deformation followed by a slow but increasing deformation until equilibrium is reached it is called creep. The other type of response is defined as a high initial stress followed by a slow decrease of the stress and this happens when a viscoelastic material is exposed to constant deformation and this is called stress relaxation.

The viscoelastic behavior during compression is mostly related to the flow of interstitial fluid but with shear it is mostly due to the motion of the collagen and PG chains. The part of the viscoelasticity that is caused by interstitial fluid is known as biphasic viscoelastic behavior and the part that is caused by the macromolecules of the matrix is known as flow independent or intrinsic behavior [24–27].

## 2.7 Permeability of Articular Cartilage

A porous material becomes permeable if the pores are interconnected thus making it possible for the fluid to flow through the material. The porosity ($\beta$) is defined as the fluid volume divided by the total volume. The permeability ($k$) is a description of how easy it is for the fluid to pass through the material and is inversely proportional to the frictional drag ($K$). So it is a measure of the force that is needed to move the fluid at a given velocity through the porous material. The frictional force is caused by the interaction of the fluid and the walls of the pores. The permeability $k$ is related to $K$ in the relationship:

$$k = \frac{\beta^2}{K}$$

There is also Darcy's law:

$$k = \frac{Qh}{A(P_1 - P_2)}$$

$Q$ is volume per unit time through the sample with the area $A$. $P_1 - P_2$ is the pressure difference between the different sides of the sample and $h$ is the height of the sample [17–19, 23, 26].

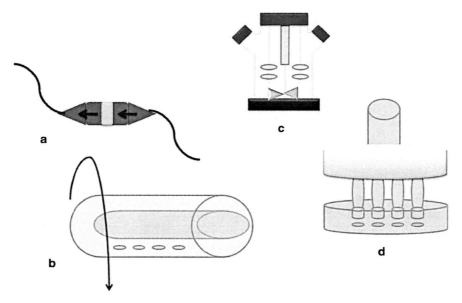

**Fig. 1** Different bioreactor systems used in tissue engineering. (**a**) Direct perfusion bioreactor. (**b**) Rotating-wall vessels. (**c**) Spinner flask. (**d**) Compression bioreactor

## 3 Bioreactor Systems

The role of a bioreactor is to mimic the in vivo conditions of a tissue. The bioreactor should be able to control the different variables that define the environment of the tissue. The pH, partial pressures ($pO_2$ and $pCO_2$), temperature, nutrient supply, and mechanical environment are a few of the parameters that need to be controlled (Fig. 1) [28–34].

### 3.1 Spinner Flasks

Spinner flasks are the most common mechanically stirred bioreactors (Fig. 1c). They use a magnetic stir bar to mix the medium. In this way the diffusion limit of 100–200 μm is overcome and the scaffolds are exposed to fresh nutrients and oxygen. Stirring is often performed at 50 rpm because this velocity inflicts no apparent physical damage to the cells, although a fibrous capsule can form on the scaffold surface. The mass transfer will not be uniform in the scaffold and there will be pH gradients and shear gradients which lead to inferior tissue formation. Some spinner-flask systems include ports and filters for gas exchange, which will lead to a more closed bioreactor but manual handling is still a must for medium exchange [35–37].

## 3.2 Rotating-Wall Vessels

The rotating-wall vessels (Fig. 1b) expose the scaffolds to an environment with low shear stresses and high mass transfer rates. The walls of the vessel rotate at a rate that balances the scaffold in the medium by evening out the forces exerted on the construct. The forces are the drag force ($F_d$), the centrifugal force ($F_c$), and the gravitational force ($F_g$) [36, 38, 39].

## 3.3 Hollow-Fiber Bioreactors

Hollow-fiber reactors use perfusion through semi-permeable fibers which increase transport of nutrients and oxygen. The cells are embedded in a gel inside the lumen of the fibers and the medium is transported from the exterior.

## 3.4 Dynamic Compression Bioreactors

By applying mechanical load, the compression reactors (Fig. 1d) try to emulate the physical environment of the tissue to generate functional tissue grafts. The dynamic compression is often control by a computer in which frequency and time interval for compression can be regulated. The dynamic strain of the scaffold is often superimposed on a static one whose purpose is to hold the scaffold still, as well as to emulate the static pressure that the cartilage is exposed to. The configuration of frequency, load, strain, and duration has varied a lot in different studies. With frequencies between 0.0001 and 3 Hz, compression stresses exerted on the scaffolds vary from 0.1 to 24 MPa and strain levels between 0.1 and 25% [40].

# 4 Perfusion Systems

Flow perfusion reactors continuously perfuse medium through the porous scaffold construct. A pump causes the medium to flow through the tubing and into the chamber, where the scaffold is confined so that the medium passes through the scaffold and not around it. This design improves the mass transfer in the interior of the three-dimensional scaffold. This enhances the nutrient delivery otherwise limited by external and internal diffusion. The fluid also exerts mechanical stimuli on the scaffold and cells in the form of fluid shear stress. The amount of shear stress that is exerted on cells in a perfusion chamber is dependent on the flow rate used. Shear stress on each individual cell will be dependent on the microstructure of the scaffold where size, porosity, and interconnectivity of the pores are of importance. An advantage of this kind of system is that the risk of contamination is much lower since the medium is not changed manually [40–43] (Fig. 2).

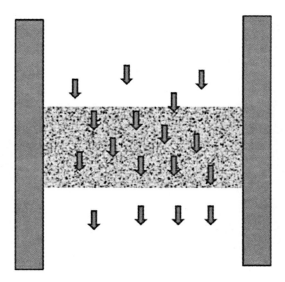

**Fig. 2** Flow through a porous scaffold

## 5 Physical Environment

There are two different kinds of flow type, laminar flow and turbulent flow. There can of course be a flow that contains both types (at either different times or positions in the flow) as well as a flow that is in a transition phase between the two flows. The physical variables that determine what kind of flow is present are the density ($\rho$), the viscosity ($\mu$), the flow velocity ($u$), and the geometry ($d$) where the liquid is present. These parameters can be combined to provide a dimensionless parameter which is called the Reynolds number ($Re$).

$$Re = \frac{\rho u d}{\mu}$$

Two flow patterns will behave the same if the $Re$ is the same even if the input variables ($\rho$, $\mu$, $u$, and $d$) are different. For small $Re$ the flow is laminar and for large $Re$ the flow is turbulent. There is no straight cut-off point but with a $Re$ of over 1,000 the flow can begin to be turbulent. The laminar flow is predictable and there exist calculation formulas. Therefore, checking the flow at the same spot at different time points will give the same result. The turbulent flow on the other hand is more chaotic and calculations need to be done using statistical computational methods.

The average flow velocity can be calculated if the flow rate is known. The flow rate can be controlled by settings on the pump. The flow rate is defined as volume per time. The average flow velocity will be known since the system is confined and all volume must continue to be in the system. It can also be seen as steady

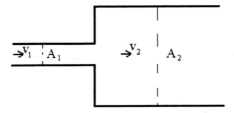

**Fig. 3** Schematic picture showing mass flow

mass flow in which mass cannot accumulate or be compressed to a smaller volume (incompressible media). This is true for this experiment when the chamber is filled (Fig. 3).

The mass that is transported through one cross-section area must go into the next cross-section area. So this will give rise to the following equation:

$$\rho_1 A_1 v_1 = \rho_2 A_2 v_2 = m$$

$A$ represents the cross-sectional area and $v$ is the velocity. The mass flow rate is represented by $m'$. The density of the media is the same in both chamber parts so no indices are necessary on the $\rho$. So the equation can be rewritten in which the mass flow rate can divide by the density.

$$v = \frac{q}{A}$$

In this simplified equation $q$ is the flow rate (volume per time). From the equation above it is possible to obtain the average velocity of the media in the different parts of the chamber.

It is not only the average velocity of the media that is interesting, but also the flow velocity profile of the medium. In this case the chamber will be like a rectangular pipe (or duct). In a laminar flow profile the velocity will always be highest in the middle of the flow. The velocity then decreases from its maximum in the middle and reaches zero at the edges of the chamber. The fact that a media particle in contact with the chamber wall stays at rest is called the no-slip condition.

The formula for the flow velocity in a rectangular pipe looks like this:

$$u(y,z) = \frac{48}{\pi^3} \frac{u_{av}}{1 - \frac{192a}{\pi^5 b} \tanh(\pi \frac{b}{2a})} \sum_i \frac{(-1)^{i+1}}{(2i-1)^3} \left[ 1 - \frac{\cosh(\frac{z\pi(2i-1)}{2a})}{\cosh(\frac{b\pi(2i-1)}{2a})} \right] \cos(\frac{y\pi(2i-1)}{2a})$$

In this formula $u_{av}$ is the average velocity; $2a$ and $2b$ are the length and the width of the chamber. The $y$ and $z$ parameters correspond to the length and width coordinate with the origin in the center of the chamber. The highest velocity will be located in the center and be twice as high as the average speed.

## 6 Perfusion Systems for Cartilage Generation

Several research groups have used perfusion systems to evaluate matrix synthesis from, for example, chondrocytes and also employed such systems for bone tissue engineering applications [43–46]. Portner et al. [4–8] developed a flow-chamber reactor system that enabled the cultivation of six cartilage-carrier-constructs in parallel. In this novel chamber a very thin medium layer enhances gas and nutrient transfer. Cartilage-carrier-constructs cultured with this chamber showed a higher degree of differentiation than static controls.

Davisson et al. [47] performed a study to compare matrix synthesis and cell content in three-dimensional cultures of chondrocytes under static and different kinds of perfusion conditions. They used both static and perfusion seeding and evaluated two different flow rates of perfusion during culturing. In the perfusion seeding 20 million cells per system were used in 30 ml of medium. The flow rate during seeding was 2.5 ml min$^{-1}$ (500 µm s$^{-1}$) for 5 h, after this the flow rate was lowered to 0.05 ml min$^{-1}$ (10 µm s$^{-1}$) and left overnight. After perfusion seeding the scaffolds were either put in a static culture, perfusion cultured at a flow rate of 0.05 ml min$^{-1}$ (10 µm s$^{-1}$), or perfusion cultured at a flow rate of 0.8 ml min$^{-1}$ (170 µm s$^{-1}$). The static-seeded scaffolds were maintained in static culture conditions. At day 1 the samples were radiolabeled with $^{35}SO_4$. Other samples were cultured for 7 days at flow rates of 0.05 ml min$^{-1}$, and were then divided into the same three groups (static, flow rate 0.05 ml min$^{-1}$, and 0.8 ml min$^{-1}$) after which they were radiolabeled. Samples were evaluated at day 3 (radiolabeled at day 1) and day 9 (radiolabeled at day 7). The cell content at day 3 was significantly higher for samples that were perfused the whole time. A flow rate of 0.05 ml min$^{-1}$ for the whole period gave the highest value. On the other hand, the static culturing gave more newly synthesized S-GAGs at day 3. At day 9 it was once again samples that had been exposed to perfusion that had the highest cell content, significantly higher than in the static culturing. The S-GAG production was significantly higher for samples cultured with perfusion at a flow rate of 0.05 ml min$^{-1}$ for the whole 9-day period as well as for 7 days followed by 2 days of perfusion at a flow rate of 0.8 ml min$^{-1}$ compared to the static control. The highest production value was achieved with a flow rate of 0.05 ml min$^{-1}$ for 7 days followed by 0.8 ml min$^{-1}$ until day 9. So flow rates as well as the time samples are exposed to perfusion are important factors in achieving the best results.

The perfusion chamber made by Bancroft et al. [30] is designed so that the flow penetrates the scaffold in a downwards motion, to avoid gas bubbles on the downside of the scaffold, whereas a flow from the other direction could perhaps create them. They used Plexiglas as the material for their chamber, which they sterilized with ethylene oxide. Plexiglas was used because it is transparent which makes it possible to follow the flow through the chamber. The scaffold is held in a cassette that is trapped between two O-rings made of neoprene with a screw on top to provide a tight seal.

Strehl et al. [43] cultured explanted cartilage pieces in a perfusion system. The system showed that the perfused pieces maintain hyaline cartilage morphology and tissue-specific content. The perfusion culturing was carried out in a serum-free environment and was compared to static cultures made with serum (fetal calf and human) and under serum-free conditions. The samples were also compared to fresh cartilage. Samples were analyzed after 14, 28, 48, and 56 days. At day 14 there were some differences but they were extremely evident at day 56. The perfused samples maintained a low mitotic activity during the whole culture time, comparable to levels seen in fresh cartilage. Immunohistochemistry levels for aggrecan, collagen I, collagen II, and COMP where much more like the fresh hyaline cartilage for the perfusion samples then for the different static samples. There was an upregulating tendency for collagen I in the static-culturing systems. The morphology of the samples changed drastically for the static samples over the 56 days of culture and an extensive weight and size loss was measured. On the other hand, the morphology for the samples that had been cultured in the perfusion chamber had a similar morphology to the fresh hyaline cartilage. The serum increased the mitotic activity, which also led to dedifferentiation of the chondrocytes. And the dedifferentiation of the cells was correlated with degeneration of the matrix.

Zhao and Ma [44] used a perfusion system for growing human mesenchymal stem cells. When growing the cells they found no difference in growth patterns between the upper and lower side of the scaffold at day 28. On the other hand, they found a much higher amount of proliferating cells on the upper surface (comparable to that found in the perfusion case) then on the lower surface in the static control case at day 21. In both the static and the perfusion culture the metabolic rates increased exponentially but the magnitude of the glucose consumption and lactate production was 1–2-times higher in the perfusion case. They found indications that the static culture had oxygen delivery limitations. This is seen in the ratio of the lactate production/glucose consumption. There is a significant effect on cell proliferation as well as on ECM secretion at oxygen tension below 40 mmHg. In the later stages of the static culturing the oxygen tension was lower than this while the perfusion samples were never close to this. Zhao and Ma developed a bioreactor system that seeded the cells and cultured the cells by perfusion. In this study they used human mesenchymal stem cells and the scaffold material was PET. It was designed in a way so that there could be different circulation loops with either the media/cell suspension being forced through the scaffold or the media/cell suspension having the possibility to flow over the scaffold surface (upper and lower surface). In the seeding process the cell suspension was forced through the scaffold in a recirculating motion for 3 h. Seeding efficiency at two different cell concentrations and at three different flow rates was evaluated. The cell distribution was also evaluated. The media could flow above and below the scaffold and through the scaffold but it was not forced through the scaffold when doing the perfusion culturing. The flow rate was 0.1 ml min$^{-1}$ for up to 40 days. The perfusion results were compared to the static culture results. In the static seeding the cell suspension was passed through the scaffolds four times and the whole process took 1 h. The samples were then put in wells and the media was changed every other day for up to 38 days. Both the static and perfusion culturing was performed in a 5% $CO_2$ incubator. Cell numbers were evaluated with DNA assays and metabolic activity was evaluated by analyzing the glucose and lactate at the inlet and outlet of the perfusion chamber. The seeding results showed an inverse correlation between flow rates and seeding efficiencies. When decreasing the cell density per chamber (each chamber containing three scaffolds) from $7.2 \times 10^5$ to $1.2 \times 10^5$ the seeding efficiency doubled. This was performed at a flow rate of 0.1 ml min$^{-1}$. For higher flow rates the difference decreases and no statistical difference is found at 0.4 ml min$^{-1}$. The seeding was uniform in the perfusion scaffolds, with no statistical difference between the upper and lower side of the scaffold and with no difference between the three scaffolds that shared a chamber. To observe the morphology of the tissue construct SEM and histological sections were used. After 7 days the static samples had a more dense cellular mass on the lower side of the scaffold indicating an uneven growth and distribution of the cells in the scaffold. On the same day the perfusion samples showed a more uniform distribution and ECM proteins were found on both sides of the scaffold. The histological sections of the central regions of the scaffolds confirmed the uniform distribution in both the static and perfusion constructs. On the other hand, when looking at the peripheral parts of the scaffolds it was seen that there were more aggregated cells and more secreted ECM proteins around the scaffolds in the static culture. This was not found in the perfusion scaffolds. At an initial cell density of $1.13 \times 10^6$ cells ml$^{-1}$ the static cultures reached a plateau after 25 days and at day 38 a 4.3-fold increase was achieved. The perfusion reactor system reached a higher final cell density even with a lower initial cell seeding density. The cell density increased from $7.91 \times 10^5$ to $4.22 \times 10^7$ cells ml$^{-1}$ after 40 days of culturing. This is a 50-fold increase. An increase in glucose consumption and lactate production was seen over time in the perfusion system especially after 20 days, which indicated a higher metabolic demand at the later stage of culturing.

In a study performed at our laboratories we evaluated the effect of confined perfusion on the ECM synthesis of human adult primary chondrocytes cultured in chitosan scaffolds and explored basic methods to monitor the metabolic activity of the cultured cells.

Cartilage was harvested from patients undergoing arthroscopy. The media used for expansion was DMEM-F12, 10% FCS, and 1% antibiotic antimycotic solution. After the first passage cells were seeded into chitosan scaffolds (thickness = 3 mm, width = 3 cm, length = 5 cm, area = 15 cm$^2$) with a fast-delivery dynamic system using a RADA 16 self-assembly peptide hydrogel (SAPH) at a density of $2 \times 10^6$ cells per cm$^2$ of chitosan scaffold. A volume of 1,500 µl of peptide hydrogel–chondrocyte mixture was statically dispersed along the scaffold and then dynamically seeded perfusing the mixture through the construct. After this step the self assembly was initiated perfusing phospate buffer solution (PBS) from the bottom to the top of the scaffold. The constructs were cultured statically for 2 days and after this period direct perfusion was initiated in the dynamic group. The dynamic culture group was cultured using a custom-made confined perfusion chamber (Fig. 4a–d). For chondrocyte cultures in 3D we used a defined media consisting of DMEM-HG supplemented with ITS-G, 5.0 µg ml$^{1-}$ linoleic acid, 1.0 mg ml$^{1-}$ human serum albumin, 2 ng ml$^{1-}$ transforming growth factor1 (TGF-$\beta_1$), 10$^{-7}$ M dexamethasone, 14 µg ml$^{1-}$ ascorbic acid, and Penicillin-Streptomycin.

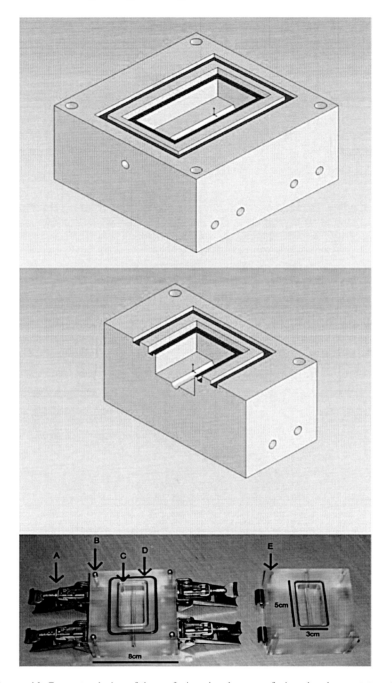

**Fig. 4 a** and **b** Computer design of the perfusion chamber, **c** perfusion chamber prototype 1

## 6.1 Flow Velocity Calculations

The mass flow was calculated using the formula $m' = p*A*v$. This is based on the fact that the same amount of mass is transported in and out of the system and that no accumulation occurs. This is true for our system when the chamber is filled up (Fig. 5).

## 6.2 Metabolic Parameter Determination

Different metabolic parameters were measured before and after perfusion. $pCO_2$, $pO_2$, bicarbonate, PH, and oxygen saturation levels were recorded and compared to determine if there are differences after perfusion of the constructs.

## 6.3 Histology

After 14 days the 3D scaffolds were fixed in Histofix™ (Histolab products AB, Sweden), dehydrated and embedded in paraffin. Five-micrometer sections were

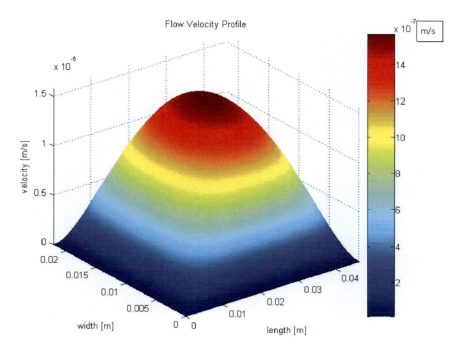

**Fig. 5** Flow velocity profile through the perfusion chamber

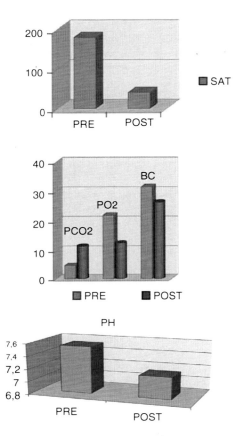

**Fig. 6** a Oxigen saturation, b pCO$_2$, pO$_2$ and Bicarbonate pre and post-perfusion, c PH values pre and post-perfusion

cut and placed onto microscope slides (Superfrost Plus, Menzel-Gläser, Germany), deparaffinized and stained with Alcian Blue.

## 6.4 Results

The flow velocity was 0.8 μm s$^{-1}$ through the scaffold. Furthermore, the flow rate along the chamber was 0.05 ml min$^{-1}$.

During the period of evaluation there was a significant difference between the different metabolic parameters evaluated. The levels of pCO$_2$, pO$_2$, bicarbonate, PH, and oxygen saturation were affected by the cell metabolism thus increasing or decreasing (Fig. 6a–c). We found considerable histologic differences between the perfused and the static group. The perfusion group showed viable cells with abundant matrix production and areas of early chondrogenic differentiation (Fig. 7). The static group showed cell necrosis and no matrix production (Fig. 8).

It was also possible to monitor GAG production in the media as well as in the scaffolds. Perfusion increased GAG: Glucosaminoglycans production over time. It was possible to detect GAG in the media supernatant as an indirect measure to evaluate matrix production nondestructively.

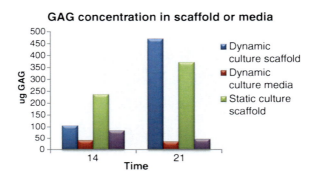

**Fig. 7** GAG concentrations (µg GAG/scaffold) in scaffolds or supernatant media under static or dynamic conditions

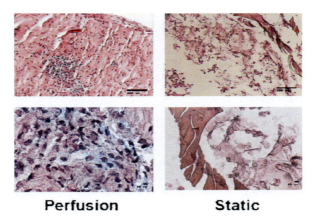

**Fig. 8** The perfusion group showed viable cells with abundant matrix production and areas of early chondrogenic differentiation. The static group showed cell necrosis and no matrix production

## 7 Conclusion

Confined perfusion increased the quality of the tissue-engineered constructs using a flow rate of 0.05 ml min$^{-1}$. A defined media with low concentrations of TGFβ was enough to start early differentiation under the described experimental conditions. It is possible to monitor the metabolic activity of the constructs using nondestructive assays. These perfusion studies indicate that perfusion culture is a powerful and successful tool to use in cell culture. It solves the diffusion limitations involved in static culturing. In static culturing the chondrocytes are more likely to start producing collagen I compared to perfusion-cultured cells. It is important not to use too high flow rates in the beginning since this can lead to removal of the cells from the scaffold. An increase of the flow rate as the tissue matures has shown better results compared to using the same flow rate through the whole culturing process.

## References

1. Brittberg M, Lindahl A, Nilsson A, Ohlsson C, Isaksson O, Peterson L (1994) Treatment of deep cartilage defects in the knee with autologous chondrocyte transplantation. N Engl J Med 331(14):889–895
2. Henderson I, Francisco R, Oakes B, Cameron J (2005) Autologous chondrocyte implantation for treatment of focal chondral defects of the knee—a clinical, arthroscopic, MRI and histologic evaluation at 2 years. Knee 12(3):209–216
3. Peterson L, Minas T, Brittberg M, Nilsson A, Sjogren-Jansson E, Lindahl A (2000) Two- to 9-year outcome after autologous chondrocyte transplantation of the knee. Clin Orthop (374):212–234
4. Nagel-Heyer S, Goepfert C, Feyerabend F, Petersen JP, Adamietz P, Meenen NM, et al. (2005) Bioreactor cultivation of three-dimensional cartilage-carrier-constructs. Bioprocess Biosyst Eng 27(4):273–280
5. Portner R, Nagel-Heyer S, Goepfert C, Adamietz P, Meenen NM (2005) Bioreactor design for tissue engineering. J Biosci Bioeng 100(3):235–245
6. Nagel-Heyer S, Goepfert C, Adamietz P, Meenen NM, Portner R (2006) Cultivation of three-dimensional cartilage-carrier-constructs under reduced oxygen tension. J Biotechnol 121(4):486–497
7. Heyland J, Wiegandt K, Goepfert C, Nagel-Heyer S, Ilinich E, Schumacher U, et al. (2006) Redifferentiation of chondrocytes and cartilage formation under intermittent hydrostatic pressure. Biotechnol Lett 28(20):1641–1648
8. Nagel-Heyer S, Goepfert C, Morlock MM, Portner R (2005) Relationship between physical, biochemical and biomechanical properties of tissue-engineered cartilage-carrier-constructs. Biotechnol Lett 27(3):187–192
9. Marlovits S, Kutscha-Lissberg F, Aldrian S, Resinger C, Singer P, Zeller P, et al. (2004) [Autologous chondrocyte transplantation for the treatment of articular cartilage defects in the knee joint. Techniques and results]. Radiologe 44(8):763–772
10. Marlovits S, Tichy B, Truppe M, Gruber D, Schlegel W (2003) Collagen expression in tissue engineered cartilage of aged human articular chondrocytes in a rotating bioreactor. Int J Artif Organs 26(4):319–330
11. Marlovits S, Striessnig G, Kutscha-Lissberg F, Resinger C, Aldrian SM, Vecsei V, et al. (2004) Early postoperative adherence of matrix-induced autologous chondrocyte implantation for the treatment of full-thickness cartilage defects of the femoral condyle. Knee Surg Sports Traumatol Arthrosc 13(6):451–457

12. Nehrer S, Domayer S, Dorotka R, Schatz K, Bindreiter U, Kotz R (2006) Three-year clinical outcome after chondrocyte transplantation using a hyaluronan matrix for cartilage repair. Eur J Radiol 57(1):3–8
13. Marcacci M, Zaffagnini S, Kon E, Visani A, Iacono F, Loreti I (2002) Arthroscopic autologous chondrocyte transplantation: technical note. Knee Surg Sports Traumatol Arthrosc 10(3):154–159
14. Marcacci M, Berruto M, Brocchetta D, Delcogliano A, Ghinelli D, Gobbi A, et al. (2005) Cartilage engineering with Hyalograft C: 3-year clinical results. Clin Orthop Relat Res (435):96–105
15. Pavesio A, Abatangelo G, Borrione A, Brocchetta D, Hollander AP, Kon E, et al. (2003) Hyaluronan-based scaffolds (Hyalograft C) in the treatment of knee cartilage defects: preliminary clinical findings. Novartis Found Symp 249:203–217; discussion 229–233, 234–208, 239–241
16. Poole AR, Kojima T, Yasuda T, Mwale F, Kobayashi M, Laverty S (2001) Composition and structure of articular cartilage: a template for tissue repair. Clin Orthop (391 Suppl):S26–S33
17. Mow V (2006) Functional tissue engineering. Conf Proc IEEE Eng Med Biol Soc 1:16–17
18. Mow VC, Ateshian GA, Spilker RL (1993) Biomechanics of diarthrodial joints: a review of twenty years of progress. J Biomech Eng 115(4B):460–467
19. Mow VC, Gibbs MC, Lai WM, Zhu WB, Athanasiou KA (1989) Biphasic indentation of articular cartilage – II. A numerical algorithm and an experimental study. J Biomech 22(8–9):853–861
20. Mow VC, Guo XE (2002) Mechano-electrochemical properties of articular cartilage: their inhomogeneities and anisotropies. Annu Rev Biomed Eng 4:175–209
21. Mow VC, Wang CC (1999) Some bioengineering considerations for tissue engineering of articular cartilage. Clin Orthop Relat Res (367 Suppl):S204–S223
22. Mow VC, Wang CC, Hung CT (1999) The extracellular matrix, interstitial fluid and ions as a mechanical signal transducer in articular cartilage. Osteoarthritis Cartilage 7(1):41–58
23. Mak AF, Lai WM, Mow VC (1987) Biphasic indentation of articular cartilage – I. Theoretical analysis. J Biomech 20(7):703–714
24. Shieh AC, Athanasiou KA (2006) Biomechanics of single zonal chondrocytes. J Biomech 39(9):1595–1602
25. Leipzig ND, Athanasiou KA (2005) Unconfined creep compression of chondrocytes. J Biomech 38(1):77–85
26. Shieh AC, Athanasiou KA (2003) Principles of cell mechanics for cartilage tissue engineering. Ann Biomed Eng 31(1):1–11
27. Shieh AC, Athanasiou KA (2002) Biomechanics of single chondrocytes and osteoarthritis. Crit Rev Biomed Eng 30(4–6):307–343
28. Marsano A, Wendt D, Raiteri R, Gottardi R, Stolz M, Wirz D, et al. (2006) Use of hydrodynamic forces to engineer cartilaginous tissues resembling the non-uniform structure and function of meniscus. Biomaterials 27(35):5927–5934
29. Wendt D, Marsano A, Jakob M, Heberer M, Martin I (2003) Oscillating perfusion of cell suspensions through three-dimensional scaffolds enhances cell seeding efficiency and uniformity. Biotechnol Bioeng 84(2):205–214
30. Bancroft GN, Sikavitsas VI, Mikos AG (2003) Design of a flow perfusion bioreactor system for bone tissue-engineering applications. Tissue Eng 9(3):549–554
31. Darling EM, Athanasiou KA (2003) Articular cartilage bioreactors and bioprocesses. Tissue Eng 9(1):9–26
32. Freed LE, Hollander AP, Martin I, Barry JR, Langer R, Vunjak-Novakovic G (1998) Chondrogenesis in a cell-polymer-bioreactor system. Exp Cell Res 240(1):58–65
33. LeBaron RG, Athanasiou KA (2000) Ex vivo synthesis of articular cartilage. Biomaterials 21(24):2575–2587
34. Mahmoudifar N, Doran PM (2005) Tissue engineering of human cartilage and osteochondral composites using recirculation bioreactors. Biomaterials 26(34):7012–7024

35. Tognana E, Chen F, Padera RF, Leddy HA, Christensen SE, Guilak F, et al. (2005) Adjacent tissues (cartilage, bone) affect the functional integration of engineered calf cartilage in vitro. Osteoarthritis Cartilage 13(2):129–138
36. Vunjak-Novakovic G, Freed LE (1998) Culture of organized cell communities. Adv Drug Deliv Rev 33(1–2):15–30
37. Freyria AM, Cortial D, Ronziere MC, Guerret S, Herbage D (2004) Influence of medium composition, static and stirred conditions on the proliferation of and matrix protein expression of bovine articular chondrocytes cultured in a 3-D collagen scaffold. Biomaterials 25(4):687–697
38. Pei M, Solchaga LA, Seidel J, Zeng L, Vunjak-Novakovic G, Caplan AI, et al. (2002) Bioreactors mediate the effectiveness of tissue engineering scaffolds. FASEB J 16(12):1691–1694
39. Vunjak-Novakovic G, Obradovic B, Martin I, Bursac PM, Langer R, Freed LE, et al. (1998) Dynamic cell seeding of polymer scaffolds for cartilage tissue engineering chondrogenesis in a cell-polymer-bioreactor system. Biotechnol Prog 14(2):193–202
40. Freyria AM, Yang Y, Chajra H, Rousseau CF, Ronziere MC, Herbage D, et al. (2005) Optimization of dynamic culture conditions: effects on biosynthetic activities of chondrocytes grown in collagen sponges. Tissue Eng 11(5–6):674–684
41. Bancroft GN, Sikavitsas VI, van den Dolder J, Sheffield TL, Ambrose CG, Jansen JA, et al. (2002) Fluid flow increases mineralized matrix deposition in 3D perfusion culture of marrow stromal osteoblasts in a dose-dependent manner. Proc Natl Acad Sci U S A 99(20):12600–12605
42. Minuth WW, Strehl R, Schumacher K, de Vries U (2001) Long term culture of epithelia in a continuous fluid gradient for biomaterial testing and tissue engineering. J Biomater Sci Polym Ed 12(3):353–365
43. Strehl R, Tallheden T, Sjogren-Jansson E, Minuth WW, Lindahl A (2005) Long-term maintenance of human articular cartilage in culture for biomaterial testing. Biomaterials 26(22):4540–4549
44. Zhao F, Ma T (2005) Perfusion bioreactor system for human mesenchymal stem cell tissue engineering: dynamic cell seeding and construct development. Biotechnol Bioeng 91(4):482–493
45. Minuth WW, Strehl R, Schumacher K (2004) Tissue factory: conceptual design of a modular system for the in vitro generation of functional tissues. Tissue Eng 10(1–2):285–294
46. Minuth WW, Schumacher K, Strehl R, Kloth S (2000) Physiological and cell biological aspects of perfusion culture technique employed to generate differentiated tissues for long term biomaterial testing and tissue engineering. J Biomater Sci Polym Ed 11(5):495–522
47. Davisson T, Sah RL, Ratcliffe A (2002) Perfusion increases cell content and matrix synthesis in chondrocyte three-dimensional cultures. Tissue Eng 8(5):807–816

# Technical Strategies to Improve Tissue Engineering of Cartilage-Carrier-Constructs

R. Pörtner, C. Goepfert, K. Wiegandt, R. Janssen, E. Ilinich, H. Paetzold, E. Eisenbarth, and M. Morlock

**Abstract** Technical aspects play an important role in tissue engineering. Especially an improved design of bioreactors is crucial for cultivation of artificial three-dimensional tissues in vitro. Here formation of cartilage-carrier-constructs is used to demonstrate that the quality of the tissue can be significantly improved by using optimized culture conditions (oxygen concentration, growth factor combination) as well as special bioreactor techniques to induce fluid-dynamic, hydrostatic or mechanical load during generation of cartilage.

**Keywords** Chondrocytes, Cartilage, Osteochondral implants, Mechanical load, Growth factors, Re-differentiation

---

R. Pörtner (✉), C. Goepfert, and K. Wiegandt
Hamburg University of Technology, Institute of Bioprocess and Biosystems Engineering, Hamburg, Germany
e-mail: poertner@tuhh.de

R. Janssen
Hamburg University of Technology, Institute of Advanced Ceramics, Hamburg, Germany

E. Ilinich
Hamburg University of Technology, Institute of Polymer Composites, Hamburg, Germany

H. Paetzold, E. Eisenbarth, and M. Morlock
Hamburg University of Technology, Institute of Biomechanics, Hamburg, Germany

**Contents**

| | |
|---|---|
| 1   Introduction | 146 |
| 2   Expansion of Cell Number | 148 |
| 3   Formation of In Vitro Cartilage on Various Types of Calcium Phosphate Carriers | 150 |
| 4   Cartilage Formation Under Reduced Oxygen Tension and Optimized Growth Factor Combination | 153 |
|         4.1   Literature Survey | 153 |
|         4.2   Experimental Results | 156 |
| 5   Cartilage Formation Under Hydrostatic and Mechanical Load | 162 |
|         5.1   Literature Survey | 162 |
|         5.2   Experimental Results for Porcine Cartilage | 171 |
| 6   Final Considerations | 173 |
| References | 175 |

# 1   Introduction

The loss and damage of tissues cause serious health problems. As the existing therapy concepts are not sufficient, new therapy concepts for practical medical applications are required [1–7]. To this end, tissue-engineered substitutes generated in vitro could open new strategies for the restoration of damaged tissues. Formation of 3D tissue substitutes in vitro requires not only a biological model (e.g., an adequate source for proliferable cells with appropriate biological functions, a protocol for proliferating cells while maintaining the tissue-specific phenotype), but also the further development of new culture strategies including bioreactor concepts.

Bioreactors for the generation of 3D tissue constructs can provide a better process control by taking into account different demands of cells during cultivation [8–13]. Furthermore, they can provide the technical means to perform controlled studies aimed at understanding specific biological, chemical or physical effects. Moreover, bioreactors enable a safe and reproducible production of tissue constructs. An overall comparison of different culture methods shows the advantages of bioreactor culture. Not only the properties of cultivated 3D tissue constructs can be improved, but also aspects such as safety of operation argue for the use of bioreactor systems. Bioreactors can be used to study effects such as shear flow and/or hydrostatic pressure on the generation of tissues. For future clinical applications, the bioreactor system should be an advantageous method in terms of low contamination risk, ease of handling and scalability.

The following considerations are intended to demonstrate how the interdisciplinary application of biological and engineering knowledge can significantly improve the properties of tissue-engineered 3D cartilage constructs. The quality of joint cartilage decreases with increasing age which often leads to orthopaedic treatments causing high costs [14–17]. Other reasons for cartilage damage are traumatic events which are the primary cause for local lesions of articular cartilage. Only an early treatment of the defect may achieve restoration of the initial joint geometry and integrity. However, for large defects the availability of autologous cartilage is not sufficient to bridge the affected area. Using heterologous transplants for missing articular cartilage bears the risk of infection and immune response. Tissue engineering offers an alternative treatment method that has the capability to overcome some of the main drawbacks of current treatments for articular defects [18–22]. Tissue engineering presently creates cartilage with poor mechanical and biochemical properties. Therefore, the improvement of properties and functionality are critical to the success of these engineered tissues. For this, new bioreactor concepts play an important role [23–35].

Our research is based on the formation of osteochondral implants which consist of a ceramic carrier as bone substitute and a layer of cultivated cartilage generated in vitro. A mosaic-like implantation [15] of these autologous cartilage-carrier-constructs may provide a reconstructed surface area inside the knee joint. The cultivation principle includes the following steps as shown in Fig. 1 [36–38]: (a) explanted chondrocytes are expanded in monolayer culture until passage 3; (b) afterwards the cells are seeded onto a solid carrier (cell coating) and cultivated for 2 weeks; (c) simultaneously expanded chondrocytes are suspended in alginate gel for 2 weeks

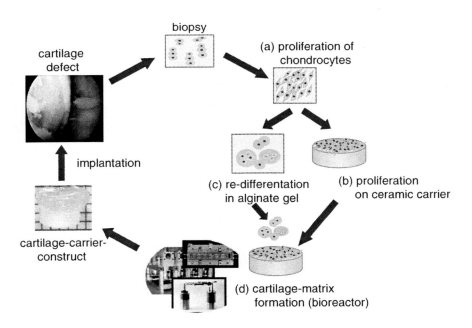

**Fig. 1** Scheme of the cultivation principle for generation of cartilage-carrier-constructs (developed by P. Adamietz, C. Goepfert and F. Feyerabend, UKE, Hamburg, compare [27, 36]): (**a**) isolated chondrocytes are expanded in monolayer culture until passage 3; (**b**) afterwards the cells are seeded onto a solid carrier (cell coating) and cultivated for 2 weeks; (**c**) simultaneously expanded chondrocytes are suspended in alginate gel for 2 weeks to induce re-differentiation; (**d**) the re-differentiated chondrocytes are then eluted out of the alginate gel and sedimented on the cell-coated carrier. These cartilage-carrier-constructs are cultivated for cartilage formation for 3 weeks (high-density cell culture)

**Fig. 2** Cartilage carrier construct from porcine chondrocytes, side view of a cartilage-carrier construct (1: newly formed cartilage by free swelling, 2: solid calcium phosphate carrier, scale 1 mm)

to induce re-differentiation; (d) the re-differentiated chondrocytes are then eluted out of the alginate gel and sedimented on the cell-coated carrier. These cartilage-carrier-constructs are cultivated for cartilage formation for 3 weeks (high-density cell culture). An example of a porcine cartilage-carrier-construct is shown in Fig. 2. The concept was applied successfully in mini-pigs [39]. Nevertheless, further improvement of the protocol is required. In the following aspects such as expansion of cell number, ceramic carrier as support, optimization of oxygen concentration and growth factor supplementation, and impact of hydrostatic and mechanical load on cartilage formation will be discussed. For all studies the same basic protocol was applied. Parameters such as content of glycosaminoglycan (GAG), DNA, height and weight of the constructs, Young's modulus to describe elastic properties, and the content of collagen type II are considered. The methods will not be discussed in detail, as they have been published elsewhere [36–38].

## 2 Expansion of Cell Number

In cell-based technologies for the treatment of articular cartilage defects, the expansion of small numbers of chondrocytes obtained from biopsies is an integral part of the manufacturing process. The initial cell number for the generation of autologous implants is limited by the small size of a biopsy which is assumed to count $0.2 \times 10^6$ to $1.2 \times 10^6$ cells. Hence, chondrocytes have to be expanded by a factor of

100–1000 until the required cell number is reached: for cartilage-carrier-constructs as presented here, an initial cell number of $1 \times 10^6$ cells is required. In order to fill a defect of about 6 cm$^2$ it is necessary to subcultivate monolayer cultures until passage 3. The propagation in T-flasks comprises a couple of trypsination steps leading to a de-differentiated phenotype of the chondrocytes (reduction of collagen type II expression and production of collagen type I) [40, 41]. Here an alternative propagation technique is presented, which may avoid these drawbacks.

Basically, microcarriers are used for proliferation instead of flasks. The term "microcarriers" comprises small beads, either solid or porous, having a diameter of approximately 100–300 μm and a density slightly higher than the growth medium density of 1.02–1.04 g cm$^{-3}$ [42]. In microcarrier culture, cells grow as monolayers on the surface of small spheres or three-dimensionally within the macroporous structures. The carriers are usually suspended in culture medium by gentle stirring. On smooth solid microcarriers cells grow on the outer surface until a monolayer is formed. In this case, each microcarrier can accumulate approximately 100–200 cells. To reach an optimal growth on all individual microcarriers, an even distribution of cells is required. For most cell lines more than seven cells per carrier are required during inoculation to ensure that the population of unoccupied microcarriers is less than 5% and the use of the available surface area is maximized [43]. In recent years, microcarrier cultures have been studied extensively in the engineering of cartilage and bone (reviewed by [44]), mostly to investigate the effect of certain culture conditions (e.g. oxygen concentration) on growth and degree of re-differentiation of proliferated chondrocytes [45].

Propagation of chondrocytes on microcarriers has the following advantages: (1) bioreactors can be used for better process control and easier handling, (2) repeated treatment with proteolytic enzymes (trypsin) can be avoided, and (3) detachment of chondrocytes after the expansion period for further use is possible, among others. Nevertheless, the common techniques for use of microcarriers bear some disadvantages. The increase in cell number is usually in the range of 10–20 during a batch culture, similar to flask culture. Therefore, a number of subculture steps are required if a certain number of cells have to be generated. During these subculture steps a detrimental de-differentiation has to be expected similar to propagation in flasks. Therefore, a culture strategy based on proliferation of chondrocytes on microcarriers with reduced subculturing steps was developed (Fig. 3). The basic idea is to increase the surface available for cell growth during the cultivation. For this purpose, a small amount of microcarriers in a watch glass is inoculated with a low number of cells. After reaching confluence, new microcarriers and fresh medium are added. Again after reaching confluence, the microcarriers are transferred to a conical bioreactor or a shake flask, where new microcarriers and fresh medium are added.

In the following the expansion and re-differentiation capacity of porcine and human articular chondrocytes (HAC) after expansion on microcarriers or in conventional monolayer culture is discussed. With the technique described above it was possible to increase the number of primary porcine chondrocyte cells by 1,400-times without sub-cultivation by trypsination.

First results with human chondrocytes showed a reduced level of cells producing collagen type I and an increased level of cells producing collagen type II compared

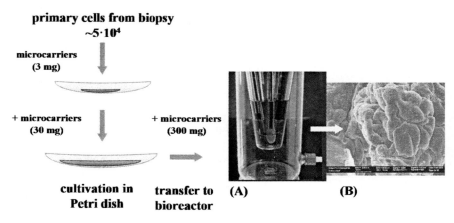

**Fig. 3** Strategy for proliferation of primary cells by bead-to-bead transfer on microcarriers without trypsinization. (**A**) Conical bioreactor (medorex) for the propagation of cells on microcarriers (working volume 100 mL); (**B**) porcine chondrocytes grown on cytodex 3 microcarriers, final cell density approximately $70 \times 10^6$ cells (expansion 1,400 times)

to chondrocytes cultivated for one passage in T-flasks [46]. To study these effects in more detail, freshly isolated human chondrocytes were expanded by factor 10 in monolayer culture and by factor 300 on the microcarriers without subcultivation, and further cultivated in alginate gel according to step (c) of the protocol shown in Fig. 1 [47]. T-flask expanded passage 1 chondrocytes (P1) formed high amounts of collagen type I during cultivation in alginate gel. The best results were obtained in a medium supplemented with IGF-I and BMP-7 (compare Sect. 4.2.2). Here, the ratio of cells producing collagen type II to those producing collagen type I was approximately 1 by aerating with 10% oxygen. For microcarrier expanded chondrocytes (P1*), the production of collagen type II was significantly increased. Here, a ratio of cells producing collagen type II to those producing collagen type I of 2.6 was reached by using IGF-I and BMP-7 and aerating with 10% oxygen.

In conclusion, expansion of human articular chondrocytes on microcarriers can be carried out starting with low cell numbers of $5*10^4$–$2*10^5$ which are obtained from biopsies. Using this method, repeated treatment with trypsin can be avoided during the expansion process. The higher rates of extracellular matrix production and better quality of matrix composition indicate, that the expansion of chondrocytes on microcarriers may improve cell-based cartilage repair.

## 3 Formation of In Vitro Cartilage on Various Types of Calcium Phosphate Carriers

The treatment of joint defects by transplantation of autologous osteochondral grafts is a well-established method. However, the application of this technique is limited due to the fact that the implants have to be obtained from non-weight bearing regions of the joint. Therefore, several methods have been developed to produce osteochondral implants consisting of a bone substitute covered by a layer of tissue-engineered cartilage. Since chondrocytes are harvested from a small biopsy and expanded in cell culture, in vitro-cultivated autologous cartilage-carrier-constructs allow for the therapy of defects avoiding the above-mentioned drawbacks of osteochondral transplantation.

In the last years, the formation of osteochondral implants by tissue-engineering methods has been addressed by various approaches in vitro and in vivo. Biphasic implants have been constructed using a mineral biomaterial such as biocoral [48], bioglass [49, 50] or calcium phosphate-based materials in combination with chondrocytes delivered to the bone substitute [51–55]. Waldman et al. [56, 57] applied the concept of high-density cultures of primary chondrocytes on top of the bone equivalent. The bone substitutes are supposed to provide repair of the subchondral bone, anchorage of the cartilage layer within the bone and initial mechanical stability to the injured joint.

Bone substitute materials closely related to the mineral component of bone are well established in reconstructive surgery. Many in vitro and in vivo studies de-monstrate the osteoconductive potential of these materials and show ingrowth and osteogenic differentiation of mesenchymal precursors. In bone surgery, the replacement materials are selected with respect to their proper sites and functions. However, little is known about the interaction between cartilage and calcium phosphate-based implant materials. Using primary bovine chondrocytes, cartilage formation in vitro could be demonstrated on sintered calcium polyphosphate [56].

In our laboratory, various types of calcium phosphate ceramics were investigated for their ability to be used as support materials for the formation of a 3D scaffold-free cartilage layer in vitro. Using the method described by Nagel-Heyer et al. [36] (Fig. 1), the carriers were covered by a layer of expanded articular chondrocytes (P3, Minipig). In parallel, chondrocytes were cultivated in alginate gel, and then released together with their cell-associated matrix formed during the re-differentiation process and cultured in high density on top of the bone replacement material (see Fig. 1). Incorporation of cell-attached matrix into the constructs resulted in cartilaginous tissue suitable for implantation, omitting artificial scaffolds for the cartilage layer [39].

The calcium phosphate carriers used as support materials included dense β-TCP and hydroxyapatite (HA) as well as the commercially available porous materials ChronOs, Cerasorb and a newly developed material, Sponceram HA (Fig. 4). The surface of the dense carrier materials was modified by ultrasonic milling. Carrier materials (2 mm in height and 4.5 mm in diameter) were kindly provided by Dr. R. Janssen (β-TCP and HA sintered as dense carriers), by Dr. M. Alini (AO Foundation, ChronOs) and by the companies Curasan (Cerasorb) and Zellwerk (Sponceram HA). For the cultivation of cartilage tissue, the carriers were inserted in tissue culture devices as described previously [27]. These biphasic constructs were cultivated for 3 weeks and then analysed for GAG and DNA content. The formation of collagen type II and type I was confirmed by immuno blotting and by immunohistochemistry.

Cartilage formation was observed on all calcium phosphate ceramics used in this study. To estimate the connection between cartilage layer and bone equivalent, a

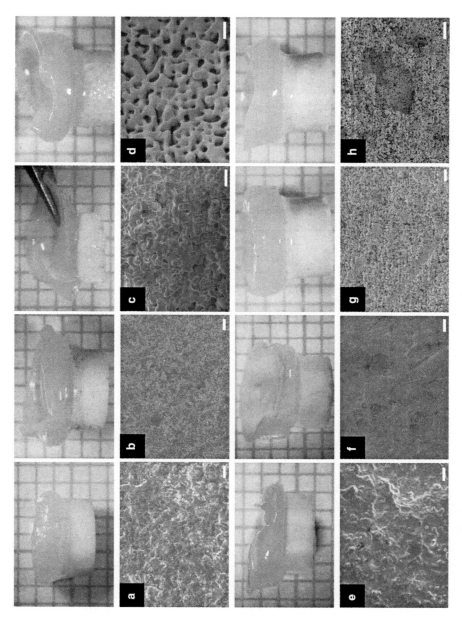

**Fig. 4** Cartilage cultivated on various types of β-TCP and HA (grid lines: 1 mm) and SEM images of Calcium Phosphate Ceramics tested (scale bar: 10 μm). (**a**) β-TCP (as sintered), (**b**) β-TCP ultrasonically milled, (**c**) Cerasorb, (**d**) ChronOs, (**e**) HA (as sintered), (**f**) HA (ultrasonically milled), (**g**) Sponceram HA (with micropores), (**h**) Sponceram HA (with macropores)

simple subjective method was applied. When tissue slipped off the carrier material upon removal from the culturing device, it was characterized as (−). When the cartilage layer could be pulled off by tearing with tweezers, the constructs were validated as (+). When cartilage layers could not be removed without destroying them, (++) was applied. Ultrasonic milling was able to improve the attachment of the cartilage layer to the carrier material consisting of β-TCP (see Table 1). On porous materials, a firm connection between carrier material and cartilage layer was achieved. Formation of collagen type II and type I was demonstrated by immunohistochemical staining of the cartilaginous tissue formed on the various substrates.

Wet weights of cartilage layers were slightly increased by roughening the surface by ultrasonic milling or by using porous materials. The macroporous material ChronOs yielded significantly higher wet weights than the other carrier materials (ANOVA, $p < 0.05$). GAG contents of the cartilaginous tissue were consistently higher on microporous and macroporous materials than on the dense materials. Furthermore, GAG/DNA ratios were calculated in order to determine the amount of cartilage matrix formed by the chondrocytes. GAG/DNA ratios were significantly increased when using carrier materials consisting of hydroxyapatite compared to carriers made of β-TCP (Tukey test for multiple comparisons, $p < 0.05$). Correspondingly, DNA contents were higher on β-TCP indicating higher growth rates during cultivation. Taken together, cartilage adhesion to the substrate appeared to be dependent on material properties as well as surface structure and porosity. High synthesis rates of cartilage matrix were favoured by using porous β-TCP and HA. On the other hand, the proportion of matrix formation and cell content (GAG/DNA ratio) was significantly improved when using HA carriers compared to the corresponding carriers made of β-TCP.

## 4 Cartilage Formation Under Reduced Oxygen Tension and Optimized Growth Factor Combination

### *4.1 Literature Survey*

A critical parameter during the cultivation of cartilage tissue is oxygen supply. The effect of oxygen during the in vitro cultivation of chondrocytes is poorly understood, and therefore it is presently a controversial issue [58]. In several studies, chondrocytes were immobilized in alginate beads and cultivated under different oxygen concentrations in the gas phase (for review see [58]). O'Driscoll et al. [59] observed a limited collagen type II production at very high (90% $O_2$) and very low (1–5% $O_2$) oxygen concentrations. Domm et al. [60] showed a stimulatory effect of a decreased oxygen tension (5% $O_2$) on matrix production. In a study of Malda et al. [61], the pellets of chondrocytes were suspended in a stirred bioreactor under different oxygen concentrations. They observed an increased production of glycosaminoglycan

Table 1 Formation of in vitro cartilage on various types of calcium phosphate carriers (adult porcine chondrocytes, passage 3)

| Carrier material/trade name | Structure | Adhesion to carrier material | GAG/ construct | GAG/mg tissue | DNA/ construct | DNA/mg tissue | GAG/ DNA | Height | Wet weight |
|---|---|---|---|---|---|---|---|---|---|
| β-TCP | Smooth | (−) | 1.00 | 1.00 | 1.00 | 1.00 | 1.00 | 1.00 | 1.00 |
| β-TCP | Roughened | (+) | 1.39 | 0.78 | 1.54 | 0.85 | 0.92 | 0.86 | 1.77 |
| β-TCP/cerasorb | Microporous | (++) | 2.10 | 1.30 | 0.94 | 0.57 | 2.29 | 0.63 | 1.55 |
| β-TCP/chronOs | Macroporous | (++) | 2.65 | 1.09 | 0.85 | 0.34 | 3.17 | 0.92 | 2.38 |
| HA | Smooth | (−) | 1.58 | 1.12 | 1.09 | 0.79 | 1.55 | 0.87 | 1.44 |
| HA | Roughened | (−) | 1.55 | 1.55 | 0.79 | 0.54 | 2.01 | 0.78 | 1.47 |
| HA/sponceram HA | Microporous | (++) | 2.56 | 1.40 | 0.74 | 0.40 | 3.56 | 0.77 | 1.76 |
| HA/sponceram HA | Macroporous | (++) | 2.21 | 1.48 | 0.57 | 0.37 | 4.00 | 1.32 | 1.43 |

Experiments were performed according to the protocol shown in Fig. 1. Results are normalized to the control condition (β-TCP, smooth surface). Medium in alginate phase and during cartilage formation was supplemented with IGF-1 + TGF-β1 at 5% and 1% $O_2$ (v/v) in comparison with aeration at 21% $O_2$ (air). The increased glycosaminoglycan production is accompanied by a decrease in collagen type I level. On the other hand, several studies of chondrocytes embedded in a 3D matrix or a scaffold have demonstrated an enhanced matrix formation, particularly proteoglycan synthesis under more aerobic conditions [62, 63]. The main difference between the applied methodologies (alginate and pellet culture vs. cartilage generation in 3D scaffolds) can be seen in oxygen gradients in the vicinity and within the formed cartilage [45, 64]. In the case of alginate and pellet culture, oxygen gradients at the surface of constructs can be neglected, and only oxygen limitations within the constructs are likely. For the cultivation of 3D scaffolds, even more significant oxygen gradients within the formed matrix are likely.

In a recent study by Wernike et al. [65] the influence of long-term confined dynamic compression and surface motion under low oxygen tension on a tissue-engineered cell scaffold was studied. Culture under reduced oxygen tension (5%) resulted in an increase in mRNA levels of type II collagen and aggrecan, whereas the expression of type I collagen was down-regulated at early time points. Higher glycosaminoglycan content was found at reduced oxygen tension. Immunohistochemical analysis showed more intense type II and weaker type I collagen staining in constructs cultivated at reduced oxygen levels. Histological results confirmed the beneficial effect of mechanical loading on chondrocyte matrix synthesis (compare Sect. 5).

Besides oxygen concentration, various growth factor supplements to the culture medium have been studied in the literature, mainly to stimulate re-differentiation of chondrocytes after de-differentiation during proliferation and subculturing (compare Sect. 2) [41, 66–69]. Research being carried out to determine the characteristics and effects of growth factors in cartilage-tissue engineering has proved that IGF-I (Insulin-like Growth Factor I) especially and TGF-β1 (Transforming Growth Factor β1) delivered positive results. From several studies, it is known that IGF-I and TGF-β1 influence differentiation of chondrocytes and cartilage formation in vitro. The varying results concerning the effects of these growth factors are attributed to the differences in culture conditions, species and age of donors, growth factor concentrations, status of cell differentiation or the addition of serum [70, 71]. To summarize, IGF-I is an important anabolic factor for chondrocytes and in vitro-engineered cartilage. It can stimulate the proliferation of cells and extracellular matrix biosynthesis, in particular the production of proteoglycans [40, 67, 71, 72]. The addition of TGF-β1 to the culture medium has shown diverse effects on the cultivation of chondrocytes. It can stimulate or inhibit cell growth and accumulation of proteoglycans and collagen type II, respectively [67]. Several reports depict a synergistical action of several growth factors or other stimuli (e.g. mechanical) [71, 73–75], which means that a combination of these factors can influence the cultivation in a different way than can each factor alone.

In conclusion low oxygen concentrations and optimized growth factor supplements can improve the quality of engineered cartilage, but they have to be adjusted for each application in cartilage-tissue engineering.

## 4.2 Experimental Results

### 4.2.1 Cultivation Under Reduced Oxygen Tension

Our work started from an engineering point of view, that is, we investigated the oxygen transport in culture systems for the generation of cartilage pellets and within the pellet itself theoretically. Two culture strategies were modelled, a static culture system with the supply of oxygen through the medium by diffusion and a perfused culture system with an oxygen-enriched medium (for details see [11]). For static culture, oxygen concentration in the vicinity of the pellet decreases very rapidly to zero within a very short time. This effect is mainly due to the very low diffusive transport of oxygen through the boundary layer between the cartilage pellet and the medium. Alternatively, a perfused system was modelled, in which oxygen-enriched medium flows over the cartilage. These calculations show that the mass transfer resistance in the boundary layer between the cartilage and the medium can be significantly decreased, but still oxygen concentration within the pellet decreases very rapidly. The main reason for this limitation is the very low rate of transport by diffusion. Similar conclusions have been drawn elsewhere [64, 76]. A flow of enriched medium, as in the perfused culture system, can overcome this problem only to some extent by decreasing the thickness of the boundary layer between the tissue matrix and the medium. These findings lead to the question of whether it is possible to predict the properties of cartilage generated in these bioreactor systems on the basis of the above conclusions. This will be addressed below by comparing experimental results from different cultivation systems (alginate, cartilage formation in static flask or in a flow chamber bioreactor) at different oxygen concentrations. The discussion will focus on porcine cartilage, human cells will be introduced only briefly. Detailed information on the methods and basic data are provided by [11, 27, 38, 77]. The results of the cultivation of porcine chondrocytes under different oxygen concentrations are summarized in Table 2, data set 1–4. The main goal of the following discussion is to identify effects due to varying culture conditions. In each set of experiments chondrocytes from one animal source were used in passage 3, but obviously the animal sources differed between the sets. To exclude variations due to different animal sources, in each set the data were based on the control experiment (set to "1" for quantitative and to "=" for qualitative parameters). Statistical significance was assessed by analysis of variance with $p < 0.05$ (ANOVA).

Data for alginate cultures (step c in the protocol shown in Fig. 1) at oxygen tensions of 21, 10 and 5% (v/v) are summarized in data set 1, Table 2. The tremendous effect of lower oxygen concentrations becomes obvious for both the amount of glycosaminoglycan per DNA (4.8x compared to the control) and the ratio of cells producing collagen type II to those producing collagen type I (3x compared to the control).

Data sets 2–4, Table 2 were performed to generate cartilage-carrier-constructs under different oxygen tensions. Data set 2 comprises experiments in a flow-chamber bioreactor especially designed for the generation of three-dimensional cartilage-carrier-constructs compared to cultures in 12-well plates as the control. A

Table 2 Summary of results for cultivation of porcine chondrocytes under different oxygen concentrations and growth factor combinations

| Data set/Reference | Culture mode | Oxygen tension | Growth factors | Culture system | GAG | DNA | GAG/DNA | Collagen II/I | Height | Weight | Young's modulus |
|---|---|---|---|---|---|---|---|---|---|---|---|
| Oxygen concentration | | | | | | | | | | | |
| 1/unpublished data | Alginate | 21% | IGF-1 + TGF-β1 | Flask | | | 1 | 1 | | | |
| | | 10% | IGF-1 + TGF-β1 | Flask | | | 1.44 | 2.41 | | | |
| | | 5% | IGF-1 + TGF-β1 | Flask | | | 4.79 | 3.03 | | | |
| | Cartilage | Alginate/cartilage | Alginate/cartilage | | | | | | | | |
| 2/[38] | | 21%/21% | IGF-1 + TGF-β1/ | Flask/flask | 1 | 1 | 1 | = | 1 | 1 | n.d. |
| | | 21%/21% | IGF-1 + TGF-β1/ | Flask/fluid | 0.59 | 1.17 | 0.52 | + | 1.45 | 1 | n.d. |
| 3/[27] | | 21%/21% | IGF-1 + TGF-β1/ | Flask/flask | 1 | 1 | 1 | = | 1 | n.d. | n.d. |
| | | 21%/10% | IGF-1 + TGF-β1/ | Flask/flask | 1.04 | 0.83 | 1.27 | + | 1.08 | n.d. | n.d. |
| | | 21%/5% | IGF-1 + TGF-β1/ | Flask/flask | 1.41 | 1.01 | 1.7 | + | 1.17 | n.d. | n.d. |
| 4/unpublished data | | 10%/10% | IGF-1 + TGF-β1/ | Flask/flask | 1 | 1 | 1 | = | 1 | 1 | 1 |
| | | 5%/5% | IGF-1 + TGF-β1/ | Flask/flask | 1.42 | 0.82 | 1.7 | + | 0.94 | 1.14 | 1.07 |
| Growth factor combination | | | | | | | | | | | |
| 5/[66] | Alginate | 5% | w/o GF | Flask | | | 1 | 1 | | | |
| | | 5% | IGF-1 | Flask | | | 1.11 | 0.65 | | | |
| | | 5% | IGF-1 + TGF-β1 | Flask | | | 0.39 | 0.68 | | | |

(continued)

**Table 2** (continued)

| Data set/ Reference | Culture mode | Oxygen tension | Growth factors | Culture system | GAG | DNA | GAG/ DNA | Collagen II/I | Height | Weight | Young's modulus |
|---|---|---|---|---|---|---|---|---|---|---|---|
| 6/[66] | Cartilage | 5%/5% | w/o GF/w/o GF | Flask/ flask | 1 | 1 | 1 | = | 1 | 1 | 1 |
| | | 5%/5% | w/o GF/IGF-1 | Flask/ flask | 1.17 | 0.51 | 2.13 | = | 2.43 | 2.28 | 1.53 |
| | | 5%/5% | IGF-1 + TGF-β1/ IGF-1 | Flask/ flask | 2.46 | 0.53 | 4.37 | + | 2.82 | 3.12 | 5.58 |
| | | 5%/5% | IGF-1 + TGF-β1/ w/o GF | Flask/ flask | 2.58 | 0.85 | 2.89 | + | 3.34 | 1.82 | 9.3 |

Experiments were performed according to the protocol shown in Fig. 1 (alginate – step c, cartilage – step d, fluid: flow chamber reactor). In each set of experiments the data were based on the control experiment (set to "1" for quantitative and to "=" for qualitative parameters). Statistical significance was assessed by analysis of variance with $p < 0.05$ (ANOVA)

specific feature of the flow chamber is a very thin medium layer for improved oxygen supply and a counter current flow of the medium and gas [27, 37]. The intention was mainly to reduce mass-transfer effects rather than to expose fluid-dynamic stress on the cartilage. The applied flow rates are much too low for this [78]. Experiments in the flow-chamber bioreactor performed at 21% $O_2$ (v/v) showed a significantly higher matrix thickness but a lower content of glycosaminoglycane than cultures in 12-well plates as the control. The appearance of the cartilage obtained in the bioreactor seemed to be closer to the native cartilage with respect to the shape of cells, distribution of cells within the matrix, and smoothness of the surface among others. The cartilage obtained from 12-well plates showed an inhomogeneous distribution of cells, an uneven surface and holes within the matrix. Another important requirement for a successful implantation is the consistency of the cartilage. In the case of bioreactor cultures, the compaction of the cartilage and the attachment between the cartilage and carrier were very good, indicating that the construct is appropriate for implantation in this respect. In contrast, the cartilage-carrier-constructs cultivated in 12-well plates were soft and the attachment between the cartilage and carrier was not sufficient. In this case, the cartilage tended to slip off the carrier with only a slight mechanical impact. Furthermore, the cultivated cartilage should contain a significant content of collagen type II. This was confirmed qualitatively by immunohistological analysis.

Experiments at different oxygen concentrations (21, 10 and 5% v/v $O_2$) were performed in 12-well plates as the control and the constructs were compared qualitatively and quantitatively. In data set 3 the oxygen tension during the alginate step was maintained at 21% $O_2$ and only during cartilage formation was the oxygen concentration varied. Here the appearance of the cartilage obtained under decreased oxygen tension seemed to be closer to the native cartilage with respect to the shape of cells, distribution of cells within the matrix, and smoothness of the surface among others. The thickness of the cartilage formed by free swelling was slightly higher under reduced oxygen concentration and in the same range as that of the native cartilage (approximately 1 mm). The amount of glycosaminoglycan per DNA was significantly higher at lower oxygen concentrations than in control experiments, but still significantly lower than that in the native cartilage (data not shown). Qualitatively, the best attachment of the cartilage on top of the carrier was found for 5% $O_2$ (v/v). Furthermore, the cultivated cartilage contained a large content of collagen type II (data not shown). In data set 4 the oxygen concentration was reduced during the alginate step as well. Here a further improvement of the cartilage properties can be seen.

From these observations, a lower oxygen concentration in the gas phase is recommended. Similar results were obtained for human cells (data not shown). Nevertheless some open questions remain, especially regarding the results obtained in the flow chamber reactor. The theoretical simulations indicate that even under ideal conditions (no mass transfer limitation in the fluid phase) a severe oxygen limitation within the engineered tissue should be expected. If oxygen supply would be the limiting factor during cartilage formation, a bioreactor system (flow chamber) with an improved oxygen supply should lead to a better quality of the engineered cartilage. On the other hand, lower oxygen concentrations in the gas phase seem to improve some matrix properties. From the results discussed above, these discrepancies can be solved only to some extent. With respect to important biochemical properties, particularly the content of GAG, the constructs from the flow chamber bioreactor showed significantly lower values than those from the 12-well plates, probably due to a higher, detrimental oxygen concentration in the matrix. On the other hand, other matrix properties, particularly the attachment between the cartilage and carrier was better for constructs from the flow chamber than for those from 12-well plates. This may be due to a better oxygen supply within the matrix close to the surface of the carrier. Further studies are required to achieve a deeper understanding of the relevant mechanisms.

### 4.2.2 Growth Factor Combination

The effects of IGF-I and TGF-β1 on porcine chondrocytes were examined during the re-differentiation in alginate beads (step c, data set 5, Table 2) and cartilage formation in high-density cultures (step d, data set 6) for the cultivation principle introduced above in order to optimize the biochemical and biomechanical properties of the cartilage-carrier-constructs in vitro [66]. All experiments were performed at 5% $O_2$ (v/v).

The absence of any growth factors during the cultivation in alginate gel resulted in a high GAG to DNA ratio, but the number of cells which produced collagen type II were fewer compared to cultures with IGF-I or IGF-I + TGF-β1. Hence, these results demonstrate that the investigated growth factors can support the re-differentiation, here identified by the collagen type II production. The generated cartilage-carrier-constructs using no growth factors during alginate culture showed an irregular shape and tissue interspersed with holes. The analysis of the alginate cultures led to the assumption that the cells cultivated with IGF-I reached the furthest state of re-differentiation compared to the cultivation without any growth factors or with IGF-I + TGF-β1. Almost 100% of the cells produced the cartilage-specific collagen type II, but unfortunately also collagen type I. However, the resulting cartilage-carrier-constructs (data not shown) were soft and not stable. It is assumed that the cells have been stimulated by IGF-I to produce large amounts of proteoglycans. This presumption corresponds to the findings of many other groups as discussed above. But possibly, the collagen network did not develop adequately to maintain the cells as a pellet. These data demonstrate that the production of large amounts of proteoglycans during the alginate culture does not necessarily lead to the formation of high quality cartilage. Thus, addition of IGF-I alone during the alginate culture is not sufficient for our application.

Cartilage-carrier-constructs which were cultivated with IGF-I + TGF-β1 during the alginate phase and afterwards during cartilage formation with IGF-I yielded the significantly highest GAG to DNA ratio (data set 6, Table 2). In contrast, the significantly highest Young's Modulus was observed for the constructs which were cultivated with IGF-I + TGF-β1 during the alginate phase and without any growth factors during the cartilage formation. The Young's Modulus achieved (0.0595 MPa) was only 15% of that of native cartilage. For both conditions, histological sections showed homogenous tissue with an intensive staining against collagen type

II. According to the presented results, the biochemical appearance of the tissue-engineered cartilage is close to native cartilage, while it is still necessary to improve the biomechanical properties.

Articular cartilage in vivo has to transmit high stress [75]. Thus, it is important that tissue-engineered cartilage can withstand these loadings. Because the Young's Modulus is an indicator for the stiffness of the extracellular matrix, we decided to proceed with our research using IGF-I and TGF-β1 as medium supplements during the re-differentiation in alginate beads and no growth factors during the cartilage formation in high-density cell cultures.

It becomes obvious that it is not suitable to investigate the phases of the cultivation principle separately as the different steps are influenced by each other. Furthermore, it can be concluded that chondrocytes react in a different way to certain growth factors during re-differentiation or cartilage formation. Nevertheless, the achievements obtained by optimizing the growth factor combination are significant (compare Fig. 5 and Fig. 6).

Re-differentiation of human articular chondrocytes during alginate culture can be improved by optimization of growth factor combination as well [47]. To prove this, re-differentiation capacity of human articular chondrocytes after expansion on microcarriers or in conventional monolayer culture will be discussed (compare Sect. 2). Freshly isolated chondrocytes were first expanded by factor 10 in monolayer culture and by factor 300 on microcarriers without subcultivation. T-flask expanded passage 1 chondrocytes (P1) formed high amounts of collagen type I during cultivation in alginate gel (Fig. 7a). The formation of collagen type II was stimulated by IGF-I (100 ng mL$^{-1}$) and TGF-β (10 ng mL$^{-1}$). GAG/DNA ratios were slightly increased by these growth factors. Microcarrier-expanded chondrocytes (P1*) displayed lower amounts of collagen type I forming cells (Fig. 7b). The formation of collagen type II was inhibited almost completely by TGF-β (10 ng mL$^{-1}$) and by a combination of IGF-I (100 ng mL$^{-1}$) and TGF-β.

Treatment with BMP-7 (10, 50 or 100 ng mL$^{-1}$) did not reduce the percentages of collagen type I positive cells in T-flask expanded chondrocytes, but collagen type II production was stimulated by high concentrations of BMP-7 (50–100 ng mL$^{-1}$) and further increased by the combined action of IGF-I and BMP-7 (Fig. 7c and 7e).

Microcarrier-expanded chondrocytes (P1*) stained only weakly for collagen type I after 3 weeks in alginate culture (Fig. 7d), when BMP-7 was added to the culture medium. The production of collagen type II was increased synergistically by IGF-I and BMP-7 (Fig. 7d and 7f). GAG/DNA ratios were highest when both, IGF-I and BMP-7, were used to stimulate the production of cartilage matrix components by microcarrier-expanded chondrocytes.

In conclusion, it was shown that BMP-7 alone or in combination with IGF-I stimulated collagen type II production by both, T-flask and microcarrier-expanded chondrocytes, but the expansion on microcarriers resulted in a higher ratio of collagen type II to type I synthesizing cells. The higher rates of extracellular matrix production and better quality of matrix composition indicate, that the expansion of chondrocytes on microcarriers may improve cell-based cartilage repair (compare Sect. 2).

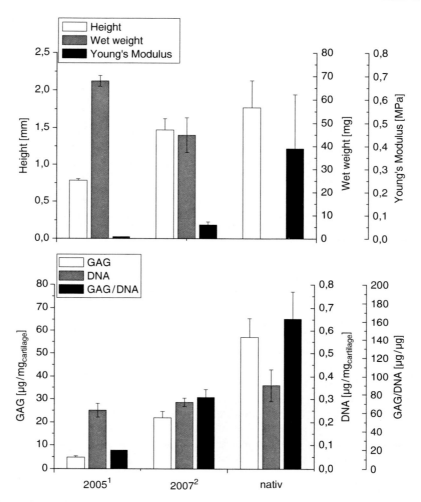

**Fig. 5** Biochemical and biomechanical data of porcine in vitro cartilage before (1) and after (2) optimization of growth factors and oxygen tension (2: data from [66])

## 5 Cartilage Formation Under Hydrostatic and Mechanical Load

### 5.1 Literature Survey

Another attempt to improve tissue-engineered cartilage is the application of mechanical force during the cultivation to produce a phenotypically appropriate tissue. Since cartilage is exposed to intermittent hydrostatic pressure, direct compression, gliding or shear due to everyday activities, it is hypothesized that mimicking mechanical load

Technical Strategies to Improve Tissue Engineering 163

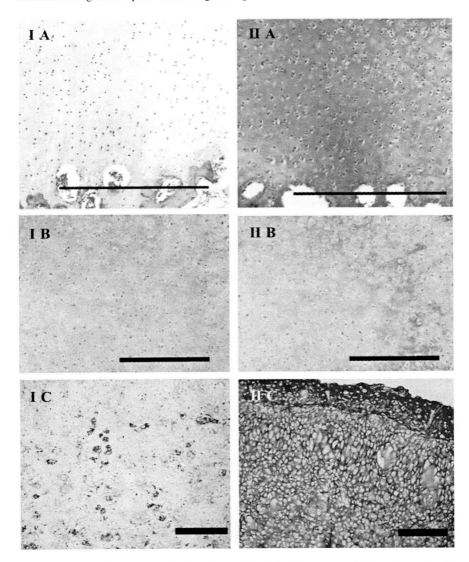

**Fig. 6** Immunohistology of porcine in vitro cartilage. I: collagen type I, II: collagen type II. (**a**) Native, scale bar 500 μm; (**b**) before optimization of growth factors and oxygen tension, scale bar 250 μm; (**c**) after optimization of growth factors and oxygen tension, scale bar 250 μm (data from [66])

might increase matrix synthesis of in vitro-engineered cartilage [79–82]. Thus, the aim of several studies was to investigate the influence of intermittent hydrostatic loading on the re-differentiation of chondrocytes during cartilage formation in vitro. However, the response of the cartilage matrix to loading is complex and involves many factors, including tissue and cell deformation, changes in hydrostatic pressure, and fluid flow [83]. Four main types of force are currently used in cartilage

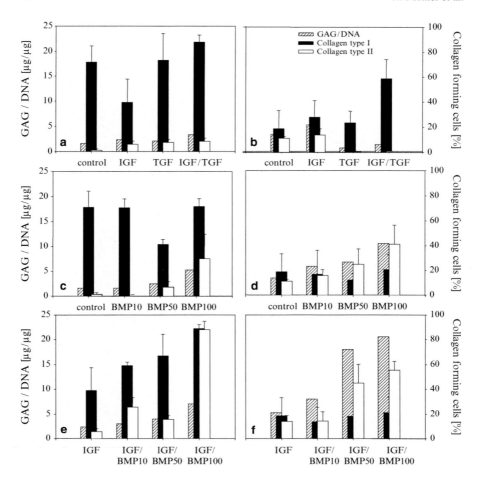

Fig. 7 Cartilage matrix formation by HAC expanded in T-flasks or on microcarriers. After expansion in T-flasks (by tenfold) or on microcarriers (by 300-fold) and subsequent re-differentiation in alginate gel, the recovered chondrocytes were analysed for cartilage-specific matrix components. GAG/DNA ratios represent glycosaminoglycan synthesis per DNA content of the cells. Percentages of collagen type I and type II positive cells were determined in each case from three images of microscopic slides. Results obtained with T-flask expanded chondrocytes are shown in (a), (c) and (e), results of microcarrier-expanded chondrocytes are represented by (b), (d) and (f), respectively. Growth factors were added to the culture medium as follows: IGF-I (100 ng mL$^{-1}$), TGF-β (10 ng mL$^{-1}$) and a combination of IGF-I and TGF-β (a and b), BMP-7 (10, 50 and 100 ng mL$^{-1}$) (c and d), IGF-I (100 ng mL$^{-1}$) and BMP-7 (10, 50 and 100 ng mL$^{-1}$) (e and f)

cultivation: hydrostatic pressure, direct compression, and "high"- and "low"-shear fluid environments. All these forces have been integrated into culturing devices used as bioreactors for articular cartilage. The individual effects have been discussed in [24], more recent data can be found in [77, 84–104]. Darling and Anthansiou [24] observed a poor comparability of the results, as experiments were performed with

# Technical Strategies to Improve Tissue Engineering

different cells (human or animal derived, primary or passaged) or different protocols used for cultivation (monolayer in flask or on microcarriers, encapsuled, free swelling cartilage, chondrocytes immobilized in three-dimensional macroporous carriers). Nevertheless, some general conclusions can be drawn.

During loading of the joint, the effect of hydrostatic pressure is mainly due to binding of water from the synovial fluid within the cartilage matrix by means of charged proteoglycanes. Swelling of the cartilage matrix is antagonized by the collagen network. Experimental results indicate best results when hydrostatic pressure within the physiological range (approximately 7–10 MPa) is applied intermittently during longer intervals. Permanent pressure seems to be inappropriate. A further important aspect of hydrostatic pressure is an impact on the solubility of oxygen and other gases (e.g. $CO_2$). Higher partial pressure of oxygen might lead to increased oxygen tension within the culture medium. This might have a negative effect, as for cartilage formation usually a lower oxygen tension is regarded as advantageous (as discussed above). An increase in carbon dioxide concentration might result in local decrease of the pH, again a negative effect. Therefore, detailed knowledge on mass transfer effects is essential for understanding the effects of hydrostatic pressure on cartilage formation. A literature survey of the effects of hydrostatic pressure is given in Table 3 and an overview of its impact is given in Table 4.

Direct compression results in a direct contact between joints, whereas different forces (normal, shear or friction) occur. For experimental simulation of these types of load, in most cases chondrocytes were cultivated under standard conditions (flask etc.) and exposed to direct compression within an external reactor for a short period of time only. In a recent publication [65] cartilage constructs were cultivated for a longer time within a bioreactor and exposed to direct compression in intervals (discussed in Sect. 4.1). Similar to hydrostatic pressure, a permanent compression is regarded as detrimental and loading should be intermittent to stimulate matrix formation. Besides a "mechano-stimulation" of the cells, improved mass transfer is regarded as important for improved cartilage properties.

Stimulation of chondrocytes by shear forces is often performed in bioreactor systems, where high shear forces (mechanically stirred systems, direct perfusion of a macroporous carrier) or low shear bioreactors (e.g. flow chambers) are distinguished. Advantages of stirred bioreactors (mostly spinner flasks), where high shear forces are exposed to cells grown on microcarriers, within macroporous carriers or immobilized in alginate beads, can be seen in an improved cell seeding rather than in enhanced matrix synthesis. Direct perfusion of cartilage tissue requires three-dimensional, macroporous, perfusable carrier, which are seeded with cells and perfused with culture medium at different flow velocities. Thus, different shear rates can be applied to the cells growing within the porous structures. This approach requires large interconnected pores within the cartilage matrix, a severe drawback. Therefore, this technique can not be used for cartilage constructs intended for later implantation.

Low shear forces can be applied within special bioreactor systems, e.g. the "rotating wall bioreactor" [105, 106]. The available results indicate that the slow rotation in this reactor induces low shear rates and a good mixing resulting in

**Table 3** Literature survey on impact of hydrostatic pressure on chondrocytes and cartilage formation

| Cells | Culture | Culture system | Amplitude | Frequency | Time | Analytics | Proteoglycan (compared to control) | Collagen type II | Total collagen | Biomechanics | Reference |
|---|---|---|---|---|---|---|---|---|---|---|---|
| Bovine, adult | Explant | A | 5 MPa<br>15 MPa<br>50 MPa<br>0.1-50 MPa | Constant | 2 h<br>5 min<br>2 h<br>20 s | B | ~1.4x<br>~1.2x<br>~0.2x<br>Maximum at 5-10 MPa ~1.5x, >40-50 MPa negative | n.d. | ~1.3x<br>~1.1x<br>~0.4x<br>Maximum at 5-10 MPa ~1.5x, >40-50 MPa negative | n.d. | [113] |
| | | | 0.1-50 MPa | | 2 h | | Maximum at 5 MPa ~1.4x, >15 MPa negative | | Maximum at 5 MPa ~1.25x, >15 MPa negative | | |
| Bovine, P0, Adult | Monolayer | A | 5 MPa | 0.0176 Hz<br>0.05 Hz<br>0.25 Hz<br>0.5 Hz<br>0.0176 Hz<br>0.05 Hz<br>0.25 Hz<br>0.5 Hz | 1.5 h<br><br><br><br>20 h | B | ~1x<br>~0.7x<br>~0.5x<br>~0.8x<br>~0.8x<br>~1x<br>~1.2x<br>~1.2x | n.d. | n.d. | n.d. | [114] |
| Bovine, P0 | Explantat | | 5 MPa | 0.0176 Hz<br>0.05 Hz<br>0.25 Hz<br>0.5 Hz | 1.5 h | | ~1x<br>~1fx<br>~1x<br>~1.17x | | | | |
| Bovine, P0, Adult | monolayer | A | 5 MPa | 0.017 Hz<br>0.25 Hz<br>0.5 Hz | 20 h | B | ~0.8x<br>~1x<br>~1.1x | n.d. | n.d. | n.d. | [115] |
| Bovine, P0, adult | Monolayer | D* | 10 MPa | 1 Hz | 4 h, +4 d outside reactor | A/B | A: ~1.3x,<br>B: ~1.7x | A: ~1.4x | n.d. | n.d. | [116] |
| Equine, P3, juveline | PGA-mesh | B | 3.44 MPa<br><br>6.87 MPa | 5/15 s on/off | 20 min every 4 h, 5 weeks | C | ~3.2x<br><br>~5.2x | n.d. | ~1x<br><br>~1.8x | n.d. | [117] |

# Technical Strategies to Improve Tissue Engineering

| | | | | | | | | |
|---|---|---|---|---|---|---|---|---|
| Equine, P3, adult | | | 3.44 MPa | | | ~2x | ~1x | Bulk modulus of elasticity, no data [117] |
| Equine, P3, juvenile | PGA-mesh | B | 6.87 MPa 3.45 MPa | 5/15 s on/off | 20 min every 4h, 5 weeks | C | ~1.3x ~4.5x | n.d. ~2x ~1x |
| Bovine, P1, adult | Monolayer Alginate gel | C | 0.2 MPa | 30/2 min on/off | 8h/days, 3 weeks | E | n.d. | 0% pos. cells 5-8% pos cells n.d. [118] |
| Bovine, P0, Adult | Pellet Monolayer | A | 10 MPa | 1 Hz | 4 h 4h/days, 4 days | A | ~20x | ~2x ~9x n.d. [119] |
| Bovine, P0, adult | Monolayer | C | 0.2 MPa | 0.1 Hz 30/2 min on/off 2/30 min on/off | 10 h/day, 6 day | D/C | n.d. | n.d. ~1x n.d. ~1.2x ~1.6x ~0.7x [82] |
| Bovine, P0, adult | Agarose gel | A | 5 MPa | 1 Hz | 4 h | B | ~1.1x | n.d. n.d. [120] |
| Bovine, P0, adult | Agarose gel | A | 5 MPa | constant | 4 h | A | ~4x | ~1.5x n.d. [120] |
| Human, P0, OA | Collagen type I/III-membrane | C | 0.2 MPa | 30/2 min on/off | 8 d | B | ~2x | n.d. ~1.1x [121] |
| Human, P0, OA | Monolayer | A | 10 MPa | 1 Hz | 4 h/day, 1day 4 h/day, 4 day 4 h/day, 1day | A | ~1.4x ~1.3x | ~1.3x ~1.1x n.d. [122] |
| Human, P0 | | | 1 MPa 5 MPa 10 MPa 1 MPa 5 MPa 10 MPa | | 4 h/day, 4 day | | ~1x ~1.3x ~1.5x ~1.4x ~1.8x ~1.9x | ~1x ~1x ~1x ~1.2x ~1.6x ~1.7x | |

(continued)

Table 3 (continued)

| Cells | Culture | Culture system | Amplitude | Frequency | Time | Analytics | Proteoglycan (compared to control) | Collagen type II | Total collagen | Biomechanics | Reference |
|---|---|---|---|---|---|---|---|---|---|---|---|
| Porcine, P0, mixture of juveline and Adult | Pellet | A* | 4 MPa | 1 Hz | 1, 5 h/d, 7 d | C/B | ~2x | n.d. | ~2.5x | n.d. | [123] |
| Bovine, P0, Juveline | Alginate gel collagen Foam | B | 2.8 MPa | 0.015 Hz constant | 15 days | C | ~1x ~2.7x ~3.1x | n.d. n.d. | ~1x n.d. | n.d. | [124] |
| Bovine, P0, juveline | agarose gel | A* | 10 MPa | 1 Hz | 4 h/day, 5 day/week, 8 weeks | C | ~1.4x | n.d. | ~1.2x | aggregate module: no sign. difference | [125] |
| Porcine, P3, adult | Alginate gel, 5 % $O_2$ | C | 0.4 MPa | 1 min on/off | 6 h/day, 2 weeks | C/E | ~1.24x (GAG/DNA) | ~1.55x (Type II/I) | n.d. | n.d. | [77] |
| | Cartilage-carrier-construct, 5 % $O_2$ | C | 0.4 MPa | 1 min on/off | 6 h/day, last week of 3 weeks total | C/E | ~1.43x | no significant diff. (Type II/I) | n.d. | Youngs modulus ~1.3x | |
| Human, P0, OA | Collagen type I-Gel | C | 40 kPa | 20/60 s on/off | 2 weeks | C | ~1.4x | n.d. | n.d. | n.d. | [126] |
| bovine, P3, < 12 month | pellet | A** | 5 MPa | 0.5 Hz | 4 h/d, 4 d | A | ~5x | ~4x | n.d. | n.d. | [127] |

# Technical Strategies to Improve Tissue Engineering 169

**culture system**

A  Cultures were performed in bags, petri dishes or culture chambers filled with media, which were placed in reactor filled with water. Load transmitted via fluid. Problem: no defined gas exchange
B  Semi-continuous perfusion system. Cartilage constructs are within a cage.
C  Pressure transmitted via incoming gas.
D  Pressure transmitted via medium (piston).

**analytics**

A  Determination of mRNA
B  Determination of inclusion of [$^{35}$S]-sulfate (Proteoglycane) or [3H]-proline (total collagen)
C  Photometrically
D  ELISA
E  Histology, Immunstayning

**abbreviations**

\*  After loading cells are taken out of the bag and cultivated in an incubator
\*\*  pO2 and pCO2 constant
OA  Osteo-arthrotic
n.d.  not done

**Table 4** Summary of results for cultivation of porcine chondrocytes under intermittent hydrostatic pressure and mechanical load

| Data set/ref. | Culture mode | Oxygen tension | Growth factors | Culture system | GAG | DNA | GAG/DNA | Collagen II/I | Height | Weight | Young's modulus |
|---|---|---|---|---|---|---|---|---|---|---|---|
| **Intermittent hydrostatic pressure** | | | | | | | | | | | |
| | Alginate | | | | | | | | | | |
| 7/[77] | | 5% | IGF-1 + TGF-β1 | Flask | 1 | 1 | 1 | 1 | | | |
| | | 5% | IGF-1 + TGF-β1 | Hydr. | | | 1.25 | 1.40 | | | |
| | Cartilage | | | | | | | | | | |
| | | Alginate/cartilage | Alginate/cartilage | Alginate/cartilage | | | | | | | |
| 8/[77] | | 5%/5% | IGF-1 + TGF-β1/IGF-1 + TGF-β1 | Flask/flask | 1 | 1 | 1 | = | 1 | 1 | 1 |
| | | 5%/5% | IGF-1 + TGF-β1/IGF-1 + TGF-β1 | Flask/hybr. | 1.11 | 1.04 | 1.08 | = | 1.3 | 0.85 | 1.01 |
| | | 5%/5% | IGF-1 + TGF-β1/IGF-1 + TGF-β1 | Hydr./hydr. | 1.43 | 1.16 | 1.22 | = | 1.22 | 0.88 | 1.31 |
| **Mechanical load** | | | | | | | | | | | |
| 9/[109] | | 5%/5% | IGF-1 + TGF-β1/IGF-1 | Flask/flask | 1 | 1 | 1 | = | 1 | 1 | 1 |
| | | 5%/5% | IGF-1 + TGF-β1/IGF-1 | Flask/compr. | 1.22 | 2 | 0.61 | + | 0.91 | n.d. | 1.45 |
| | | 5%/5% | IGF-1 + TGF-β1/IGF-1 | Flask/shear | 1.36 | 1 | 1.36 | + | 1,01 | n.d. | 1.28 |
| Hydr. | Hydrostatic load (4 bar abs., 1 min on/off, 6 h per day) | | | | | | | | | | |
| Compr. | Compression 5% | | | | | | | | | | |
| Shear | Shear stress | | | | | | | | | | |

Experiments were performed according to the protocol shown in Fig. 1 (alginate – step c, cartilage – step d, fluid: flow chamber reactor)
In each set of experiments the data were based on the control experiment (set to "1" for quantitative and to "=" for qualitative parameters). Statistical significance was assessed by analysis of variance with $p < 0.05$ (ANOVA)

improved cartilage properties. But it has to be stated, that most of these promising results were obtained with primary cells rather than with passaged cells. The main disadvantage of these systems lies in an inhomogeneous flow pattern and therefore heterogeneous distribution of the forces acting on the cartilage constructs.

The above considerations underline, that the effects of mechanical stimulation on cartilage formation have not been investigated sufficiently and are not yet understood. Some of the applied systems favour synthesis of collagen, mainly collagen type II (e.g. direct compression), some induce matrix synthesis (proteoglycans, e.g. hydrostatic pressure). Furthermore, the optimal conditions have to be adjusted for each application in cartilage-tissue engineering.

## 5.2 Experimental Results for Porcine Cartilage

Three-dimensional cartilage-carrier-constructs were produced according to the protocol described in Fig. 1 using porcine chondrocytes. The alginate step c and/or the final step d for cartilage formation of this protocol were performed either in bioreactors (intermittent hydrostatic pressure or direct mechanical load) or in 12-well plates as the control. The main results of these experiments are summarized in Table 4.

### 5.2.1 Hydrostatic Pressure

Heyland et al. [77] used a bioreactor system to study the response of porcine chondrocytes to intermittent hydrostatic pressure realized by gassing with overpressure at low oxygen concentrations of 5% (v/v) $O_2$. The hydrostatic pressure was transduced via the uncompressible medium to the cells. For 6 h per day, intermittent hydrostatic pressure (0.4 MPa abs.) was applied during the cultivation in the alginate gel (step c) or during the last week of cartilage formation (step d) with a frequency 1 min on/off. The pressure amplitude of 0.3 MPa used here was within the range of amplitudes found to have a stimulating effect in other systems [24]. The results of this study show that intermittent loading can influence matrix synthesis during re-differentiation of chondrocytes in alginate beads and during cartilage formation.

During alginate culture (data set 7, Table 4) a significant increase in the ratio of glycosaminoglycan to DNA was found, 25% higher compared to the corresponding static control. A 40% higher ratio of collagen type II to I compared to the static control was observed. Furthermore, the immunostaining of the generated cartilage-carrier-constructs indicated a superior intensity of collagen type II and a lower intensity of collagen type I compared to the static control. It does, however, seem obvious that intermittent loading can introduce matrix synthesis, although it should be pointed out that little is known about the effects of loading stress. The results indicate that applying intermittent loading during re-differentiation on chondrocytes embedded in three-dimensional alginate beads can induce higher rates of synthesis of the matrix components glycosaminoglycan and collagen type II at low oxygen concentrations.

For chondrocytes cultivated in alginate beads at intermittent hydrostatic loading enhanced chondrogenesis during cartilage formation was observed (data set 8, Table 4). One interesting result is that a better Young's modulus was achieved for loaded cartilage-carrier-constructs, indicating a stiffer matrix when applying intermittent loading during both cultivation steps. Obviously, a more stable collagen network with a higher stability is formed when cartilage is cultivated under hydrostatic load. Promising results were found for the ratio of GAG/DNA, where an increase of more than 8% in cartilages with intermittent hydrostatic pressure only during cartilage formation and an increase of more than 22% in cartilage, applying intermittent hydrostatic loading during both cultivation steps, compared to the static controls was determined.

The stimulation observed when intermittent hydrostatic pressure was exerted underlines the positive effects of mimicking in vivo conditions. Although previous reports have also described the stimulating effects of intermittent loading on glycosaminoglycan synthesis of bovine chondrocytes [24, 82], no advice in the literature has been found reporting this stimulation for porcine chondrocytes. However, our results demonstrated that intermittent loading applied during re-differentiation in the alginate culture and later on the cartilage-carrier-constructs had positive effects on the characteristics of the matrix. Although no increase in collagen type II expression of the cartilage-carrier-constructs was observed, a higher Young's modulus was achieved. Moreover, glycosaminoglycan production was enhanced throughout intermittent hydrostatic loading, although it was also shown that a combination of applied loading during alginate culture and again on cartilage culture yielded the highest values.

### 5.2.2 Mechanical Load

Existing bioreactors designed for mechanical stimulation mostly enable the stimulation of the tissue but not its analysis. In order to determine the influence of the load regimes on the cartilage's development, it is necessary to remove the samples from the cultivation process for investigation [107]. Either several tests are done, each interrupting the cultivation process at a different stage [108], or parallel cultivation of many samples allows for subsequent removal of samples [20]. Both methods require a high number of samples since the specimen has to be removed from the cultivation process for testing. A bioreactor that facilitates cultivation of cartilage under mechanical stimuli as well as measuring biomechanical properties would be of great value to determine and possibly optimize the cartilage's properties throughout the cultivation process. The bioreactor introduced by Ilinich [109] enables cultivation of cartilage under shear loading as well as static and dynamic compression and biomechanical testing without interrupting the cultivation process. The thickness and the Young's modulus of the cartilage-carrier-constructs can be determined by means of an indentation or unconfined compression test. All mechanical stimulations and tests are performed within the cultivation chamber without discontinuation of the culturing process. This allows permanent monitoring of the cartilage development throughout the process of cartilage synthesis. The reactor consists of eight cultivation chambers, made of polyetheretherketone (PEEK), each containing one cartilage construct (Fig. 8).

# Technical Strategies to Improve Tissue Engineering

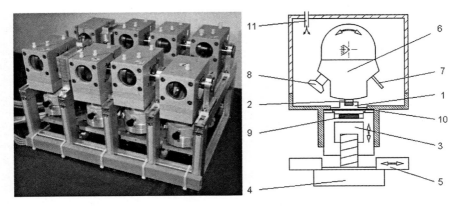

**Fig. 8** Bioreactor with eight chambers for cultivation of eight cartilage samples under load application [109]). Thickness and Young's modulus can be measured within the chamber during cultivation (*left*); *right*: Cross section through a bioreactor chamber with sample support (1), cartilage-carrier-construct (2), step motor for horizontal displacement (3), common plate (4), shear displacement step motor (5), steel segment (6), indenter (7), punch (8), load cell (9), membrane (10), gas exchange port (11)

This bioreactor was used for cultivation of cartilage-carrier-constructs (step d of the protocol shown in Fig. 1). The cartilage-carrier-constructs within the bioreactor were cultivated at 5% $O_2$ and loaded 10-times per day. Each load interval consisted of ten load cycles with 5% strain amplitude and a compression rate of 0.05 mm $s^{-1}$, followed by a break of 30 min. In case of compression, starting from a pre-load force of 0.05 N a compression ratio of 5% was applied, in case of shear loading a compression ratio of 5% and a displacement of 100 µm was applied. The thickness of the cartilage-carrier-constructs was measured prior to each load interval at a contact force of 0.05 N. The contact load was required in order to guarantee flush contact between probe and indenter. For further details compare Ilinich [109].

Data set 9 in Table 4 summarizes the final properties of cartilage-carrier-constructs cultivated under compression or shear loading. Compared to the control, the content of GAG/DNA was lower in case of compression and higher in case of shear loading compared to the control. The height of the construct was for all conditions in a similar range; the Young's modulus, however, was significantly up to 45% (in case of compression) increased. For both loading conditions, histological sections showed a homogenous tissue with a more intensive staining against collagen type II as the control.

## 6 Final Considerations

The above examples underline the philosophy that engineering knowledge can be used to improve cultivation systems and applied strategies for cartilage-tissue engineering. On the other hand it has to be realized, that both engineering and fundamental studies on cell biology are required to further clarify the observed effects. Since for all studies presented the same basic protocol was used, the impact of isolated parameters can be compared (even so it has to be expected that certain cross talk effects could not be fully eliminated). With respect to oxygen tension cultivation at 5% $O_2$ in the gas phase gave the best results. This finding is in agreement with several other literature studies and might be satisfactory for setting up a cultivation protocol, but the actual concentration within the tissue-engineered cartilage will be much lower. Therefore, more sophisticated theoretical or experimental analysis of the oxygen profile within the cartilage matrix is required. Furthermore, concentration profiles for $CO_2$ are essential, as insufficient removal of $CO_2$ might lead to local pH-gradients.

Studies on growth factor combinations underlined the tremendous potential of this parameter. In general the conclusions drawn from our studies are in agreement with others. It should be emphasized that chondrocytes react in a different way to certain growth factors during re-differentiation or cartilage formation. Therefore, a growth factor optimization is required for each individual protocol used.

From the studies on hydrodynamic and mechanical load, it can be assumed that mimicking in vivo conditions such as loading stress during cultivation might be a useful tool in cartilage-tissue engineering and might lead to optimized culture conditions. But again—some questions remain. The promising results of cartilage-carrier-constructs cultivated under intermittent hydrostatic pressure or mechanical load were obtained at a decreased oxygen concentration in the gas phase. Initially, this phenomenon is difficult to understand, especially in the case of hydrostatic pressure. As the partial pressure of oxygen in the gas phase depends on total pressure, a higher pressure should even increase oxygen concentration significantly, leading to even worse matrix properties. However, further investigations by our group (data not shown) indicated that due to the very low mass transfer between the static culture medium and the gas phase, oxygen concentration in the culture medium did not increase during intervals with a higher pressure. Therefore, other factors that explain the significantly improved matrix properties of constructs cultivated under intermittent hydrostatic pressure compared with static cultures in 12-well plates should be identified. Probably, the mechanical stimulation of chondrocytes is the main effect. The results obtained under mechanical stimulation underline this hypothesis. The results of the first cultivation of porcine cartilage under mechanical compression with the new bioreactor design showed an increase in stiffness. This is very promising. However, the mean stiffness is still too low compared to native cartilage (approximately 1.1 MPa [109]). Systematically varying the parameters of the loading protocol such as the load's degrees of freedom, the load amplitude and the amount of stimuli will show whether the properties can be adapted to the native cartilage to a degree that is sufficient for implantation. As the shown experiments were performed before optimization of growth factor combination, a further increase in the biomechanical properties of the generated cartilage can be expected.

Furthermore, to clarify the biological effects, further biochemical and cell biological studies combined with analysis of flow and mass-transfer effects and more advanced bioreactor systems are required [78, 110–112]. To date the goals and expectations of bioreactor development have been fulfilled only to some extent, as bioreactor design in tissue engineering is very complex and still at an early stage of development. In the future, a very intimate collaboration between engineers and biologists will lead to an increased fundamental understanding of complex issues that can have an impact on tissue formation in bioreactors. On one hand, devices are required with a well-described microenvironment of cells for fundamental studies. On the other hand, a transition from a laboratory scale to an industrial scale will require a high adaptability of specialized bioreactors in a standardized production process. These advances are the prerequisite for ensuring that tissue engineering will fulfil the expectations for revolutionizing medical care.

**Acknowledgement** Financial support by the BMBF, FHH Hamburg and DFG is gratefully acknowledged.

# References

1. Godbey WT, Atala A (2002) In vitro systems for tissue engineering. Ann N Y Acad Sci 961: 10–26
2. Griffith LG, Naughton G (2002) Tissue engineering-current challenges and expanding opportunities. Science 295: 1009–1014
3. Koch RJ, Gorti GK (2002) Tissue engineering with chondrocytes. Facial Plast Surg 18: 59–68
4. Lalan S, Pomerantseva I, Vacanti JP (2001) Tissue engineering and its potential impact on surgery. World J Surg 25: 1458–1466
5. Langer R (2000) Tissue engineering. Mol Ther 1: 12–15
6. Naughton GK (2002) From lab bench to market: critical issues in tissue engineering. Ann N Y Acad Sci 961: 372–385
7. Stoltz JF, Bensoussan D, Decot V, Ciree A, Netter P, Gillet P (2006) Cell and tissue engineering and clinical applications: an overview. Biomed Mater Eng 16: 3–18
8. Chen HC, Hu YC (2006) Bioreactors for tissue engineering. Biotechnol Lett 28: 1415–1423
9. Martin I, Wendt D, Heberer M (2004) The role of bioreactors in tissue engineering. Trends Biotechnol 22: 80–86
10. Pei M, Solchaga LA, Seidel J, Zeng L, Vunjak-Novakovic G, Caplan AI, Freed LE (2002) Bioreactors mediate the effectiveness of tissue engineering scaffolds. FASEB J 16: 1691–1694
11. Pörtner R, Nagel-Heyer St, Goepfert Ch, Adamietz P, Meenen NM (2005) Bioreactor design for tissue engineering. J Bioeng Biosci 100: 235–245
12. Pörtner R, Giese Ch (2007) Overview on bioreactor design, prototyping and in process controls for reproducible three dimensional tissue culture. In: Marx U, Sandig V (eds.) Drug testing in vitro. WILEY-VCH Verlag, Weinheim
13. Ratcliffe A, Niklason LE (2002) Bioreactors and bioprocessing for tissue engineering. Ann N Y Acad Sci 961: 210–215
14. Risbud MV, Sittinger M (2002) Tissue engineering: advances in in vitro cartilage generation. Trends Biotechnol 20: 351–356
15. Petersen JP, Rücker A, von Stechow D, Adamietz P, Pörtner R, Rueger JM, Meenen NM (2003) Present and future therapies of articular cartilage defects. Eur J Trauma 1: 1–10
16. Meenen NM, Ueblacker P, Pörtner R, Göpfert Ch, Nagel-Heyer St, Petersen JP, Adamietz P (2005) Knorpel aus dem Labor-eine Sackgasse? Arthroskopie 18: 245–252

17. Kuo CK, Li WJ, Mauck RL, Tuan RS (2006) Cartilage tissue engineering: its potential and uses. Curr Opin Rheumatol 18(1): 64–73
18. Glowacki J (2000) In vitro engineering of cartilage. J Rehabil Res Dev 37: 171–177
19. Kim HW, Han CD (2000) An overview of cartilage tissue engineering. Yonsei Med J 41: 766–773
20. Lima EG, Mauck RL, Shelley HH, Park S, Ng KW, Ateshian GA, Hung CT (2004) Functional tissue engineering of chondral and osteochondral constructs. Biorheology 41: 577–590
21. Nesic D, Whiteside R, Brittberg M, Wendt D, Martin I, Mainil-Varlet P (2006) Cartilage tissue engineering for degenerative joint disease. Adv Drug Deliv Rev 58: 300–322
22. Martin I, Miot S, Barbero A, Jakob M, Wendt D (2007) Osteochondral tissue engineering. J Biomech 40: 750–765
23. Obradovic B, Martin I, Freed LE, Vunjak-Novakovic G (2001) Bioreactor studies of natural and tissue engineered cartilage. Ortop Traumatol Rehabil 3: 181–189
24. Darling EM, Athanasiou KA (2003) Articular cartilage bioreactors and bioprocesses. Tissue Eng 9: 9–26 (Erratum in: Tissue Eng 2003 9: 565)
25. Chen HC, Lee HP, Sung ML, Liao CJ, Hu YC (2004) A novel rotating-shaft bioreactor for two-phase cultivation of tissue-engineered cartilage. Biotechnol Prog 20: 1802–1809
26. Lee DA, Martin I (2004) Bioreactor culture techniques for cartilage-tissue engineering. Methods Mol Biol 238: 159–170
27. Nagel-Heyer St, Goepfert Ch, Adamietz P, Meenen NM, Petersen JP, Pörtner R (2005) Flow-chamber bioreactor culture for generation of three-dimensional cartilage-carrier-constructs. Bioprocess Biosyst Eng 27: 273–280
28. Vunjak-Novakovic G, Meinel L, Altman G, Kaplan D (2005) Bioreactor cultivation of osteochondral grafts. Orthod Craniofac Res 8: 209–218
29. Wendt D, Jakob M, Martin I (2005) Bioreactor-based engineering of osteochondral grafts: from model systems to tissue manufacturing. J Biosci Bioeng 100: 489–494
30. Akmal M, Anand A, Anand B, Wiseman M, Goodship AE, Bentley G (2006) The culture of articular chondrocytes in hydrogel constructs within a bioreactor enhances cell proliferation and matrix synthesis. J Bone Joint Surg Br 88: 544–553
31. Bilgen B, Sucosky P, Neitzel GP, Barabino GA (2006) Flow characterization of a wavy-walled bioreactor for cartilage tissue engineering. Biotechnol Bioeng 95: 1009–1022
32. Freed LE, Guilak F, Guo XE, Gray ML, Tranquillo R, Holmes JW, Radisic M, Sefton MV, Kaplan D, Vunjak-Novakovic G (2006) Advanced tools for tissue engineering: scaffolds, bioreactors, and signaling. Tissue Eng 12: 3285–3305
33. Cooper JA Jr, Li WJ, Bailey LO, Hudson SD, Lin-Gibson S, Anseth KS, Tuan RS, Washburn NR (2007) Encapsulated chondrocyte response in a pulsatile flow bioreactor. Acta Biomater 3: 13–21
34. Schulz RM, Bader A (2007) Cartilage tissue engineering and bioreactor systems for the cultivation and stimulation of chondrocytes. Eur Biophys J 36: 539–568
35. Yoshioka T, Mishima H, Ohyabu Y, Sakai S, Akaogi H, Ishii T, Kojima H, Tanaka J, Ochiai N, Uemura T (2007) Repair of large osteochondral defects with allogeneic cartilaginous aggregates formed from bone marrow-derived cells using RWV bioreactor. J Orthop Res 25: 1291–1298
36. Nagel-Heyer St, Goepfert Ch, Morlock MM, Pörtner R (2005) Relationship between gross morphological and biochemical data of tissue engineered cartilage-carrier-constructs. Biotechnol Lett 27: 187–192
37. Nagel-Heyer St (2004) Engineering aspects for generation of three dimensional cartilage-carrier-constructs. Books on Demand GmbH, Norderstedt, Germany, ISBN 3-8334-1478-2
38. Nagel-Heyer St, Goepfert Ch, Adamietz P, Meenen NM, Pörtner R (2006) Cultivation of three-dimensional cartilage-carrier-constructs under reduced oxygen tension. J Biotechnol 121: 486–497
39. Petersen JP, Uebelacker P, Goepfert Ch, Adamietz P, Stork A, Rueger JM, Pörtner R, Amling M, Meenen NM. Long term results after implantation of tissue engineered cartilage for the treatment of osteochondral lesions in a minipig model. J Mat Sci Mater Med, DOI: 10.1007/s10856-007-3291-3

40. Pei M, Seidel J, Vunjak-Novakovic G, Freed LE (2002) Growth factors for sequential cellular de- and re-differentiation in tissue engineering. Biochem Biophys Res Commun 294: 149–154
41. Jakob M, Demarteau O, Schäfer D, Hintermann B, Dick W, Heberer M, Martin I (2001) Specific growth factors during the expansion and redifferentiation of adult human articular chondrocytes enhances chondrogenesis and cartilaginous tissue formation. J Cell Biochem 81: 368–377
42. Lundgren B, Blüml G (1998) Microcarriers in cell culture production. In: Subramanian G (ed.) Bioseparation and bioprocessing – a handbook. WILEY-VCH, Weinheim
43. Butler M (2004) Animal cell culture and technology – the basics. Oxford University Press, Oxford
44. Malda J, Frondoza CG (2006) Microcarriers in the engineering of cartilage and bone. Trends Biotechnol 24: 299–304
45. Malda J, van den Brink P, Meeuwse P, Grojec M, Martens DE, Tramper J, Riesle J, van Blitterswijk CA (2004) Effect of oxygen tension on adult articular chondrocytes in microcarrier bioreactor culture. Tissue Eng 10: 987–994
46. Nagel-Heyer St, Leist Ch, Lünse S, Goepfert Ch, Pörtner R (2005) From biopsy to cartilage-carrier constructs by using microcarrier cultures as sub-process. In: Proceedings of 19th ESACT meeting, Harrogate, UK, p. 139
47. Goepfert Ch, Lutz V, Lünse S, Kittel S, Wiegandt K, Pörtner R, Kammal M, Püschel K (2007) Expansion of human articular chondrocytes on microcarriers enhances the production of cartilage specific matrix components. 20th Meeting of the European Society for Animal Cell Technology, 17–20th June 2007, Dresden
48. Kreklau B, Sittinger M, Mensing M, Voigt C, Berger G, Burmester G, Rahmanzadeh R, Gross U (1999) Tissue engineering of biphasic joint cartilage implants. Biomaterials 20: 1743–1749
49. Suominen E, Vedel A, Knagasniemi I, Uusipaikka E, Yli-Urpo A (1996) Subchondral bone and cartilage repair with bioactive glasses, hydroxyapatite, and hydroxyapatite-glass composite. J Biomed Mater Res 32: 143–551
50. Asselin A, Hattar S, Oboef M, Greenspan D, Berdal A, Sautier J (2004). The modulation of tissue-specific gene expression in rat nasal chondrocyte cultures by bioactive glasses. Biomaterials 25: 2621–5630
51. Van Susante J, Buma P, Schuman L, Homminga G, Van den Berg W, Veth R (1999) Resurfacing potential of heterologous chondrocytes suspended in fibrin glue in large full-thickness defects on femoral articular cartilage: an experimental study in the goat. Biomaterials 20: 1167–1175
52. Janssen, R.; S. Nagel-Heyer; Ch. Goepfert; R. Pörtner; D. Toykan; O. Krummhauer; M.M. Morlock; P. Adamietz; N.M. Meenen; W.M. Kriven; D.-K. Kim; A. Tampieri, G. Celotti (2004) Calcium phosphate ceramics as substrate for cartilage cultivation. Ceramic Eng Sci Proc 25(4): 523–528
53. Chang CH, Lin FH, Lin CC, Chou CH, Liu HC (2004) Cartilage tissue engineering on the surface of a novel gelatin-calcium-phosphate biphasic scaffold in a double-chamber bioreactor. J Biomed Mater Res B Appl Biomater 71(2): 313–21
54. Tanaka T, Komaki H, Chazono M, Fuji K (2005) Use of a biphasic graft constructed with chondrocytes overlying a -tricalcium phosphate block in the treatment of rabbit osteochondral defects. Tissue Eng 11: 331–339
55. Kandel RA, Grynpas M, Pilliar R, Lee J, Wang J, Waldman S, Zalzal P, Hurtig M (2006) Repair of osteochondral defects with biphasic cartilage-calcium polyphosphate constructs in a sheep model. Biomaterials (22): 4120–4131
56. Waldman SD, Grypnas M, Pilliar R. Kandel R (2002) Characterization of cartilagenous tissue formed on calcium polyphosphate substrates in vitro. J Biomed Mater Res 62: 323–330
57. Waldman SD, Grypnas M, Pilliar R, Kandel R (2003) The use of specific chondrocyte populations to modulate the properties of tissue-engineered cartilage. J Orthop Res 21: 132–138
58. Malda J, Martens DE, Tramper J, van Blitterswijk CA, Riesle J (2003) Cartilage tissue engineering: controversy in the effect of oxygen. Crit Rev Biotechnol 23: 175–194

59. O'Driscoll SW, Fitzsimmons JS, Commisso CN (1977) Role of oxygen tension during cartilage formation by periosteum. J Orthop Res 15: 682–687
60. Domm C, Schünke M, Christesen K, Kurz B (2002) Redifferentiation of dedifferentiated bovine articular chondrocytes in alginate culture under low oxygen tension. Osteoarthr Cartil 10: 13–22
61. Malda J, Rouwkema J, Martens DE, Le Comte EP, Kooy FK, Tramper J, van Blitterswijk CA, Riesle J (2004) Oxygen gradients in tissue-engineered PEGT/PBT cartilaginous constructs: measurement and modelling. Biotechnol Bioeng 86: 9–18
62. Obradovic B, Carrier RL, Vunjak-Novakovic G, Freed LE (1999) Gas exchange is essential for bioreactor cultivation of tissue engineered cartilage. Biotechnol Bioeng 63: 197–205
63. Ysart GE, Mason RM (1994) Responses of articular cartilage explant cultures to different oxygen tensions. Biochim Biophys Acta 1221: 15–20
64. Nehring D, Adamietz P, Meenen NM, Pörtner R (1999) Perfusion cultures and modelling of oxygen uptake with three-dimensional chondrocyte pellets. Biotechnol Technol 13: 701–706
65. Wernike E, Li Z, Alini M, Grad S (2008) Effect of reduced oxygen tension and long-term mechanical stimulation on chondrocyte-polymer constructs. Cell Tissue Res 331: 473–483
66. Wiegandt K, Goepfert Ch, Pörtner R (2007) Improving in vitro generated cartilage-carrier-constructs by optimizing growth factor combination. Open Biomed Eng J 1: 85–90
67. Blunk T, Sieminski AL, Gooch KJ, Courter DL, Hollander AP, Nahir AM, Langer R, Vunjak-Novakovic G, Freed LE (2002) Differential effects of growth factors on tissue-engineered cartilage. Tissue Eng 8: 73–84
68. Benz K, Breit S, Lukoschek M, Mau H, Richter W (2002) Molecular analysis of expansion, differentiation, and growth factor treatment of human chondrocytes identifies differentiation markers and growth-related genes. Biochem Biophys Res Commun 293: 284–292
69. Francioli SE, Martin I, Sie CP, Hagg R, Tommasini R, Candrian C, Heberer M, Barbero A (2007) Growth factors for clinical-scale expansion of human articular chondrocytes: relevance for automated bioreactor systems. Tissue Eng 13: 1227–1234
70. Guerne PA, Sublet A, Lotz M (1994) Growth factor responsiveness of human articular chondrocytes: distinct profiles in primary chondrocytes, subcultured chondrocytes, and fibroblasts. J Cell Physiol 158: 476–484
71. Yaeger PC, Masi T, Buck de Ortiz JL, Binette F, Tubo R, Mc Pherson JM (1997) Synergistic action of transforming growth factor-$\beta$ and insulin like growth factor-I induces expression of type II collagen and aggrecan genes in adult human articular chondrocytes. Exp Cell Res 237: 318–325
72. Van Osch GJ, Van den Berg WB, Hunziker EB, Häuselmann HJ (1998) Differential effects of IGF-1 and TGF beta-2 on the assembly of proteoglycans in pericellular and territorial matrix by cultured bovine articular chondrocytes. Osteoarthr Cartil 6: 187–195
73. Tsukazaki T, Usa T, Matsumoto T, Enomoto H, Ohtsuru A, Namba H, Iwasaki K, Yamashita S (1994) Effect of transforming growth factor-$\beta$ on the insulin-like growth factor-I autocrine/paracrine axis in cultured rat articular chondrocytes. Exp Cell Res 215: 9–16
74. Gooch KJ, Blunk T, Courter DL, Sieminski AL, Bursac PM, Vunjak-Novakovic G, Freed LE (2001) IGF-I and mechanical environment interact to modulate engineered cartilage development. Biochem Biophys Res Commun 286: 909–915
75. Mauck RL, Nicoll SB, Seyhan SL, Athesian G, Hung CT (2003) Synergistic action of growth factors and dynamic loading for articular cartilage tissue engineering. Tissue Eng 9: 597–611
76. Malda J, van Blitterswijk CA, van Geffen M, Martens DE, Tramper J, Riesle J (2004) Low oxygen tension stimulates the redifferentiation of dedifferentiated adult human nasal chondrocytes. Osteoarthr Cartil 12: 306–313
77. Heyland J, Wiegandt K, Goepfert Ch, Nagel-Heyer St., Ilinich E, Schumacher U, Pörtner R (2006) Redifferentiation of chondrocytes and cartilage formation under intermittent hydrostatic pressure. Biotechnol Lett 28: 1641–1648
78. Hussein MA, Esterl S, Pörtner R, Wiegandt K, Becker T (2008) On the Lattice Boltzmann method simulation of a two phase flow bioreactor for artificially grown cartilage cells. Bioscience (in press)

79. Altman GH, Horan RL, Martin I, Farhadi J, Stark PR, Volloch V, Richmond JC, Vunjak-Novakovic G, Kaplan DL (2002) Cell differentiation by mechanical stress. FASEB J 16: 270–272
80. Butler DL, Goldstein SA, Guilak F (2000) Functional tissue engineering: the role of biomechanics. J Biomech Eng 122: 570–575
81. Mauck RL, Soltz MA, Wang CC, Wong DD, Chao PH, Valhmu WB, Hung CT, Ateshian GA (2000) Functional tissue engineering of articular cartilage through dynamic loading of chondrocyte-seeded agarose gels. J Biomech Eng 122: 252–260
82. Hansen U, Schunke M, Domm C, Ioannidis N, Hassenpflug J, Gehrke T, Kurz B (2001) Combination of reduced oxygen tension and intermittent hydrostatic pressure: a useful tool in articular cartilage tissue engineering. J Biomech 7: 941–949
83. Weightman B, Isherwood DP, Swanson SA (1979) The fracture of ultrahigh molecular weight polyethylene in the human body. J Biomed Mater Res 4: 669–672
84. Connelly JT, Vanderploeg EJ, Levenston ME (2004) The influence of cyclic tension amplitude on chondrocyte matrix synthesis: experimental and finite element analyses. Biorheology 41: 377–387
85. Gemmiti CV, Guldberg RE (2006) Fluid flow increases type II collagen deposition and tensile mechanical properties in bioreactor-grown tissue-engineered cartilage. Tissue Eng 12: 469–479
86. Gokorsch S, Nehring D, Grottke C, Czermak P (2004) Hydrodynamic stimulation and long term cultivation of nucleus pulposus cells: a new bioreactor system to induce extracellular matrix synthesis by nucleus pulposus cells dependent on intermittent hydrostatic pressure. Int J Artif Organs 27: 962–970
87. Gokorsch S, Weber C, Wedler T, Czermak P (2005) A stimulation unit for the application of mechanical strain on tissue engineered anulus fibrosus cells: a new system to induce extracellular matrix synthesis by anulus fibrosus cells dependent on cyclic mechanical strain. Int J Artif Organs 28: 1242–1250
88. Hsu SH, Kuo CC, Whu SW, Lin CH, Tsai CL (2006) The effect of ultrasound stimulation versus bioreactors on neocartilage formation in tissue engineering scaffolds seeded with human chondrocytes in vitro. Biomol Eng 23: 259–264
89. Lappa M (2003) Organic tissues in rotating bioreactors: fluid-mechanical aspects, dynamic growth models, and morphological evolution. Biotechnol Bioeng 84: 518–532
90. Lee DA, Knight MM (2004) Mechanical loading of chondrocytes embedded in 3D constructs: in vitro methods for assessment of morphological and metabolic response to compressive strain. Methods Mol Med 100: 307–324
91. Li KW, Klein TJ, Chawla K, Nugent GE, Bae WC, Sah RL (2004) In vitro physical stimulation of tissue-engineered and native cartilage. Methods Mol Med 100: 325–352
92. Li Z, Yao S, Alini M, Grad S (2007) Different response of articular chondrocyte subpopulations to surface motion. Osteoarthr Cartil 15: 1034–1041
93. Mauck RL, Byers BA, Yuan X, Tuan RS (2007) Regulation of cartilaginous ECM gene transcription by chondrocytes and MSCs in 3D culture in response to dynamic loading. Biomech Model Mechanobiol 6: 113–125
94. Meyer U, Büchter A, Nazer N, Wiesmann HP (2006) Design and performance of a bioreactor system for mechanically promoted three-dimensional tissue engineering. Br J Oral Maxillofac Surg 44: 134–140
95. Raimondi MT, Boschetti F, Falcone L, Migliavacca F, Remuzzi A, Dubini G (2004) The effect of media perfusion on three-dimensional cultures of human chondrocytes: integration of experimental and computational approaches. Biorheology 41: 401–410
96. Raimondi MT, Moretti M, Cioffi M, Giordano C, Boschetti F, Laganà K, Pietrabissa R (2006) The effect of hydrodynamic shear on 3D engineered chondrocyte systems subject to direct perfusion. Biorheology 43: 215–222
97. Schmidt O, Mizrahi J, Elisseeff J, Seliktar D (2006) Immobilized fibrinogen in PEG hydrogels does not improve chondrocyte-mediated matrix deposition in response to mechanical stimulation. Biotechnol Bioeng 95: 1061–1069

98. Seidel JO, Pei M, Gray ML, Langer R, Freed LE, Vunjak-Novakovic G (2004) Long-term culture of tissue engineered cartilage in a perfused chamber with mechanical stimulation. Biorheology 41: 445–458
99. Stoddart MJ, Ettinger L, Häuselmann HJ (2006) Enhanced matrix synthesis in de novo, scaffold free cartilage-like tissue subjected to compression and shear. Biotechnol Bioeng 95: 1043–1051
100. Terraciano V, Hwang N, Moroni L, Park HB, Zhang Z, Mizrahi J, Seliktar D, Elisseeff J (2007) Differential response of adult and embryonic mesenchymal progenitor cells to mechanical compression in hydrogels. Stem Cells 25: 2730–2738
101. Waldman SD, Couto DC, Grynpas MD, Pilliar RM, Kandel RA (2007) Multi-axial mechanical stimulation of tissue engineered cartilage: review. Eur Cell Mater 13: 66–73 (discussion 73–74)
102. Waldman SD, Couto DC, Grynpas MD, Pilliar RM, Kandel RA (2006) A single application of cyclic loading can accelerate matrix deposition and enhance the properties of tissue-engineered cartilage. Osteoarthr Cartil 14: 323–330
103. Waldman SD, Spiteri CG, Grynpas MD, Pilliar RM, Kandel RA (2004) Long-term intermittent compressive stimulation improves the composition and mechanical properties of tissue-engineered cartilage. Tissue Eng 10: 1323–1331
104. Wenger R, Hans MG, Welter JF, Solchaga LA, Sheu YR, Malemud CJ (2006) Hydrostatic pressure increases apoptosis in cartilage-constructs produced from human osteoarthritic chondrocytes. Front Biosci 11: 1690–1695
105. Begley CM, Kleis SJ (2000) The fluid dynamic and shear environment in the NASA/JSC rotating-wall perfused-vessel bioreactor. Biotechnol Bioeng 70: 32–40
106. Freed LE, Langer R, Martin I, Pellis NR, Vunjak-Novakovic G (1997) Tissue engineering of cartilage in space. Proc Natl Acad Sci U S A 94: 13885–13890
107. Korhonen RK, Laasanen MS, Toyras J, Rieppo J, Hirvonen J, Helminen HJ, Jurvelin JS (2002) Comparison of the equilibrium response of articular cartilage in unconfined compression, confined compression and indentation. J Biomech 35: 903–909
108. Demarteau O, Wendt D, Graccini A, Jakob M, Schäfer D, Heber M, Martin I (2003) Dynamic compression of cartilage constructs engineered from expanded human articular chondrocytes. Biochem Biophys Res Commun 310: 580
109. Ilinich, E (2007) Kultivierung und Analyse von Knorpel-Träger-Konstrukten in einem neuartigen Bioreaktor. Technisch wissenschaftliche Schriftenreihe, Bd. 5, ISBN 978-3-930400-98-0
110. Sengers BG, Oomens CWJ, Baaijens FPT (2004) An integrated finite-element approach to mechanics, transport and biosynthesis in tissue engineering. J Biomech Eng 126: 83–91
111. Williams KA, Saini S, Wick TM (2002) Computational fluid dynamics modelling of steady-state momentum and mass transport in a bioreactor for cartilage tissue engineering. Biotechnol Prog 18: 951–963
112. Kallemeyn NA, Grosland NM, Pedersen DR, Martin JA, Brown TD (2006) Loading and boundary condition influences in a poroelastic finite element model of cartilage stresses in a triaxial compression bioreactor. Iowa Orthop J 26: 5–16
113. Hall AC, Urban JPG, Gehl KA (1991) The effects of hydrostatic pressure on matrix synthesis in articular cartilage. J Orthop Res 9: 1–10
114. Parkkinen JJ, Ikonen J, Lammi MJ, Laakkonen J, Tammi M, Helminen HJ (1993) Effects of hydrostatic pressure on protoglycan synthesis in cultured chondrocytes and articular cartilage explants. Arch Biochem Biophys 300: 458–465
115. Lammi MJ, Inkinen R, Parkkinen JJ, Häkkinen T, Jortikka M, Nelimarkka LO et al. (1994) Expression of reduced amounts of structurally altered aggrecan in articular cartilage chondrocytes exposed to high hydrostatic pressure. Biochem J 304: 723–730
116. Smith RL, Rusk SF, Ellison BE, Wessells P, Tsuchiya K, Carter DR et al. (1996) In vitro stimulation of articular chondrocyte mRNA and extracellular matrix synthesis by hydrostatic pressure. J Orthop Res 14: 53–60
117. Carver SE, Heath C (1999) Increasing extracellular matrix production in regenerative cartilage with intermittent physiological pressure. Biotechnol Bioeng 62: 166–174

118. Domm C, Fay J, Schünke M, Kurz B (2000) Die Redifferenzierung von dedifferenzierten Gelenkknorpelzellen in Alginatkultur. Orthopäde 29: 91–99
119. Smith RL, Lin J, Trinidade MCD, Shida J, Kajiyama BS, Vu T et al. (2000) Time-dependent effects of intermittent hydrostatic pressure on articular chondrocytes type II collagen and aggrecan mRNA expression. JRRD 37(2)
120. Toyoda T, Seedhom BB, Yoa JQ, Kirkham J, Brookes S, Bonass WA (2003) Hydrostatic pressure modulates proteoglycans metabolism in chondrocytes seeded in agarose. Arthritis Rheum 48: 2865–2872
121. Scherer K, Schünke M, Sellckau R, Hassenpflug J, Kurz B (2004) The influence of oxygen and hydrostatic pressure on articular chondrocytes and adherent bone marrow cells in vitro. Biorheology 41: 323–333
122. Smith RL, Carter DR, Schurman DJ (2004) Pressure and shear differentially alter human articular chondrocytes metabolism. Clin Orthop Rel Res 427: 89–95
123. Elder SH, Sanders SW, McCulley WR, Marr ML, Shim JW, Hasty KA (2006) Chondrocyte response to cyclic hydrostatic pressure in alginate versus pellet culture. J Orthop Res 24: 740–747
124. Mizuno S (2005) A novel method for assessing effects of hydrostatic fluid pressure on intracellular calcium: a study with bovine articular chondrocytes. Am J Physiol Cell Physiol 288: 329–337
125. Hu JC, Athanasiou KA (2006) The effects of intermittent hydrostatic pressure on self-assembled articular cartilage constructs. Tissue Eng 12: 1337–1344
126. Gavenis K, Kremer A, von Walter M, Hollander DA, Schneider U, Schmidt-Rohlfing B (2007) Effects of cyclic hydrostatic pressure on the metabolism of human osteoarthritic chondrocytes cultivated in a collagen gel. Artif Organs 31: 91–98
127. Kawanishi M, Oura A, Furukawa K, Fukubayashi T, Nakamura K, Tateishi T, Ushida T (2007) Redifferentiation of dedifferentiated bovine articular chondrocytes enhanced by cyclic hydrostatic pressure under a gas-controlled system. Tissue Eng 13: 957–964

# Application of Disposable Bag Bioreactors in Tissue Engineering and for the Production of Therapeutic Agents

**R. Eibl and D. Eibl**

**Abstract** In order to increase process efficiency, many pharmaceutical and biotechnology companies have introduced disposable bag technology over the last 10 years. Because this technology also greatly reduces the risk of cross-contamination, disposable bags are preferred in applications in which an absolute or improved process safety is a necessity, namely the production of functional tissue for implantation (tissue engineering), the production of human cells for the treatment of cancer and immune system diseases (cellular therapy), the production of viruses for gene therapies, the production of therapeutic proteins, and veterinary as well as human vaccines.

Bioreactors with a pre-sterile cultivation bag made of plastic material are currently used in both development and manufacturing processes primarily operating with animal and human cells at small- and middle-volume scale. Because of their scalability, hydrodynamic expertise and the convincing results of oxygen transport efficiency studies, wave-mixed bioreactors are the most used, together with stirred bag bioreactors and static bags, which have the longest tradition.

Starting with a general overview of disposable bag bioreactors and their main applications, this chapter summarizes the working principles and engineering aspects of bag bioreactors suitable for cell expansion, formation of functional tissue and production of therapeutic agents. Furthermore, results from selected cultivation studies are presented and discussed.

**Keywords** Cell expansion, Disposable bag bioreactor, Functional tissue, Stem cells, Therapeutic agents

---

R. Eibl (✉), D. Eibl
Zurich University of Applied Sciences, School of Life Sciences and Facility Management, Institute of Biotechnology, Campus Grüntal, P.O. Box 8820, Wädenswil, Switzerland

## Contents

| | |
|---|---|
| 1 Introduction | 186 |
| 2 Disposable Bag Bioreactors: General Overview and Categorization | 186 |
| 2.1 Static Bag Bioreactors | 189 |
| 2.2 Stirred Bag Bioreactors: Working Principle and Engineering Aspects | 189 |
| 2.3 Bag Bioreactor with Vibromixer | 192 |
| 2.4 Wave-Mixed Bioreactors | 192 |
| 2.5 Hybrid Bag Bioreactor | 195 |
| 3 Use of Bag Bioreactors in Expansion of Functional Cells, Tissue Formation and Biomanufacturing | 196 |
| 3.1 Bioengineered Functional Tissue | 197 |
| 3.2 Haematopoietic Cell Expansion | 198 |
| 3.3 Production of Therapeutic Proteins and Viruses | 199 |
| 4 Conclusion | 203 |
| References | 204 |

## Abbreviations

| | |
|---|---|
| AAV | adeno-associated virus |
| ACD | aseptic connection device |
| AD | Aujeszky's disease |
| ADV | Aujeszky's disease virus |
| B cells | lymphocytes which produce antibodies against soluble antigens |
| BEV | Baculovirus expression vector |
| BHK cells | Baby hamster kidney cells |
| BHV | bovine herpes virus |
| CHO cells | Chinese hamster ovary cells |
| dhfr⁻ | dihydrofolate reductase deficient |
| E-FL cells | embryogenic feline lung fibroblast cells |
| E. coli | *Escherichia coli* |
| FDA | Food and Drug Administration |
| G. max | *Glycine max* |
| GMP | Good Manufacturing Practice |
| GS-NS0 | glutamine synthethase deficient mouse cell line |
| HEK cells | human embryogenic kidney cells |
| HSC | haematopoietic stem cells |
| H. muticus | *Hyoscyamus muticus* |
| H. procumbens | *Harpagophytum procumbens* |
| IL-2 | recombinant interleukin-2 |
| ISO | International Organization for Standardization |
| IgG | immunoglobulin |
| $k_L \times a$ | gas–liquid mass transfer coefficient |

| | |
|---|---|
| M | motor |
| MDBK cells | Madin–Darby bovine kidney cells |
| MDCK cells | Madin–Darby canine kidney cells |
| MEV | mink enteritis virus |
| M. domesticus | *Malus domesticus* |
| mAb | monoclonal antibody |
| NK cells | natural killer cells |
| MOI | multiplicity of infection or optimal ratio of virus particles per cell |
| NS0 cells | mouse myeloma cells |
| N. tabacum | *Nicotiana tabacum* |
| PEI | polyethyleneimine |
| PER.C6™ cells | human embryogenic retinoblast cells |
| PGA | polyglycolic acid |
| PLA | polylactic acid |
| P/V | power input per volume |
| P. ginseng | *Panax ginseng* |
| P. pastoris | *Pichia pastoris* |
| pDNA | plasmid DNA |
| Re | Reynolds number |
| RV | rabies virus |
| r | recombinant |
| rpm | revolution per minute |
| SEAP | secreted alkaline phosphatase |
| SeMet | selenomethionine |
| S. cerevisiae | *Saccharomyces cerevisiae* |
| Sf | *Spodoptera frugiperda* |
| T cells | thymus cells, belonging to the group of lymphocytes |
| $TCID_{50}$ | tissue culture infectious dose |
| TIB | temporary immersion bioreactor |
| T. baccata | *Taxus baccata* |
| Tn5 cells a ni | cells from *Trichoplusi* (insect cells, also called High Five® cells) |
| TOI | optimal density of cells at infection |
| tu | transducing units |
| VLPs | virus-like particles |
| Vero cells | kidney epithelial cells from African green monkey |
| V. vinifera | *Vitis vinifera* |
| vvm | volume per volume per minute |
| WIM | wave-induced motion |
| USP | United States Pharmacopeial Convention |
| VM | Vibromixer |
| 3D | three-dimensional |

# 1 Introduction

Disposable bioreactors are intended for single use only. They consist of a cultivation container which is made of FDA-approved polymeric materials, predominantly polyethylene, polystyrene, polytetrafluorethylene and polypropylene, and meet USP Class VI as well as ISO 10993 specifications. The cultivation container is typically sterilized by gamma radiation, customized and validated. After filling with culture medium, subsequent inoculation with cells and finishing harvest, it is discarded. Directly after harvest, a new cultivation in a plastic container can be started.

Compared with their counterparts made of glass or stainless steel, disposable bioreactors have many advantages such as short set-up times, no sterilization and cleaning, reduced contamination levels, high simplicity and flexibility, and shorter production turnaround times [1, 2]. Therefore, their use in processes where it is necessary to work under GMP (Good Manufacturing Practice) compliant standards will result in minimized process costs, reduced development time, and reduced time-to-market for new products. As a direct consequence of the advantageous features listed as well as increasing cost pressure in pharmacy and medicine, the acceptance of disposable bioreactors has grown and culminated in numerous disposable bioreactor developments over the last 10 years [3–5].

Given that a bioreactor is a closed system in which production organisms are converted into the desired product, there are three main categories of disposable bioreactors: (1) traditional small-scale culture systems and their modifications, (2) hollow fibre bioreactors and (3) bag bioreactors, which will be focused on in the following.

# 2 Disposable Bag Bioreactors: General Overview and Categorization

In the case of bag bioreactors, the disposable cultivation container is designed as a flexible bag. As shown in Table 1 the currently available systems are preferred for the cultivation of various kinds of animal and human cells at laboratory and pilot scale. Cultivations performed in batch-, feeding- or perfusion mode are aimed at cell expansion, r-protein-, mAB- and virus productions in the first instance. Both suspension cells and adherent cells can be grown. For strictly adherent cells, the application of microcarriers is stringently required, with the exception of static bags characterized by unenforced power input.

In static bag bioreactors cell growth and/or product formation are exclusively caused by conduction and reaction processes within the culture bag and its interaction with its environment. For energy and mass transfer, and thereby cell growth and/or product expression in dynamic bag bioreactors, the power input generated is responsible. Taking mass and energy transfer as well as power input into account, the most widely used bag bioreactors for animal and human cell cultivations can be categorized into static bag bioreactors, mechanically driven bag bioreactors with stirrer, Vibromixer or wave-induced motion (WIM), and hybrid bag bioreactors where mechanical and pneumatic power input are combined (Fig. 1).

**Table 1** Current bag bioreactors and their applications

| Reactor | Culture volume | Cells | Product | References |
|---|---|---|---|---|
| Static cell bags from various manufacturers (e.g. CulturSil bag, LAMPIRE™ Cell Culture Bag, LIFECELL Tissue Culture Bag, VectraCell™ Single-Use Bioreactor, VueLife® culture bag) | 30 mL–2 L | Animal cell culture lines (e.g. CHO, hybridomas) | Seed inoculum, r-proteins, mAbs | [6, 7, 81] |
|  |  | Haematopoietic cells (stem cells, progenitor cells, T cells) | Expanded and differentiated cells for transplantation | [60, 64–69] |
| Multiple bag bioreactor | c | Fibroblasts | Temporary skin substitute | [8, 9, 51] |
| Static bag bioreactor with temporary immersion ("Box-in Bags" disposable TIB) | 10 L[a] | Embryonic plant cells (Arabusta and Robusta clones) | Mass propagated microplants | [82, 83] |
| Stirred bag bioreactor (Artelis-ATMI LifeSciences'Pad-Drive™ disposable bioreactor, S.U.B. Single-Use Bioreactor, XDR™-Disposable Stirred Tank Bioreactor) | 10–1,000 L | Animal cell culture lines (e.g. CHO, hybridomas, MDBK[b], PER.C6™, Vero[b]) | r-proteins[c], mAbs (e.g. IgG), viruses for veterinary vaccines (e.g. BHV-1, influenza virus) | [10–14, 16–19, 84] |
| Bag bioreactor with Vibromixer (bio-t® bag) | 2–10 L | Animal cell culture lines (e.g. CHO, Sf9) | Seed inoculum, r-proteins | [20, 21] |
| Wave-mixed bag bioreactors (AppliFlex, BioWave®, BIOSTAT®CultiBagRM, CELL-tainer®, Optima-mini™, Tsunami®Bioreactor, Wave Bioreactor, WUB-Wave and Undertow Bioreactor) | 1–500 L | Microbial cells (e.g. E. coli, S. cerevisiae) | Seed inoculum, r-proteins, pDNA for gene therapies | [4, 85–87] |
|  |  | Plant cell and tissue cultures (e.g. suspension cells from G. max, M. domesticus, N. tabacum, T. baccata, V. vinifera) and hairy roots from H. muticus, H. procumbens, P. ginseng) | Biomass, secondary metabolites (e.g. baccatin III, ginsenosides, hyoscyamine, paclitaxel, scopolamine), mAbs (e.g. anti-RV mAbs) | [42, 88–96] |
|  |  | Animal cell cultures (e.g. BHK, CHO, Drosophila S2, E-FL[b], HEK[e], Tn5, MDCK[e],NS0, Sf9, Sf21) | Seed inoculum[d], r-proteins (e.g. resistin, SEAP), mAbs (e.g. IgG), viruses and VLPs for vaccines (e.g. MEV, rAAV, influenza) | [22–38, 45–48, 79, 80, 97–99] |
|  |  | T cells[f] | Expanded cells for cancer- and immunotherapy | [39–41] |

(continued)

Table 1 (continued)

| Reactor | Culture volume | Cells | Product | References |
|---|---|---|---|---|
| Bubble column bag bioreactor (SBB-Slug Bubble Bioreactor, Plastic-lined Bioreactor) | 10–150 L | Plant cell and tissue cultures (e.g. suspension cells from *G. max*, *H. muticus*, *N. tabacum*) and hairy roots from *H. muticus* | Biomass, isoflavones | [93, 100] |
| Airlift bag bioreactor (LifeReactor™) | 0.8–7 L | Organ cultures of plant origin (e.g. meristematic clusters and somatic embryos of banana, fern, gladiolus, orchid, pineapple, potato) | Micropropagated biomass | [101–104] |
| (CellMaker Lite2™) |  | Microbial cells (*E. coli*, *P. pastoris*) | Seed inoculum | [49] |
| Hybrid bag bioreactor (CellMaker Plus, combined stirred and airlift bioreactor) | 1–7 L | Animal cell cultures (e.g. CHO, insect) | Seed inoculum | [49, 50] |

[a] Total volume
[b] Cultivated on microcarriers
[c] No detailed specification
[d] Cryopreservation possible
[e] With and without microcarriers
[f] Cultivated on microbeads

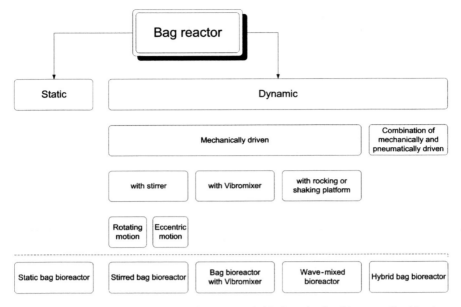

**Fig. 1** Categorization of disposable bag bioreactors suitable for animal and human cell cultivations

## 2.1 Static Bag Bioreactors

Among disposable bag bioreactors, static gas-permeable culture bags represent the oldest and most simple closed cultivation devices [6–8]. The existing systems differ in size and polymer type, and are ideally suited for the expansion, differentiation and partial freezing of cells.

Despite their widespread application at laboratory and clinical scale, static bags are limited to relatively low cell densities with low total cell output. Likewise in t-flasks or well-plates, limitations arise from the restricted surface for cell expansion, the inability of the user to readily monitor and control the culture microenvironment ($CO_2$ incubator is essential), and the necessary manual feeding steps with increased risk of contamination [9].

## 2.2 Stirred Bag Bioreactors: Working Principle and Engineering Aspects

Stirred bag bioreactors with their own measurement and control unit overcome these limitations by providing better control of culture parameters, automated feeding strategies, and the ability to support higher cell densities. They are basically equipped with an aeration device (microsparger or sparger ring) and a

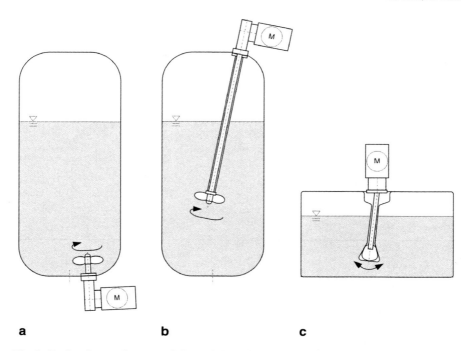

**Fig. 2** Basic scheme of commercially available stirred bag bioreactors: (**a**) XDR™-Disposable Stirred Tank Bioreactor, (**b**) S.U.B. Single-Use Bioreactor, (**c**) Artelis-ATMI Life-Sciences'Pad-Drive™ disposable bioreactor

rotating axial flow impeller (Fig. 2a, b) or tumbling impeller (Fig. 2c), ensuring mass and temperature homogeneity as well as gas dispersion inside the bag [10–14].

Beside sparger and disposable impeller assembly, the cylindrical or cubical bag includes gas filters and ports for integration of sensor probes and line sets. As exemplified for the S.U.B. (single-use bioreactor) in Fig. 3, the bag is generally shaped and fixed in a customized steel support container with heater jacket. Using sterile couplers [e.g. ACDs (aseptic connection devices) from Pall], all three stirred bag bioreactors allow sterile insertion of standard sensors for pH and dissolved oxygen.

Whereas the bottom-driven XDR™-Disposable Stirred Tank Bioreactor operates with a magnetically coupled impeller, the S.U.B. Single-Use Bioreactor is top-driven and has a mechanical seal. Therefore, it is necessary to penetrate the S.U.B.'s driveshaft through the mixing drive und to lock it into the disposable impeller assembly during installation. Similar to S.U.B.'s driveshaft, the mixing paddle of the Artelis-ATMI Life-Sciences'Pad-Drive™ disposable bioreactor is protected by a film (identical to bag material) to avoid product contact.

In order to assess cultivation results and to compare stirred bag bioreactors to traditional stirred steel bioreactors, which represent the golden standard in animal

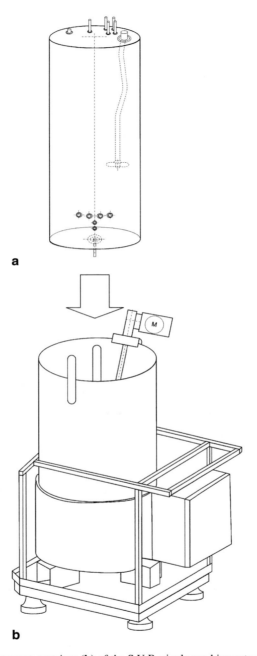

**Fig. 3** Bag (**a**) and support container (**b**) of the S.U.B. single-use bioreactor

**Table 2** Hydrodynamic and oxygen transfer efficiency data of a 250-L S.U.B. Single-Use Bioreactor [15, modified]

| Parameter | Result | | |
|---|---|---|---|
| Stirring speed [rpm] | 50 | 100 | 200 |
| Tip speed [m s$^{-1}$] | 0.53 | 1.06 | 2.13 |
| P/V [W m$^{-3}$] | 1.6 | 13.4 | 106.6 |
| Re | 34,000 | 69,000 | 137,000 |
| Mixing time [s] | 90 | 60 | 45 |
| $k_L \times a$ [h$^{-1}$] in the range between 0.5 and 2 L min$^{-1}$ | 7–11 | 7–15 | n.d. |

*n.d.* not determined

cell culture-based bioprocessing, the hydrodynamic characteristics (fluid flow, fluid mixing, shear stress pattern, temperature profile etc.) and oxygen transfer efficiency have to be determined. To date, only mixing characterization results and scale-up parameters of the 250-L S.U.B. Single-Use Bioreactor have been published. The values summarized in Table 2 as well as achieved cell densities and product titres from batch and perfusion experiments with mABs secreting CHO and PER.C6™ cells support the conclusion that the S.U.B. Single-Use Bioreactor is a good alternative to stirred steel bioreactors [10, 11, 15–19].

## 2.3 Bag Bioreactor with Vibromixer

In contrast to stirred bag bioreactors, where the power input is significantly influenced by the tip speed, the power input of the bag bioreactor with Vibromixer (bio-t bag) can be regulated by the amplitude and frequency. Here, the key element is the vertically oscillating hollow shaft with a perforated disk, which is heat-sealed to the flexible plastic bag and contains disposable sensors for the measurement of pH and/or dissolved oxygen content. The movement of the conically perforated disk induces an axial flow in the bag, which mixes and aerates the culture medium and the cells (Fig. 4). Depending on the position of the conical drill holes on the disk, an upward flow ("riser" flow) or downward flow ("downcomer" flow) results. This may eliminate vortex formation.

It was recently demonstrated that high gas–liquid mass transfer coefficients ranging between 26 and 82 h$^{-1}$ can be achieved in a 2-L bag bioreactor with Vibromixer. The investigations were performed at maximum power input and aeration rates between 0.05 and 1 vvm. Running and subsequent investigations focus on fluid flow modelling [20, 21].

## 2.4 Wave-Mixed Bioreactors

It is an indisputable fact that the early version of a wave-mixed bag bioreactor, which was introduced in 1998, and its success story has promoted the development of disposable bioreactors. Various types of wave-mixed bag bioreactors with rock-

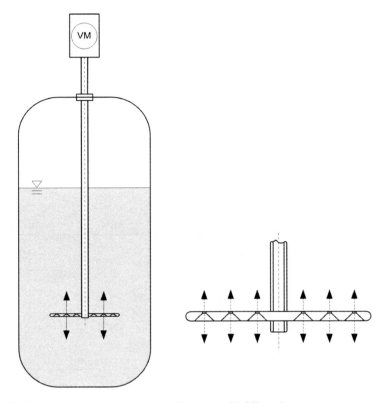

**Fig. 4** Basic scheme of the single-use bag bioreactor with Vibromixer

ing or shaking platforms have been tested for growing animal and human cells, but BioWave® and Wave Reactor (Fig. 5a) are the most widely used [22–41]. Differing in bag shape and sensor probes from equally well-suited AppliFlex (Fig. 5b), both systems are based on the first prototype of a wave-mixed bioreactor. Rocking the platform induces a wave in the culture bag, which contains culture medium and cells. In this way, mixing occurs while the surface of the medium is continuously renewed, and bubble-free surface aeration takes place. Mechanical power input produced by the rocking platform facilitates mixing and improves mass and energy transfer within BioWave®.

Hydrodynamic and oxygen transfer efficiency studies on the BioWave® with 1 L, 10 L and 100-L culture volumes were carried out for comparison with traditional stirred cell culture bioreactors with surface and membrane aeration. It was found that fluid flow, mixing time, residence time distribution, specific power input and oxygen transfer efficiency were dependent on rocking angle, rocking rate, bag type and its geometry, and culture volume. In this context, a modified Reynolds number was established to describe the fluid flow in the culture bag of BioWave® [42].

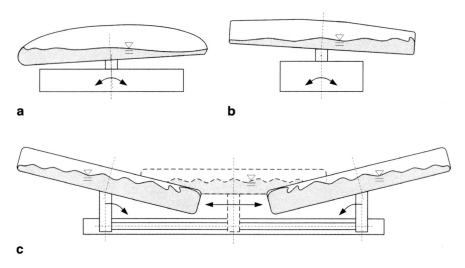

**Fig. 5** Basic scheme of wave-mixed bag bioreactors: (**a**) BioWave® and Wave Bioreactor, (**b**) AppliFlex, (**c**) CELL-tainer®

Mixing times based on 40 and 50% filling level lie between 10 and 1,400 s and can be considered as satisfactory values for cell culture bioreactors. As expected, the most ineffective mixing was observed at the lowest possible rocking rate, rocking angle and maximum filling level of 50%. A reduction of the mixing time is possible by increasing the rocking rate and/or the rocking angle, which results in a more intensive wave movement. Whereas the most effective mixing was obtained in the 2 L bag (9–264 s), the most ineffective mixing was found in the 20 L bag (40–1,402 s) [43].

By considering filling level, rocking rate and rocking angle, power input analysis for a 2-L culture bag was conducted [44]. A minimum filling level, maximum rocking rate and rocking angle result in maximum specific power input, which is one decimal power higher than operation with maximum filling level (50% culture volume) (Fig. 6).

Assuming a constant rocking angle and culture volume with rocking rates of up to 20 rpm, the specific power input is directly proportional to the rocking rate. In other words, a stepwise increase in the rocking rate up to 20 rpm raises the power input. If the rocking rate is increased further, the power input levels out and may even be followed by a slight decrease. The last observation can be explained by the occurring phase shift of the wave towards rocking movement. Experiments confirmed our hypothesis that, in this case, the subsequent increase in the rocking rate causes lower hydrodynamic cell stress but improves nutrient and oxygen transfer efficiency, which in turn promote cell growth. In addition, residence time distribution experiments have evidenced that a continuously operating BioWave® in perfusion mode can be described by the ideally mixed stirred tank model [42].

Oxygen transfer coefficients provided by the BioWave® reach comparable or higher values than those which have been reported for stirred cell bioreactors

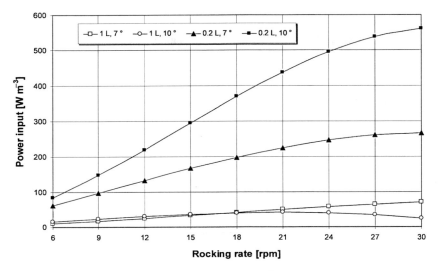

**Fig. 6** Specific power input course in 2 L bag of BioWave® [42, modified]

with membrane aeration or surface aeration [42]. For instance, in a 2 L bag operating with 1-L culture volume at rocking angles between 6 and 10°, rocking rate of 30 rpm and aeration rate of 0.25 vvm, $k_L \times a$ values between 4 and 10 $h^{-1}$ were obtained. High oxygen transfer efficiency can be ensured by increased rocking rate, rocking angle and aeration rate. Because of increased surface area, a decreased filling level in the bag increases $k_L \times a$ at constant parameters. However, even small changes in the rocking rate and rocking angle increase the $k_L \times a$ significantly. From our experience, oxygen transfer coefficients in ranges above 11 $h^{-1}$ can only be achieved by aeration rates over 0.5 vvm or aeration with pure oxygen in BioWave® [44].

More than 70-fold higher gas–liquid oxygen transfer coefficients than in BioWave® and Wave Reactor are provided by the CELL-tainer, which is a new wave-mixed bag bioreactor system (Fig. 5c) [45–47]. From its two-dimensional movement (a combination of rocking and horizontal displacement), higher cell densities can be achieved [48]. Contrariwise, it is supposed that specific power input and therefore hydrodynamic stress for cells is higher in CELL-tainer than in BioWave® and Wave Reactor.

## 2.5 Hybrid Bag Bioreactor

In the hybrid bag bioreactor mixing and aeration is achieved by using the air flow from a sparger tube and two magnetically driven propellers, all of which are integral to the bag (Fig. 7). The cell broth is aerated from the bag base, where the low

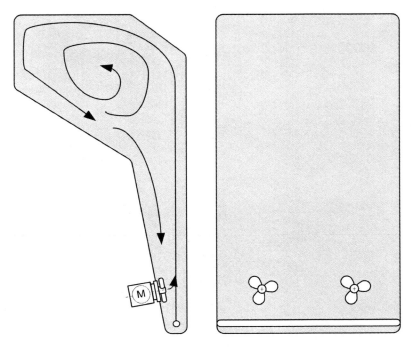

**Fig. 7** Working principle of the hybrid CellMaker Plus

density mixture of aqueous solution and air generates the "riser" flow. Because of the unique asymmetric shape of the bioreactor, the upper surface pushes this flow to the front of the top part of the bag. This transverse movement guarantees gas exchange with the bag headspace, which can even be independently aerated with gas mixtures. Finally, the "riser" flow is displaced by the "downcomer" flow, which has a higher density than the "riser" flow. By applying pressure to the bag headspace, excessive foam formation (typically representing a serious problem in pneumatically driven bioreactors) can be reduced while mass transfer, in particular oxygen mass transfer rates, are increased. Again, oxygen and pH are non-invasively measured [49, 50].

## 3 Use of Bag Bioreactors in Expansion of Functional Cells, Tissue Formation and Biomanufacturing

As already mentioned, the bag bioreactors described in the preceding sections are generally suitable for in vitro cultivation of animal and human cells from mL up to 1,000 L scale. The cultivated cell material includes primary cells derived from a patient's own healthy tissue or from healthy donors with the same and various cell lines. Quantity and quality of the products aimed for (which include expanded functional cells and 3D tissue for patient-specific therapies, proteins for biotherapeutics, and viruses for vaccines and gene therapies) are closely linked to bioreactor design and especially to the resulting hydrodynamic cell stress [51–53].

A further factor, which mainly influences product formation in disposable bag bioreactors, is the possibility of undesired interactions between the bag material, culture medium, cultivated cells and secreted product. For example, chemical exposures (so-called leachables and extractables) arising from product-contact with the side of the bag can change cell metabolism, inhibit cell growth and/or contaminate the product [54–57]. This may be a problem, especially for patient-specific indications, where functional cells and tissues are directly transplanted, as in the case of tissue regeneration, cancer- and immunotherapies.

## 3.1 Bioengineered Functional Tissue

In vitro generation of tissue implies the consecutive realization of a well-known three-step procedure involving cell proliferation, cell adhesion on a porous scaffold, and finally tissue formation. A bioreactor which is able to grow functional tissue constructs has to provide an in vivo-like biochemical and biomechanical environment supporting cell and tissue density over weeks [51, 58, 59]. Normally, a bioreactor at the mL-scale supplies enough tissue for the implantation.

With the exception of cell proliferation, for which static bags are commonly used, bag bioreactors are relatively unimportant for the production of functional tissue. Referring to the literature, there is only one report on the generation of 3D tissue in a disposable bag bioreactor. Halberstadt et al. [9] describe the superior manufacturing of a human dermal replacement in a bag bioreactor system operating in perfusion mode. This multiple bioreactor system contains 16 Teflon™ bags, a 16-channel peristaltic pump ensuring medium supply, two medium reservoir bags refrigerated at 4°C and one 100-L waste bag. All 120 mL bags are attached in parallel by connecting them to silicone tubing via polypropylene connectors. In order to provide the necessary template for cell growth and dermal tissue development, each Teflon™ bag houses a free-floating 3D mesh scaffold of biodegradable Vicryl™ (PGA/PLA). The Teflon™ bags were inoculated with human neonatal dermal fibroblasts ($6 \times 10^5$ cells mL$^{-1}$) which had previously been harvested from roller flasks. After realization of an optimized seeding and attachment procedure (24 h slow rotation by using a roller flask apparatus, 24 h laid flat on incubator shelf) in serum-supported growth medium, the bioreactor system was assembled, placed into a standard cell incubator, and the medium pump was turned on. During cultivations of 16 days, the perfusion flow rate was adjusted by controlling glucose concentrations and lactate concentrations. This method resulted in a tissue comparable in histology with that obtained after 22 days of growth in a single static Teflon™ bag with a Vicryl™ mesh scaffold. It was shown that deposition of collagen type I, decorin and fibronectin were virtually identical [9] in both continuously perfused and fed static culture bags. As these results were encouraging, this perfusion bioreactor was modified to manufacture the commercially available tissue-engineered skin replacement Dermagraft® in disposable bags [51]. Each bag (composed of fibroblasts, extracellular matrix and the mesh scaffold) is sealed and stored frozen at −70°C before it is taken to the patient, where it is opened for direct application.

## 3.2 Haematopoietic Cell Expansion

Haematopoietic stem cells (HSC) derived for clinical use from bone marrow, peripheral blood and umbilical cord blood are the most extensively studied adult stem cells. Typically, they generate an intermediate cell type (called progenitor cells) before they give rise to all the blood cell types, including myeloid (monocytes, macrophages, neutrophils, basophils, eosinophils, erythrocytes, megakaryocytes/platelets, dendritic cells) and lymphoid lineages (T cells, B cells, NK cells) [60, 61]. Compared to ex vivo manufacturing of functional tissue (discussed in Sect. 3.1), haematopoietic cell expansion normally requires larger cell amounts. According to Tzanakakis and Verfaillie [62] and Carswell et al. [63], bioreactors with a capacity between 1 and 1,000 L are necessary in order to produce clinically relevant cell amounts for haematopoietic cells destined for immuno- or cancer therapies. Indeed, the majority of clinical trial protocols make use of static gas-permeable culture bags for haematopoietic cell expansions [60, 64–69].

Consequently, the development and implementation of the first wave-mixed bag bioreactor-based process for GMP manufacture of biologically active T cells (Xcellerated T Cells™) for the treatment of chronic lymphatic leukaemia can be regarded as a great achievement. The starting material for this process (known as the Wave Bioreactor-based Xcellerate III process) is an autologous peripheral blood mononuclear cell leukapheresis product from healthy donors [39]. It is thawed and $CD3^+$ T cells are magnetically concentrated using custom Xcyte™Dynabeads® M-450 CD3/CD28 T. This occurs in the presence of anti($\alpha$)-human CD3 and $\alpha$–human CD28 mAbs that are co-immobilized on the beads. By binding the immobilized mAbs to the CD3 and CD28 receptors on T cells, the necessary signals are delivered to induce T cell activation [70]. Subsequently, the mixture of T cells and beads is transferred to a 20-L culture bag on the Wave Bioreactor platform in serum-free culture medium containing recombinant IL-2. While undergoing perfusion, the cells are activated and expanded at 37°C in 10-L culture volume. After 13 days of cultivation, the beads are removed with a magnet and activated T cells are harvested, formulated and cryopreserved. $137 \pm 34.3 \times 10^9$ Xcellerated T Cells™ at $98.5 \pm 1\%$ $CD3^+$ T cell purity were manufactured. These cells were characterized by high biological activity and the restoring of T cell receptor repertoire to a normal diversity. In this way, the treatment of 17 patients was feasible [40, 41].

## 3.3 Production of Therapeutic Proteins and Viruses

Whereas the use of animal and human cells for patient-specific cellular therapies or even tissue replacement therapies is a rather new field, the production of proteins and viruses by classical animal cell technology has a long tradition. The number of applications referred to in the literature in both fields underscores this statement.

Over the past 50 years, bioreactors up to 25 m$^3$ have been developed for growing various animal cell lines (e.g. CHO cells, BHK cells, MDBK cells, MDCK cells, NS0 cells, HEK cells, PER.C6™ cells, Vero cells, hybridomas and insect cells). Conventional stirred bioreactors made of glass or stainless steel have become the system of choice in commercial biomanufacturing and R&D processes [71, 72]. As a result of increasing demands for shorter processing times and reduced production costs along with high process safety and facility flexibility in pharmaceutical production processes, various scalable disposable bag bioreactors have been designed. The studies currently available confirm the capability and superiority of wave-mixed and stirred bag bioreactors for animal cell-based therapeutic agents at small- and middle-volume scale [10, 11, 16–19, 22–41, 73, 74]. Although newer systems such as CELL-tainer®, bag bioreactor with eccentric stirrer, bag bioreactor with Vibromixer and the hybrid CellMaker have also showed promising cultivation results, there is a lack of important data to confirm these findings, unlike for BioWave® and Wave Bioreactor (and in parts S.U.B. Single-Use Bioreactor), for which data have already been collected.

### 3.3.1 Therapeutic Proteins

Animal cell-based therapeutic r-proteins and mAbs are biotechnological products which are characterized by a tremendous growth. So it comes as no surprise that a good deal of research focuses on their efficient production using novel technology, which explains the numerous papers and application reports concerned with the cultivation of industry-relevant animal cell lines in disposable bag bioreactors. When cultivation studies with animal model cell lines are not taken into account [4, 43], three main working fields for BioWave®, Wave Bioreactor and S.U.B. Single-Use Bioreactor in therapeutic protein production become apparent.

Expression of r-proteins in BEV/Insect Cell-Culture Systems

In 1999 Singh [33] published the first short manual on scale up of *Sf*9 suspension cells and baculovirus production in the Wave Bioreactor and 10-L culture volume. In order to start a 20 L bag containing 1-L serum-free culture medium with a cell density of $5 \times 10^5$ cells mL$^{-1}$, the previous harvest of a 2 L bag was used. After 5 days of cultivation at a rocking rate of 20 rpm and 27°C, $4.75 \times 10^6$ cells mL$^{-1}$ were determined and 4 L fresh medium fed. Three days ($3 \times 10^6$ cells mL$^{-1}$) later, a further 4 L fresh medium along with the virus (MOI 0.5) were added. Harvest was performed 2-days post infection. This procedure resulted in virus titres which were comparable to 100-mL shake flask experiments.

Since baculovirus-based production processes usually take 4–7 days, Weber et al. [36] developed an improved cultivation protocol for a more rapid *Sf*9 cell-based protein generation in a 20 L bag. In order to avoid mass transfer limitations in the growth phase, they recommended an adjustment of the agitation intensity (rising rocking rate and rocking angle) to increase cell density and particular attention being paid to an adequate oxygen supply in the production phase (averaging 2–3 days). With regard to efficient r-protein production, an increase in specific oxygen uptake by 50% has to be guaranteed after virus infection [75]. Weber and Fussenegger [76] therefore suggested an appropriate increase in the oxygen transfer rate by aeration with pure oxygen.

Oxygen enrichment up to 40% at aeration flow rates between 0.033 and 0.066 vvm and a stepwise rocking rate rising (15–20 rpm) at a constant rocking angle of 7° were applied by Schlaeppi et al. [32] in parallel BioWave® fermentations. They developed a semi-automated large-scale process for the production of recombinant tagged proteins in a BEV/*Sf*21 cell culture. Besides cell harvesting, lysis by tangential flow filtration and automated purification by chromatography, the instrumental process platform included insect cell fermentation in up to eight BioWave® systems with 10-L culture volume. It was possible to process four proteins in less than 24 h with final yields between 1 and 100 mg, and purities between 50 and 95%.

Finally, in baculovirus-infected *Sf*9 and High Five® cells which have been grown in Wave Bioreactors (10-L culture volume), selenomethionyl-derivatized proteins with an SeMet occupance of ~75% were successfully produced by Cronin et al. [23]. The data presented by the authors support implementation of BioWave® and Wave Bioreactor for the rapid production of milligram quantities of r-proteins with BEV/insect cell-culture systems.

Transient Transfection in Mammalian Cells

The majority of r-proteins produced for preclinical and clinical trials are normally expressed in stably transfected CHO cell lines. However, stable expression associated with generating a highly productive cell line requires a good deal of time and resources, and incurs high costs. Transient transfection in mammalian cells represents an innovative approach to producing a large number of functional human proteins within a short time and thereby streamlining the process from research to development [77, 78]. The use of disposable bioreactors in combination with this method can lead to additional savings in time and costs.

Haldankar et al. developed a protocol for large-scale transient fed batch transfections in a Wave Bioreactor with 10-L working volume. The transient transfections based on CHO suspension cells and polyethyleneimine (PEI) were performed with the chimeric human/mouse IgG1 isotype antibody molecule. The seed train was produced in shake flasks and inoculation was carried out at a cell density of 1

× $10^6$ cells mL$^{-1}$. For the Wave Bioreactor parameters, the reader is referred to [79]. After inoculation, the transfection complex was introduced into the culture (DNA/PEI ratio 1:3). Production runs were carried out in fed batch mode (transfected culture fed on days 2, 4 and 6 post transfection) and harvesting was performed 6 days or 8 days post transfection. The harvested cultures were characterized by cell densities between 5.1 × $10^6$ cells mL$^{-1}$ and 1.22 × $10^7$ cells mL$^{-1}$ and viabilities between 95.6 and 97%. Within 10 days, 100 mg of r-protein for further study was produced.

mAb Production with Stable Hybridomas and Established Mammalian Cells

Acceptable, reproducible and comparable (to traditional stirred cell culture bioreactors) mAb production was achieved in studies with stable hybridomas, CHO-, GS-NS0 and PER.C6™ cells in a Wave Bioreactor and S.U.B. Single-Use Bioreactor [15–17, 19, 24, 31, 33, 35].

Oashi et al. [31] and Tang et al. [35] considered mAb (IgG) production of a hybridoma cell line, which had been grown in a Wave Bioreactor (1-L culture volume) operating in fed batch (feeding) and perfusion mode. The perfusion mode was ensured by a specially designed culture bag with a floating membrane cell-retention filter (7 μm pore size) inside. A weight-based perfusion controller balanced the medium renewal rate and the harvest rate during perfusion. Operating parameters of the Wave Bioreactor running with inoculum from pooled t-flasks are shown in [31] and [35]. The perfusion was initiated at cell density >1 × $10^6$ cells mL$^{-1}$ with an initial dilution rate of 0.1 day$^{-1}$ and 0.2 day$^{-1}$. To keep glucose/lactate concentrations constant, the dilution rate was adjusted during cultivation. The maximum living cell density was between 7- and 10-times higher and the total mAb between 8- and 13-times higher in perfusion mode than in batch operation.

Fries et al. studied growth, metabolism and mAb production of GS-NS0 cells and dhfr⁻ CHO cells in a traditional stirred cell culture bioreactor (2-L culture volume) and the Wave Bioreactor (10-L culture volume). For both bioreactor systems, which differed in pH control and agitation, the seed train was generated in shake flasks. Parameters of realized batch cultivations are shown in [24]. In GS-NS0 cell-derived experiments, average maximum living cell densities of 1 × $10^6$ cells mL$^{-1}$ in the Wave Bioreactor and 0.9 × $10^6$ cells mL$^{-1}$ in the stirred bioreactor were reached. The average maximum mAb concentration was 83 mg L$^{-1}$ in the Wave Bioreactor and 64 mg L$^{-1}$ in the stirred bioreactor. In the case of dhfr⁻ CHO cell cultivation in the Wave Bioreactor and stirred bioreactor, larger differences in the extent of biomass production, metabolic patterns and antibody formation were observed. The stirred bioreactor guaranteed approximately two-times higher cell densities and mAb concentrations than the Wave bioreactor. However, an excellent reproducibility of the Wave Bioreactor's data was shown. Moreover, volumetric product formation and mAb quality for both cell lines were found to be quite comparable in both bioreactor types.

In addition, the first application notes and scientific publications have reported the indisputable suitability of the S.U.B. Single-Use Bioreactor for hybridoma-, PER.C6™- and CHO cell-based mAb productions up to 250-L culture volume [15–17, 19].

### 3.3.2 Viruses for Vaccines and Gene Therapies

The potential of BioWave® for vaccine virus production was systematically investigated using BHK 21, E-FL and MDCK cells. Its comparability with and advantages over traditional vaccine production bioreactors (mainly represented by roller flasks, Cell Factories and stirred microcarrier bioreactors) [73, 74] were evident.

In BioWave® with 450-mL culture volume, Slivac et al. [34] demonstrated its suitability for the production of Aujeszky's disease virus (ADV), which is important for the vaccination of pigs. BHK 21 C13 suspension cells were cultivated during a 3-day growth period at a rocking angle of 6°, a rocking rate of 10 rpm, an air flow rate of 0.44 vvm (0.2 L min$^{-1}$) and 5% $CO_2$. The inoculum production had previously been executed in spinner flasks. Starting with $5.5 \times 10^5$ cells mL$^{-1}$, a maximum cell density of $1.82 \times 10^6$ cells mL$^{-1}$ and viability of 99% were achieved during cell growth. In order to prevent nutrient limitations, a partial medium exchange of 65% was realized on the second day of the growth phase. The production phase was introduced by infecting the cells with gE$^-$ Bartha K-61 strain virus suspension ($10^{5.9}$ TCID$_{50}$) with MOI of 0.01. After 144 h incubation in the BioWave®, 400 mL of ADV harvest was obtained with a titre of $10^{7.0}$ TCID$_{50}$ mL$^{-1}$. This corresponds to 40,000 doses of vaccine against Aujeszky's disease (AD).

Genzel et al. [26] showed that an identical wave-mixed bioreactor with 1-L culture volume was superior to a stirred bioreactor for equine influenza virus production. MDCK cells were grown in serum-supported culture medium at Cytodex™ 1 microcarrier concentrations of 2 and 4 g L$^{-1}$. Subsequent to cell and microcarrier transfer, and cell attachment, the bioreactor's parameters were set up at 37°C, a rocking angle of 7°, a rocking rate of 15 rpm, an aeration rate of 0.1 vvm and $CO_2$ amounts between 2 and 5%. Washing, medium exchange and virus addition (MOI 0.04–0.05) were performed as described in [26] after 4 days of cultivation. The final cell density on the microcarriers (observed after 99 h) was about double that of the stirred bioreactor. Peak titres of $10^{7.7}$ TCID$_{50}$ mL$^{-1}$ were reached 20-h post infection.

Hundt et al. [27] studied the possibility of transferring a roller bottle-based manufacturing process for a mink enteritis virus (MEV) vaccine ("FEBRIVAC 3-Plus") to a BioWave®-derived Cytodex™ 1 microcarrier process (1 L and 10-L culture volume). After 5 days preculture of feline lung fibroblasts (E-FL) in roller bottles, $2 \times 10^5$ cells mL$^{-1}$ and 2 g L$^{-1}$ prepared microcarriers were fed to 1 L and 10-L serum-supported culture medium in the bag. Depending on the experimental setup, the MEV was added at 0 h to the medium (MOI 0.01–0.1). In addition, three to four medium changes and virus harvests were performed. Compared to cultivations in roller bottles, an increase in virus titres by a factor of approximately 10 was found in the BioWave®. The virus titres ranged between $10^{6.6}$ and $10^{6.8}$ TCID$_{50}$ mL$^{-1}$.

Further known developments deal with VLP influenza vaccine production using the BEV system in *Sf9* cells, which are grown in a Wave Bioreactor up to 500-L culture volume [80]. Moreover, it was reported that the S.U.B Single-Use Bioreactor and the ATMI Life-Sciences'Pad-Drive™ disposable bioreactor were successfully tested for the production of veterinary vaccines with MDBK- as well as Vero cells and microcarriers (HyQ®Sphere™, Cytodex™ 1) [10, 13, 14]. To date, more detailed information about these processes has not been published.

In addition to using viruses for veterinary or human vaccine production, their use in therapeutic gene transfer is another promising field in which disposable bioreactors are attractive candidates for manufacturing. Recently, a method for producing recombinant adeno-associated virus vectors (rAAV) exploiting the BEV system in insect cells was presented [30]. A Wave Bioreactor with 5 L or 25-L culture volume was operated at 27°C, a rocking angle of 9°, a rocking rate of 25 rpm and an air flow rate of 0.012–0.06 vvm (0.3 L min$^{-1}$). Oxygen enrichment in the air supply was 30%. The cells were infected with three different BEVs (MOI 3) at TOI of $2 \times 10^6$ cells mL$^{-1}$. The culture was harvested 72-h post infection. The yield of transducing units (tu), which describe the biological activity (crucial for gene therapies) of the virus, reached a value of ~2E + 13 particles L$^{-1}$ of cell culture. This yield was comparable to those obtained in 200-mL shake flasks, and in 10 L and 40-L stirred bioreactors. Therefore, it is expected that Wave Bioreactor technology will promote the production of rAAV and further virus vectors, which are important for gene therapy treatments.

# 4 Conclusion

In summary, this review indicates that bag bioreactors from mL to 1,000 L scale are well-established for animal and human cell-based processes wherever simple, safe, fast and flexible production receives priority. While static bag bioreactors are common devices at mL-scale, wave-mixed and stirred bag bioreactors are preferred when higher cell amounts are desired. The increasing tendency to produce therapeutic r-proteins, mAbs and viruses is obvious in the fully automated BioWave®, Wave Bioreactor and S.U.B. Single-Use Bioreactor. This increase applies mainly to R&D and GMP manufacturing in classical cell culture processes.

In personalized medicine, where ex vivo generated biologically active cells in clinically relevant numbers are required, the single-use of the bioreactor is not merely advantageous. Where the approval of functional cells (including stem cells, their intermediate cell types and lineages) for cancer-, immuno- or tissue replacement therapies is concerned, it becomes a necessity. Thus, the successful implementation of the Wave Bioreactor-based GMP manufacturing of active T cells for the treatment of chronic lymphatic leukaemia represented a breakthrough in haematopoietic cell expansion.

With the exception of gas-permeable blood bags for the cell proliferation stage, bag bioreactors are of less importance for ex vivo formation of functional tissue (such as cartilage, bones, liver tissue and cardiac tissue). There is only one approach to manufacturing human dermal replacement, this being in a multiple bag bioreactor system containing a mesh scaffold and operating in perfusion mode.

In view of current industrial activities (focusing on new disposable bioreactor concepts and including intensive efforts to overcome the limitations of existing disposable bag bioreactors) as well as the rapid knowledge increase in cell culture technology and modern medicine, it is assumed that disposable bag bioreactors will become increasingly influential.

## References

1. DePalma A (2006) GEN 26:60
2. Morrow KJ (2007) GEN 28:37
3. Flanagan N (2007) GEN 27:38
4. Eibl R, Eibl D (2006) Disposable bioreactors for pharmaceutical research and manufacturing. Proceedings 2nd international conference on bioreactor technology in cell, tissue culture and biomedical applications. Saariselkä, Finland
5. Eibl R, Eibl D (2007) PROCESS PharmaTEC 4:14
6. Stang BV, Wood PA, Reddington JJ, Reddington GM, Heidel JR (1997) Monoclonal antibody production in gas-permeable bags using serum-free media. Monoclonal antibody workshop. Baltimore, California
7. Daley LP, Gagliardo LF, Duffy MS, Smith MC, Appleton JA (2005) Clin Diagn Lab Immunol 12:380
8. Purdue, GF, Hunt JL, Still JM, Law EJ, Herndon DN, Goldfarb IW, Schiller WR, Hansbrough JF, Hickerson WL, Himel HN, Kealey GP, Twomey J, Missavage AE, Solem LD, Davis M, Totoritis M, Gentzkow GD (1997) J Burn Care Rehab 18:52
9. Halberstadt, CR, Hardin R, Bezverkov K, Snyder D, Allen L, Landeen L (1994) Biotechnol Bioeng 43:740
10. Card, C, Smith T (2006) Application report draft – SUB050601
11. Thermo Fisher Scientific (2007) Application note: AN003 Rev 1
12. Galliher P (2007) Case study: scale up to a 1,000 L perfusion in a disposable stirred tank bioreactor. BioProduction. Berlin, Germany
13. Castillo J, Vanhamel S (2007) GEN 27:40
14. Zambaux JP (2007) How synergy answers the biotech industry needs. BioProduction 2007. Berlin, Germany
15. Kunas KT, Keating J (2005) Stirred tank-single-use bioreactor: comparison to traditional stirred tank bioreactor. bioLOGIC Europe, Geneva, Switzerland
16. Card C (2007) Large scale, animal free production of human monoclonal antibody using PERC.6™ cells in a disposable, stirred-tank bioreactor. 20th ESACT meeting 2007, Dresden, Germany (poster)
17. Zijlstra, G (2007) Scale-up of a PER.C6® fed-batch process in 50 and 250 L Hyclone single use bioreactors compared to 50 and 250 L stainless steel bioreactors. 20th ESACT meeting 2007, Dresden, Germany (poster)
18. Brecht R (2007) Disposable bioreactor technologies: challenges and trends in cGMP manufacturing. BioProduction 2007. Berlin, Germany

19. Ozturk SS (2007) Comparison of product quality: disposable and stainless steel bioreactor. BioProduction 2007. Berlin, Germany
20. Zurich University of Applied Sciences, Department for Life Sciences and Facility Management, IBT, Cell cultivation techniques and biochemical engineering (2006–2007) Protocols of experiments, unpublished
21. Werner S, Nägeli M (2007) BioTechnology 3:22
22. Amanullah A, Burden E, Jug-Dujakovic M, Mikola M, Pearre C, Herber W (2004) Development of a large-scale cell bank in cryobags for the production of biologics. http://www.wavebiotech.com/pdfs/literature/Merck_Cancun-2004.pfd. Cited November 4, 2007
23. Cronin CN, Lim KB, Rogers J (2007) Protein Sci 16:2023
24. Fries, S, Glazomitsky K, Woods A, Forrest G, Hsu A, Olewinski R, Robinson D, Chartrain M (2005) BioProcess Int 3:36
25. Genzel Y, Behrendt I, Koenig S, Sann H, Reichl U (2004) Vaccine 22:2202
26. Genzel Y, Olmer RM, Schaefer B, Reichl U (2006) Vaccine 24:6074
27. Hundt B, Best C, Schlawin N, Kassner H, Genzel Y, Reichl U (2007) Vaccine 25:3987
28. Knevelman C, Hearle DC, Osman JJ, Khan M, Dean M, Smith M, Aiyedebinu Cheung K (2002) Characterization and operation of a disposable bioreactor as a replacement for conventional steam-in-place inoculum bioreactors for mammalian cell culture processes. 224th National Meeting of the American Chemical Society, Boston, MA; American Chemical Society, Washington DC; BIOT 210 (poster)
29. Matthews T, Wolk B (2005) The use of disposable technologies in antibody manufacturing processes. http://www.wavebiotech.com/pdfs/literature/IBCDisposables_2005.pdf. Cited November 4, 2007
30. Negrete A, Kotin RM (2007) J Virol 145:155
31. Ohashi R, Singh V, Hammel JF (2001) Perfusion cell culture in disposable bioreactors. 17th ESACT meeting 2001, Tylösand, Sweden
32. Schlaeppi JM, Henke M, Mahnke M, Hartmann S, Schmitz R, Pouliquen Y, Kerins B, Weber E, Kolbinger F, Kocher HP (2006) Protein Expr Purif 50:185
33. Singh V (1999) Cytotechnology 30:149
34. Slivac I, Srček VG, Radoševic K, Kmetič I, Kniewald Z (2006) J Biosci 3:363
35. Tang YJ, Ohashi R, Hamel JP (2007) Biotechnol Prog 23:255
36. Weber W, Weber E, Geisse S, Memmert K (2002) Cytotechnology 38:77
37. Weber W, Bacchus W, Daoud-El Baba M, Fussenegger M (2007) Nucleic Acids Res 35:e116
38. Weber W, Stelling J, Rimann M, Keller B, Daoud-El Baba M, Weber CC, Aubel D, Fussenegger M (2007) PNAS 104:2643
39. Hami LS, Chana H, Yuan V, Craig S (2003) BioProc J 2:23
40. Hamis LS, Green C, Leshinsky N, Markham E, Miller K, Craig S (2004) Cytotherapy 6:554
41. Levine B (2007) Making waves in cell therapy: the Wave bioreactor for the generation of adherent and non-adherent cells for clinical use. http://www.wavebiotech.com/pdf/literature/ISCT_2007_Levine_Final.pdf. Cited November 4, 2007
42. Eibl R, Eibl D (2006) Design and use of the Wave Bioreactor for plant cell culture. In: Dutta Gupta S, Ibaraki Y (eds.) Plant tissue culture engineering, series: focus on biotechnology, vol 6. Springer, Dordrecht, p. 203
43. Eibl R, Eibl D, Pechmann G, Ducommun C, Lisica L, Lisica S, Blum P, Schär M, Wolfram L, Rhiel M, Emmerling M, Röll M, Lettenbauer C, Rothmaier M, Flükiger M (2003) Produktion pharmazeutischer Wirkstoffe in disposable Systemen bis zum 100 L Massstab, Teil 1. KTI-Projekt 5844.2 FHS, Final Report, University of Applied Sciences Wädenswil, Switzerland, unpublished
44. Lisica S (2004) Energieeintrag in Wave-Bioreaktoren. Modelling approaches, University of Applied Sciences Wädenswil, Switzerland, unpublished
45. CeLLution Biotech BV (2007) Mass transfer in the CELL-tainer®disposable bioreactor. http://www.cellutionbiotech.com. Cited October 20, 2007
46. CeLLution Biotech BV (2007) Cultivation of PER.C6®-cells in the CELL-tainer® disposable bioreactor. www.cellutionbiotech.com. Cited October 20, 2007

47. CeLLution Biotech BV (2007) Cultivation of CHO-cells in the CELL-tainer® disposable bioreactor. www.cellutionbiotech.com. Cited October 20, 2007
48. Zijlstra G, Oosterhuis N (2007) Cultivation of PERC.$^{6®cells}$ in the novel CELL-tainer™ high-performance disposable bioreactor. 20th ESACT meeting 2007, Dresden, Germany (poster)
49. Taylor, I (2007) The CellMaker Plus™ single-use bioreactor: a new bioreactor capable of culturing bacteria, yeast, insect and mammalian cells. Biotechnica, Hannover, Germany
50. Auton KA, Bick JA, Taylor IM (2007) GEN 27:42
51. Ratcliffe A, Niklason L (2002) Bioreactors and bioprocessing for tissue engineering. Ann N Y Acad Sci 961:210
52. Nienow AW (2006) Cytotechnology 50:9
53. Martin I, Wendt D, Heberer M (2004) Trends Biotechnol 22:80
54. Altaras GM, Eklund C, Ranucci C, Maheswari G (2007) Biotechnol Bioeng 96:999
55. Jenke D (2007) J Pharm Sci 96:2566
56. Okonkowski J, Balasubramanian U, Seamans C, Fischrogen S, Zhang J, Lachs P, Robinson D, Chartrain M (2007) J Biosci Bioeng 103:50
57. van Tienhoven EAE, Korbee D, Schipper L, Verharen HW, Jong De WH (2006) J Biomed Mater Res A78:175
58. Pörtner R, Nagel-Heyer St, Goepfert C, Adamietz P, Meenen NM (2005) J Bioeng Biosci 100:235
59. Ye H, Xia Z, Ferguson DJP, Triffitt JT, Cui Z (2007) J Mater Sci: Mater Med 18:641
60. Safinia L, Panoskaltsis N, Mantalaris A (2005) Haematopoietic culture systems. In: Chaudhuri JB, Al-Rubeai M (eds.) Bioreactors for tissue engineering. Springer, Dordrecht, p. 309
61. Cabrita GJM, Ferreira BS, da Silva CL, Goncales R, Almeida-Porada G, Cabral JMS (2003) TIBTECH 21:233
62. Tzanakakis ES, Verfaillie CM (2006) Advances in adult stem cell culture. In: Ozturk SS, Hu WS (eds.) Cell culture technology for pharmaceutical and cell-based therapies. CRC Press, New York, p. 693
63. Carswell KS, Papoutsakis ET (2000) Biotechnol Bioeng 68:328
64. Purdy MH, Hogan CJ, Hami L, McNiece I, Franklin W, Jones RB, Bearman SI, Berenson RI, Cagnoni BI, Heimfeld S, Shpall EJ (1995) J Hematother 4:515
65. Robinet E, Certoux JM, Ferrand C, Maples P, Hardwick A, Cahn JY, Reynolds CW, Jacob W, Hervé, Tiberghien P (1998) J Hematother 7:205
66. Andrews RG, Briddell RA, Hill R, Gough M, McNiece IK (1999) Stem Cells 17:210
67. CorCell Inc. (2003) Expansion of umbilical cord blood stem cells. http://www.corcell.com/healthcare/expansion.html. Cited August 30, 2008
68. Mu LJ, Gaudernack G, Saeboe-Larssen S, Hammerstad H, Tierens A, Kvalheim G (2003) Scand J Immunol 58:578
69. Mu LJ, Lazarova P, Gaudernack G, Saeboe-Larssen S, Kvalheim G (2004) Int J Immunopathol Pharmacol 17:255
70. June CH, Ledbetter JA, Linsley PS, Thompson CB (1990) Immunol Today 11:211
71. Wurm FM (2004) Nat Biotechnol 22:1393
72. Wurm FM (2007) Novel technologies for rapid and low cost provisioning of antibodies and process details in mammalian cell culture based biomanufacturing. BioProduction, Berlin, Germany
73. Schwander E, Rasmusen H (2005) GEN 25:29
74. DePalma A (2002) GEN 22:58
75. Vaughn J (1999) Insect cell culture, protein expression. In: Flickinger MC, Drew SW (eds.) Encyclopedia of bioprocess technology, vol 3. Wiley & Sons, New York, p. 1444
76. Weber W, Fussenegger M (2005) Baculovirus-based production of biopharmaceuticals using insect cell culture processes. In: Knäblein J (ed.) Modern biopharmaceuticals. Wiley VCH, Weinheim, Germany, p. 1045
77. Durocher Y, Perret S, Kamen A (2002) Nucleic Acids Res 30:e9

78. Wurm FM, Bernard A (1999) Curr Opin Biotechnol 313:156
79. Haldankar R, Li D, Saremi Z, Baikalov C, Deshpande R (2006) Bioreactors Mol Biotechnol 34:191
80. Rios M (2006) PharmaTech 4:1
81. Sambrook J, Fritsch EF, Maniatis T (1989) Molecular cloning. A laboratory manual, 2nd edn. Cold Spring Harbor Laboratory Press, New York
82. Ducos JP, Lambot C, Pétiard V (2007) Int J Dev Biol 1:1
83. Ducos JP, Chantanumat P, Vuong P, Lambot C, Pètiard V (2007) Acta Horticulturae 764:33
84. Collignon F, Gelbras V, Havelange N, Drugmand JC, Debras F, Mathieu E, Halloin V, Castillo J (2007) CHO cell cultivation and antibody production in a new disposable bioreactor based on magnetic driven centrifugal pump. http://www.artelis.be. Cited October 20, 2007
85. Hallmann S, Bertelsen HP, Scheffler U, Luttmann R (2007) Einsatz von Massflow-Controllern zur Steuerung von Bioreaktionsprozessen. Biotechnica, Hannover, Germany (poster)
86. Mikola M, Seto J, Amanullah, A (2007) Bioprocess Biosyst Eng 30:231
87. Laderman K, Quezada V, Dunphy N, Anderson J, Derecho J, McMahom R, Hsu D, Couture L (2007) DNA production in the Wave Bioreactor under cGMP conditions. http://www.wavebiotech.com/pdfs/press/pDNA_Poster_COH2007.pfd. Cited November 6, 2007
88. Eibl R, Eibl D (2002) Bioreactors for plant cell and tissue cultures. In: Oksman-Caldentey KM, Barz WH (eds.) Plant biotechnology and transgenic plants. Marcel Dekker, New York, p. 163
89. Palazón J, Mallol A, Eibl R, Lettenbauer C, Cusidó RM Piñol MT (2003) Planta Med 69: 344
90. Bentebibel S, Moyano E, Palazón J, Cusidó RM, Bonfill M, Eibl R, Piñol MT (2005) Biotechnol Bioeng 89:647
91. Girard LS, Fabis MJ, Bastin M, Courtois D, Pétiard V, Koprowski H (2006) Biochem Biophys Res Commun 345:602
92. Kilani J, Lebeaut JM (2006) Appl Microbiol Biotechnol 74:324
93. Terrier B, Courtois D, Hénault N, Cuvier A, Bastin M, Aknin A, Dubreuil J, Pétiard V (2006) Biotechnol Bioeng 96:914
94. Eibl R, Eibl D (2007) Phytochem Rev DOI: 10.1007/s 11101-007-9083-z
95. Cuperus S, Eibl R, Hühn T, Amado R (2007) BioForum Europe 6:2
96. Bonfill M, Bentebibel S, Moyano E, Palazón J, Cusidó RM, Eibl R, Piñol MT (2007) BIOL PLANT 51:647
97. DÀquino R (2006) Chem Eng Prog 102:8
98. Houtzager E, van der Linden R, de Roo G, Huurman S, Priem P, Sijmons C (2005) BioProcess Int 3:60
99. Pierce LN, Sabraham PW (2004) Bioprocessing J 3:51
100. Curtis WR (2004) United States Patent, 6,709,862 B2
101. Ziv M, Ronen G, Raviv M (1998) Dev Biol-Plant 34:152
102. Harrell RC, Bienek M, Hood CF, Munilla R, Cantliffe DY (1994) Plant Cell Tissue Org Cult 39:171
103. Fukui H, Tanaka M (1995) Plant Cell Tissue Org Cult 41:17
104. Escalona M, Lorenzo JC, Gonzalez BL, Daquinta M, Gonzalez JL, Desjardine Y, Borroto CG (1999) Plant Cell Rep 18:743

# Methodology for Optimal In Vitro Cell Expansion in Tissue Engineering

**J.M. Melero-Martin, S. Santhalingam, and M. Al-Rubeai**

**Abstract** Expansion of the cell population in vitro has become an essential step in the process of tissue engineering and also the systematic optimization of culture conditions is now a fundamental problem that needs to be addressed. Herein, we provide a rational methodology for searching culture conditions that optimize the acquisition of large quantities of cells following a sequential expansion process. In particular, the analysis of both seeding density and passage length was considered crucial, and their correct selection should be taken as a requisite to establish culture conditions for monolayer systems. This methodology also introduces additional considerations concerning the running cost of the expansion process. The selection of culture conditions will be a compromise between optimal cell expansion and acceptable running cost. This compromise will normally translate into an increase of passage length further away from the optimal value dictated by the growth kinetic of the cells. Finally, the importance of incorporating functional assays to validate the phenotypical and functional characteristics of the expanded cells has been highlighted. The optimization approach presented will contribute to the development of feasible large scale expansion of cells required by the tissue engineering industry.

**Keywords** Cell expansion, Progenitor cells, Regenerative medicine, Stem cells, Tissue engineering.

---

J.M. Melero-Martin
Vascular Biology Program and Department of Surgery, Children's Hospital Boston, Harvard Medical School, Boston, MA, USA

S. Santhalingam
Department of Chemical Engineering, School of Engineering, University of Birmingham, Birmingham, B15 2TT, UK

M. Al-Rubeai (✉)
School of Chemical and Bioprocess Engineering and Centre for Synthesis and Chemical Biology, University College Dublin, Belfield, Dublin 4, Ireland

**Contents**

| | | |
|---|---|---|
| 1 | Introduction | 210 |
| | 1.1 Tissue Engineering in Regenerative Medicine | 210 |
| | 1.2 Cell Sources for Tissue Engineering | 211 |
| | 1.3 In Vitro Expansion of Autologous Cells | 211 |
| 2 | Definitions | 213 |
| | 2.1 Parameters Related to Cell Growth in a Single Passage | 213 |
| | 2.2 Parameters Related to a Sequential Monolayer Expansion | 214 |
| 3 | Determination of Growth Curves | 215 |
| 4 | Optimal Cell Expansion | 218 |
| | 4.1 Unaltered Growth Kinetics | 219 |
| | 4.2 Altered Growth Kinetics | 220 |
| 5 | Exponential Growth Kinetics | 222 |
| 6 | Running Cost of the Expansion Process | 224 |
| 7 | Preservation of Cell Phenotype | 226 |
| 8 | Conclusions | 227 |
| References | | 228 |

# 1 Introduction

## 1.1 Tissue Engineering in Regenerative Medicine

Organs and tissues often necessitate reconstruction or replacement due to damage produced from congenital disorders, cancer, and trauma, among other conditions [1]. These defects are normally treated by either replacement with autologous tissue or by allogeneic organ transplantation. However, both approaches present important constrains: (1) autologous treatment imposes serious problems of morbidity for most tissues, and (2) there is a severe shortage of donor organs, which is worsening with aging of the world population. In addition, any of the mentioned approaches rarely replace the entire function of the original organ, and tissues used for reconstruction often lead to complications due to their inherent different functional parameters.

An alternative therapy for the repair of damaged tissue resides in the tissue engineering approach. Tissue engineering is an interdisciplinary field that applies principles and methods of engineering and the life sciences toward the development of biological substitutes that restore, maintain, and improve the function of damaged tissues and organs [2]. Such a tissue reconstitution process can be conducted either entirely in vitro or partially in vitro and then completed in vivo. Tissue engineering strategies based on autologous cells are normally initiated from a small piece of donor tissue (biopsy), from which individual cells are isolated. These cells are expanded in culture, attached to a support matrix (scaffold), and re-implanted into the host. Major advances have been achieved in this field within the past decade, resulting in the creation of functional tissues such as small diameter vascular grafts [3], heart valves [4, 5], and urinary bladder [6] among others.

## 1.2 Cell Sources for Tissue Engineering

Establishing a reliable source of cells is a principal priority for tissue engineers [7]. Cells used in tissue engineering may be drawn from a variety of sources, including primary tissues and cell lines. Primary tissues may be xenogeneic (from different species), allogeneic (from different members of the same species), syngeneic (from a genetically identical individual), or autologous (from the same individual). Ideally, both structural and functional tissue replacement will occur with minimal complications. Although animal cells are a possibility, ensuring that they are safe remains a concern, as does the high likelihood of their rejection by the immune system [1, 8]. Currently, the clinical use of allogeneic cells is still limited by the need for host immunosuppression. However, with the advent of techniques to render cells immunologically "transparent," the use of banked allogeneic cells may become a clinical reality [7]. An alternative cell source for bioengineering of tissues and organs is therapeutic cloning, wherein patient-specific embryonic stem cells (ESCs) can be derived from pre-implantation stage embryos produced by somatic cell nuclear transfer, therefore obtaining histocompatible cells for engraftment [9]. Stem cells derived from this source might have the potential to replace and regenerate damaged tissues; however, the mechanisms controlling their differentiation must be fully understood, and ethical issues surrounding their use must be resolved prior to their implementation in therapeutic strategies.

Until further advances allow other cell sources to become a clinical reality, autologous cells are the preferred cells to use in regenerative medicine. To acquire autologous cells, a biopsy of tissue is obtained from the host, the cells are dissociated and expanded in culture, and the expanded cells are implanted into the same host. The use of autologous cells, although it may cause an inflammatory response, avoids rejection, and thus the side effects of immunosuppressive medications can be avoided. In addition, the use of autologous adult stem cells is ethically sound and accepted worldwide [1].

## 1.3 In Vitro Expansion of Autologous Cells

Most adult human cells have a limited lifespan, and after repeated divisions, they eventually enter replicative senescence, a state in which they are still viable, yet no longer divide and display reduced functionality. This presents a challenge to using differentiated autologous cells as a cell source for tissue engineering. Moreover, engineered tissues must contain a sufficient amount of cells to remain functional over clinically relevant time periods [10]. Therefore, one of the major limitations of applying cell-based regenerative medicine techniques to organ replacement is the inherent difficulty of growing specific cell types in large quantities [1]. This problem affects the majority of differentiated human cells. For example, articular chondrocytes in culture rapidly undergo dedifferentiation [11]. As a result, chondrocytes isolated from their tissue-specific extracellular matrix fail to produce cartilage matrix after extensive expansion in monolayer culture, a phenomenon that limits their availability in sufficient quantities for tissue engineering applications [12]. Similar constrains are found in the field of vascular and urinary bladder tissue engineering, where the acquisition of large quantities of endothelial [13] and urothelial [1] cells from differentiated functional tissues is enormously challenging.

To overcome this expansion limitation, researchers have been exploring alternatives to obtain sufficiently large, functional autologous cell populations for tissue engineering and regenerative medicine applications. One area of interest is the use of autologous adult stem or progenitor cells. By studying the sites for stem or progenitor cells in specific organs, as well as exploring the conditions that promote their differentiation, it may be possible to overcome the obstacles that limit cell expansion in vitro. For example, the identification of chondroprogenitor cells in the superficial zones of articular cartilages [14], has provided an alternative avenue to obtain chondrocytes that retain their ability to form cartilage after extensive expansion in culture [15, 12]. For vascular tissue engineering, the identification of endothelial progenitor cells (EPCs) in blood has offered an opportunity to noninvasively obtain large quantities of functional endothelial cells [16, 17, 18, 13]. Similar advantages are found with the identification of urothelial progenitor cells for urinary bladder tissue engineering [19]. All these studies indicated that it is possible to collect autologous cells from human patients, expand them in culture, and return them to the donor in sufficient quantities for reconstructive purposes. Major advances have been achieved within the past decade in the possible expansion of a variety of progenitor cells and adult stem cells, with specific techniques that make the use of autologous cells possible for clinical application.

One of the challenges that tissue engineers will have to address in the near future is the development of feasible large-scale cell-expansion processes. Routine tissue culturing methodologies can hardly cope with the scale of cell production required for the clinical generation of tissue-engineered products. Expansion of the cell population in vitro has become an essential step in the process of tissue engineering, and optimization of the culture conditions and expansion protocols are fundamental issues that need to be addressed. In fact, the enhanced expansion potential of stem and progenitor cells in culture opens up the possibility for more intense expansion processes that may enable the generation of large cell banks for use in regenerative medicine. The aim of this article is to provide a rational methodology for searching culture conditions that optimize the acquisition of large quantities of cells following a sequential expansion process. The proposed methodology uses mathematical expressions that relate the growth curve of the cells with expansion process parameters, and it facilitates the optimal selection of routine culture conditions irrespective of the source of autologous cells under investigation.

## 2 Definitions

### 2.1 Parameters Related to Cell Growth in a Single Passage

Several parameters can be defined concerning cell expansion in a single monolayer passage (see Fig. 1):

$I_p$ (cells): initial cell number available at the beginning of the passage.
$X0_p$ (cell cm$^{-2}$): initial cell density used at inoculation.

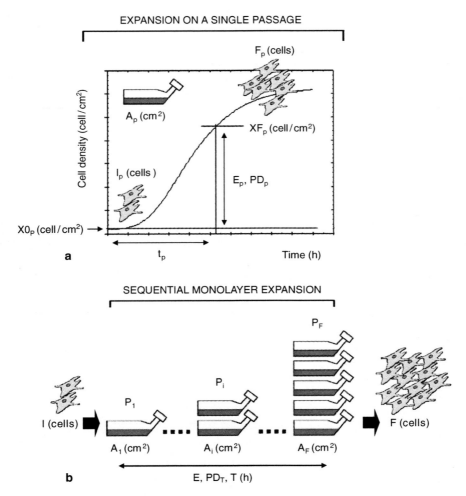

**Fig. 1** Schematic diagrams of cell expansion in a monolayer. (**a**) Growth of the cell population during a single monolayer passage. (**b**) The overall expansion process is constituted by sequential monolayer expansion steps

$XF_p$ (cell cm$^{-2}$): final cell density observed at the time of harvesting.

$PD_p$ (–): passage population doubling corresponds to the number of doublings that cells undergo from inoculation to harvesting.

$t_p$ (h): passage length corresponds to the duration of the passage, from inoculation to harvesting.

$E_p$ (–): passage expansion factor corresponds to the factor by which the viable cell density is multiplied from inoculation to harvesting.

Given these definitions of single passage parameters, the following equations apply:

$$E_p = \frac{XF_p}{X0_p} \tag{1}$$

$$2^{PD_p} = \frac{XF_p}{X0_p} \tag{2}$$

$$E_p = 2^{PD_p} \tag{3}$$

## 2.2 Parameters Related to a Sequential Monolayer Expansion

Additionally, the following parameters can be defined concerning a sequential monolayer expansion process (see Fig. 1):

$I$ (cells): initial cell number available at the beginning of the expansion process.

$F$ (cells): final cell number available at the end of the expansion process.

$PD_T$ (–): total population doubling corresponds to the final number of doublings that cells undergo during the expansion process.

$T$ (h): total expansion time corresponds to the duration of the expansion process.

$N$ (–): total number of passages that the cells undergo during the expansion process.

$E$ (–): total expansion factor corresponds to the factor by which the viable cell density is multiplied during the expansion process.

Given these definitions of sequential expansion parameters, the following equations apply:

$$PD_T = \sum_{p=1}^{N} PD_p \tag{4}$$

$$T = \sum_{p=1}^{N} t_p \tag{5}$$

$$E = \prod_{p=1}^{N} E_p \tag{6}$$

$$2^{PD_T} = \frac{F}{I} \tag{7}$$

$$E = 2^{PD_T} \tag{8}$$

$$E = \frac{F}{I} \tag{9}$$

## 3  Determination of Growth Curves

Optimization of cell expansion relies on the analysis of cell growth curves at different culture conditions. Elaboration of detailed cell growth curves (evolution of cell density over time) is normally done in the laboratory by using small-scale experiments in which cells are plated in sufficient replicated tissue culture wells, and manually counted at intervals using a hemocytometer. This microscopy technique is limited by inter user variability, laborious and often time-consuming data collection, user fatigue, and small sampling size (i.e., a few hundred cells) [20]. There is interest in the potential benefits of replacing standard cell counting methods with an automated systems. In this regard, several direct methods to monitor cell concentration have been proposed. Such methods include optical techniques based on light absorbance and or scattering, real-time imaging, particle size analysis, and techniques to measure culture fluid density [21, 22, 23, 24]. However, most of these techniques are not applicable to monitor anchorage-dependent cell concentration and viability accurately at all the stages of the culture [25]. Alternative automated systems such as Microcyte flow cytometer [26], NucleoCounter [27], and Guava PCA are being introduced into laboratories for rapid and accurate determination of viable cell numbers. However, these automated systems are mainly designed for suspended cell cultures, and not for routine evaluations of anchorage-dependent primary cells. For these reasons, trypan blue dye exclusion assay, DNA content, MTT assay, and nuclei counting are still the most commonly utilized methods for the determination of cell density of primary cells.

Each type of primary cell presents a specific growth curve that needs to be determined using the culture conditions under evaluation. Although some authors still use the exponential growth approach to characterize the behavior of proliferative cells in culture, this assumption is not accurate for most primary cells. The concept of exponential growth by mammalian cells in culture is based upon the apparent linearity of semi logarithmic data plots. Nevertheless, this method of graphical analysis is known to be an unreliable test of the exponential hypothesis. The question of exponential growth was re-examined more than two decades ago by using the more sensitive method of Smith plots, in which specific growth rate is plotted against time. With exponential growth, data points should fall on a horizontal straight line when specific growth rate is plotted against time, but after analyzing the growth of 125 different mammalian and avian cell lines, it was found that only 11 exhibited an exponential phase while the remaining cell lines all had nonexponential growth patterns [28]. The most common of these consisted of an initial period of growth acceleration followed by a later phase of deceleratory growth. Examples of decelerator growth kinetics can be found from very diverse cell sources, including chondroprogenitor cells isolated from the superficial zone of articular cartilages [29], and umbilical cord blood-derived endothelial progenitor cells [13]. As long as essential nutrients and cytokines are provided in sufficient amounts (achieved by adequate feeding), anchorage-dependent primary cell proliferation in culture is controlled by cell–cell contact inhibition. Since this mechanism of control is essentially density-dependent [30], cultures are characterized by a progressive generation of inhibitory contacts that ultimately lead to the appearance of the stationary phase or plateau. The kinetics of decelerating growth can be described with high accuracy by diverse equations such as Gompertz, logistic, inverse cube root and power functions [31, 32, 33]. In particular, Gompertz functions have been proposed to model the growth of a diversity of biological systems, including tumors [34, 35, 36], and normal primary cells [29]. The use of mathematical expressions can be instrumental for modeling the proliferative behavior of the cells in culture, but is not mandatory for the optimization analysis, which can be developed from appropriate experimental data.

The analysis of growth curves may seem straight forward at first examination, however, the election of an optimal growth curve requires careful analysis of the data. To illustrate this, we analyzed the growth curves of chondroprogenitor cells obtained at five different seeding densities (Fig. 2a) [29]. In this example, all culture parameters, with the exception of the seeding density, were kept constant. From Fig. 2a it can be observed how the initial cell density played a central role in the proliferation performance of the cultures, affecting both the kinetics of cell growth and the final cell density achieved. These results were expected since the proliferation of anchorage-dependent primary cells is controlled by cell–cell contact inhibition [15], a mechanism that is essentially density-dependent [30]. Therefore, it could be expected that higher seeding densities will necessarily lead to a more rapid generation of inhibitory contacts and consequently to a more rapid appearance of the stationary phase. While cultures with seeding densities of $10^5$ cell $cm^{-2}$ reached the stationary phase at day 5, the cultures with seeding density of $10^4$ cell $cm^{-2}$ required 8 days. Although higher initial cell densities had faster growth

# Methodology for Optimal In Vitro Cell Expansion in Tissue Engineering

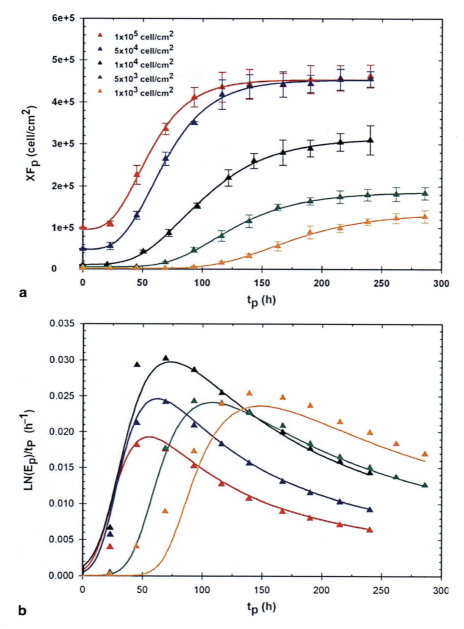

**Fig. 2** Expansion of chondroprogenitor cells in a monolayer. (**a**) The effect of initial cell density on cell proliferation was evaluated at five different seeding densities corresponding to $10^5$ cells cm$^{-2}$, $5 \times 10^4$ cells cm$^{-2}$, $10^4$ cells cm$^{-2}$, $5 \times 10^3$ cells cm$^{-2}$, and $10^3$ cells cm$^{-2}$. All the cultures were supplied with 0.3 ml cm$^{-2}$ of DMEM supplemented with 40% FCS and 1 ng ml$^{-1}$ of TGF-β1. Viable cell densities were evaluated at intervals for each of the different seeding densities tested. Each data point represents the mean of three separate cultures ±SD. (**b**) Optimization of passage length at different initial cell densities was determined by the maximum of (ln $(E_p/E_p)$). This figure was adapted from [29] with permission from the editor

kinetics (shorter lag phases and sharper growth curves) attention should also be paid to the expansion factor ($E_p$) achieved for each seeding density. The expansion factors achieved for each of the cultures examined corresponded to 5, 9, 30, 36, and 103 as the initial cell density was increased ($10^5$, $5 \times 10^4$, $10^4$, $5 \times 10^3$, and $10^3$ cell cm$^{-2}$, respectively). Consequently, while higher initial cell densities produced faster growth kinetics, it also imposed a lower value of the expansion factor achievable in a single passage. Moreover, since chondroprogenitor cells (and many progenitor cells) have the ability to form colonies even at very low seeding densities [14] the value of the expansion factor could be potentially improved by further reduction in the initial cell density of the cultures. The problem resides in the detrimental effect that a very low seeding density will have on the growth kinetics (and consequently on the time necessary to achieve the desired expansion factor). From the above example, it is clear that the determination of an optimal seeding density necessarily involves a compromise between the growth kinetic (i.e., time) and the expansion factor achievable, and that it cannot be easily done by simple direct observation of experimental growth curves.

## 4  Optimal Cell Expansion

The optimal culture conditions for cell expansion correspond to those culture conditions that would allow us to achieve the desired final cell number in the minimal time and with the minimal cost. Additionally, these optimally expanded cells need to maintain the required phenotype for their final therapeutic use. Despite the fact that the definition of optimal culture conditions is straight forward, the number of parameters affecting the growth of cells in culture is enormous and therefore their simultaneous optimization may seem like an unbearable task for most cell-culture investigators. However, from a practical stand point, the decisions that investigators need to make when culturing primary cells are often restricted to four categories: (1) media formulation and feeding strategy, (2) coating of culture plate, (3) seeding density, and (4) length of each passage (e.g., time of harvest and subculture). In fact, out of these four categories, two questions are considered essential for optimal expansion: firstly, it is necessary to determine the seeding density ($X0$). Secondly, for any selected seeding density, it is necessary to establish the value of the passage length ($t_p$) that makes the expansion process optimal in a serial operation. Once the methodology for optimal selection of both seeding density and passage length is established, it can be used to study the effect that different media formulations, feeding strategies, and/or coating materials have on the expansion performance of the cells under investigation.

Starting with a constant number of cells ($I$), the expansion optimization can be performed by maximizing the final cell number ($F$) for a given total expansion time ($T$)

$$\text{Max}(F) = \text{Max}(E \times I) \tag{10}$$

To ease the analysis and further develop this equation, we can distinguish between unaltered or altered growth kinetics as cells are expanded over a number of passages.

## 4.1 Unaltered Growth Kinetics

We can initially assume that the growth kinetic of the cells is not significantly altered during all the passages required for the expansion process. From this assumption one obtains the following equations:

$$PD_T = \sum_{i=1}^{N} PD_i = N \times PD_p \tag{11}$$

$$T = \sum_{i=1}^{N} t_i = N \times t_p \tag{12}$$

$$E = \prod_{i=1}^{N} E_i = (E_p)^N \tag{13}$$

Consequently, Eq. (10) can be written as:

$$\text{Max}(F) = \text{Max}(E \times I) = \text{Max}(E_p)^N \tag{14}$$

$$\text{Max}(F) = \text{Max}(E_p)^N = \text{Max}\left[(E_p)^{\frac{T}{t_p}}\right] \tag{15}$$

$$\text{Max}(F) = \text{Max}\left[(E_p)^{\frac{T}{t_p}}\right] = \text{Max}\left[\frac{T}{t_p} \times \ln(E_p)\right] \tag{16}$$

$$\text{Max}(F) = \text{Max}\left[\frac{\ln(E_p)}{t_p}\right] \tag{17}$$

Therefore, the optimal passage length ($t_p$) for a given initial cell number ($I$) and total expansion time ($T$) can be found by plotting ($\ln(E_p)/t_p$) against $t_p$ and determining the maximum.

Continuing with the example of chondroprogenitor cells at different seeding densities, and according to Eq. (17), the optimal passage length ($t_p$) was found by plotting $\ln(E_p)/t_p$ against $t_p$ and determining the maximum of the resulting curve for each of the seeding densities investigated (Fig. 2b). The optimal seeding density for chondroprogenitor cells corresponded to $10^4$ cell cm$^{-2}$ as reflected by the maximum of the curves. In addition, the optimal passage length for this seeding density corresponded to 73 h (time at which the maximum of the curve was reached). The selection of the adequate passage length is crucial. While longer passage length will lead to higher expansion factor in a single passage (as reflected in the growth curves, Fig. 2a), it would be detrimental in a serial operation process as it would impose longer total expansion time (or lower final expansion factor if the process time is constant). Consequently, although each seeding density is associated with an optimal value of passage length, $10^4$ cell cm$^{-2}$ for seeding density with 73 h of passage length is found to be the optimal condition overall for cell expansion.

Although the analysis just presented is suitable for a single passage, the validation of the estimated seeding density and passage length during a sequential expansion relies on the veracity of the unaltered growth kinetic assumption. Only in those situations where cell growth curves remain fairly constant during passaging can the optimization analysis be reduced to a single passage. This is the case of the chondroprogenitor cell growth kinetic, which has been shown to be reasonably unaltered up to seven passages [29].

## 4.2 Altered Growth Kinetics

Despite that on some occasions the assumption of an unaltered growth kinetic is valid for a number of passages, in some cell types the growth kinetic varies significantly as cells are expanded in culture. One example of cells that present an altered growth kinetic in culture is endothelial progenitor cells. The identification of EPCs in blood a decade ago presented an opportunity to noninvasively obtain endothelial cells for therapeutic applications [16, 17, 18]. For example, blood-derived EPCs have been used to endothelialize small-diameter blood vessels [3], and to form functional vascular networks in vivo [37, 13, 38]. For therapeutic application, extensive expansion of EPCs in vitro is likely to be required, and therefore changes in growth kinetics over time will influence the search for optimal expansion conditions.

To illustrate this concept, an analysis of the growth curves of cord blood-derived EPCs (cbEPCs) obtained at five different passages is made (Fig. 3a) [13]. In this example, all culture parameters were kept constant. From Fig. 3a it can be observed how the passage number played a central role in the proliferation performance of the cultures, affecting both the kinetics of cell growth and the final cell density achieved. In fact, it has been reported that as cbEPCs are expanded in culture, their morphology, growth kinetics, and proliferative responses toward angiogenic factors progressively resemble those of mature microvascular endothelial cells [13]. Since cbEPC growth curves are significantly different at each passage, we cannot assume that the optimal culture conditions found for one passage will remain constant during the sequential expansion of cbEPCs. For example, optimal seeding density may be different at each passage. In addition, the optimal passage length can vary as we expand the cells.

Methodology for Optimal In Vitro Cell Expansion in Tissue Engineering

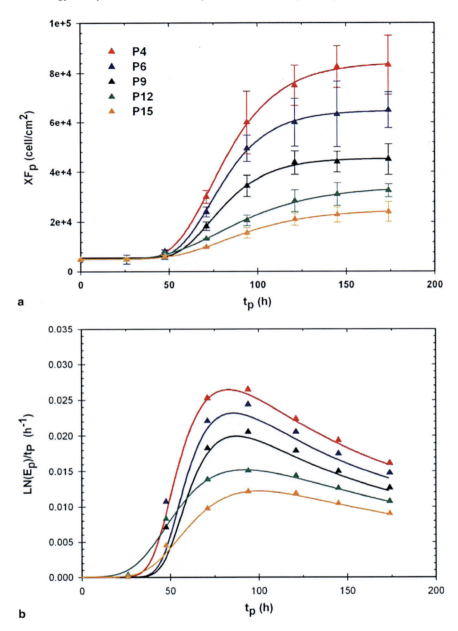

**Fig. 3** Expansion of cord blood-derived endothelial progenitor cells in a monolayer. (**a**) Growth curves of cbEPCs at different passage numbers (P4, P6, P9, P12, and P15). Each data point represents the mean of three separate cultures ±SD. All the cultures were supplied with 0.3 ml cm$^{-2}$ of EGM-2 supplemented with 20% FCS. (**b**) Optimization of passage length at different initial cell densities was determined by the maximum of (ln $(E_p)/t_p$). Panel (**a**) of this figure was adapted from [13] with permission from the editor

Therefore, in cells where the growth kinetic changes (such as cbEPCs), both the optimal seeding density ($X0$) and passage length ($t_p$) need to be found at different passages by plotting $\ln(E_p)/t_p$ against $t_p$ [Eq. (17)] and determining the maximum of the resulting curve for each of the seeding densities investigated. In the example depicted in Fig. 3, the seeding density of cbEPCs was kept constant ($5 \times 10^3$ cell cm$^{-2}$), and the optimal passage length determined at different passages. The optimal passage lengths ($t_p$) corresponded to 83, 86, 87, 92, and 100 h for EPCs at passages 4, 6, 9, 12, and 15, respectively, illustrating the effect of prolonged culture on expansion optimization (Fig. 3b). As a result, if cbEPCs are to be grown at $5 \times 10^3$ cell cm$^{-2}$, the recommended passage length for optimal sequential expansion needs to be progressively increased as cells are being passaged. Importantly, this example also illustrates that even though the growth kinetic varies dramatically as cells are expanded in vitro, the resulting optimal parameters can eventually oscillate around a narrow range of values. Therefore, some cell types may present important differences in growth kinetic as they are expanded, but their optimal seeding density and/or passage length may remain essentially unchanged. However, this may not be the case for all primary cells, and as a general rule, optimal seeding density and passage length need to be found at each passage (or at least every few passages) for each type of cell under consideration.

## 5 Exponential Growth Kinetics

Even though most anchorage-dependent primary cells present deceleratory growth patterns, some cell lines that are routinely used by the biopharmaceutical industry proliferate following exponential growth kinetics. This is the case of CHO 320, a gamma-interferon producing cell line. Growth curves of CHO 320 in suspension cultures have been modeled with high accuracy by an exponential growth function [39], given by the following simplified version of the Monod equation [40]:

$$XF = X0 \times \exp\left(\mu \times t_p\right) \qquad (18)$$

where $\mu$ (h$^{-1}$) corresponds to the specific growth rate, and remains essentially constant during the growth phase of the culture. The Monod equation [Eq. (18)] has been used extensively for the description of the growth of animal cells [40]. This model assumes an exponential cell growth until the amount of an essential substrate becomes limiting. This simplified version of the Monod equation does not take into consideration death rate by accumulation of toxic compounds, and therefore it can only be applied to stages of the culture where cell viability is high. The confirmation of an exponential growth pattern will have important implications for the expansion optimization of a cell line, and it will be necessarily different from that shown for primary cells with a deceleratory growth kinetic [29].

Using the exponential growth Eq. (18), the expansion factor of each passage can be estimated by (19)

$$E_p = \frac{XF}{X0} = \frac{X0 \times \exp(\mu \times t_p)}{X0} = \exp(\mu \times t_p) \qquad (19)$$

and the evaluation of $(\ln(E_p)/t_p)$ results in

$$\frac{\ln(E_p)}{t_p} = \frac{\mu \times t_p}{t_p} = \mu \qquad (20)$$

Therefore, determining the maximum of $(\ln(E_p)/t_p)$ in a culture system that follows a true exponential growth is equivalent to determining the maximum specific growth rate. This means that for any given initial amount of cells, the culture conditions (including seeding density) that maximize the specific growth rate of the cells will be the optimal conditions to achieve the highest number of cells in a given constant culture time. Even though this analysis may seem obvious at first examination, it is important to remember that not all mammalian cell-culture systems will lead to the same conclusion. For instance, during the evaluation of primary chondroprogenitor cells (Fig. 2) [29], the values of specific growth rate were far from constant throughout the growth period of the culture. As a result, the optimal seeding density was found to be a compromise between the growth kinetic and the expansion factor achievable (i.e., not always the highest specific growth rate corresponded to the optimal). Only when a culture follows a true exponential growth can the value of specific growth rate be taken as the sole parameter for cell expansion optimization.

In the study of CHO 320 [39], the maximum specific growth rate (and therefore optimal expansion) was found for the lowest seeding density evaluated ($0.5 \times 10^5$ cell ml$^{-1}$). This finding led us to speculate that lowering the initial cell density beyond the optimal $0.5 \times 10^5$ cell ml$^{-1}$ could be beneficial for the expansion process. However, this should be evaluated carefully; by lowering the seeding density of the culture, we could eventually reach a threshold seeding density that results in an inferior specific growth rate due to a very sparse distribution of the suspended cells. Also, very low initial cell densities are known to result in nonexponential growth patterns [41], and therefore the straight forward correlation between specific growth rate and optimal expansion condition will no longer be valid.

Another consequence of an exponential growth kinetic is found in the determination of optimal passage lengths. From Eq. (20), the answer to this question seems straight forward: as long as we can maintain the value of specific growth rate constant, it should not matter at what point we passage the cells. However, this statement should be looked at more carefully by taking into account additional considerations concerning the cost of the expansion process. Independently of the seeding density, the selection of the passage length will have a direct influence on the final cost of the expansion process (discussed below). Although the optimal cell expansion will be identical as long as the specific growth rate remains constant, longer values of passage length will eventually result in lower total process cost by reducing the total number of passages required (each passage involves evident costs in labor, materials, and services). Therefore, from costing considerations, it will be beneficial to maintain the cells in culture as long as its specific growth rate remains constant. To optimize transfer times between passages (passage length) with a minimal cost, it is recommended to maintain the cells in culture until the end of the exponential growth phase.

## 6 Running Cost of the Expansion Process

Independently of the seeding density, the selection of the passage length will have a direct influence on the final cost of the expansion process. Although the optimal passage length will always give the desired expansion factor in a minimal process time, longer values of the passage length could eventually create a situation where the same desired expansion factor is achieved with a lower total process cost (in detriment of the process time) by reducing for example the number of passages required. To illustrate this concept, we can define the following costing parameters and equations:

$C_{MPA}$ ($ cm$^{-2}$): corresponds to the cost of culture medium per unit of area necessary to carry out one single passage. For a given cell type, the concentration of medium used (expressed in ml cm$^{-2}$) and the cost of the medium ($ ml$^{-1}$) can be considered as constants, independent of the passage length. The value of $C_{MPA}$ ($ cm$^{-2}$) depends directly on the number of medium changes performed during the extent of the passage. However, if the feeding strategy is constantly maintained (typically medium is replenished every 2–3 days), the value of $C_{MPA}$ ($ cm$^{-2}$) can be considered as a constant for every passage.

$C_{FPA}$ ($ cm$^{-2}$): corresponds to the cost of tissue culture flasks expressed per unit of area. For a given kind of tissue culture flasks, this value can be considered as a constant, independent of the passage length.

$C_{XPA}$ ($ cm$^{-2}$): corresponds to the rest of the running cost of any single passage expressed per unit of area (excluding the cost considered in $C_{MPA}$ and $C_{FPA}$). This cost includes concepts such as PBS buffer and trypsin-EDTA solution required for cell harvesting, labor cost, and all the other operating costs. By its own definition, it is clear that the value of this parameter depends on the length of the passage. Nevertheless, to ease the analysis, this value was considered as a constant. The reasons for such hypothesis are based on the fact that some of the costs included in this parameter are independent of the passage length (e.g., the cost of PBS buffer and trypsin-EDTA solution required for cell harvesting, and the labor cost attributed to the inoculation and cell harvesting). Taking also into consideration that $C_{XPA}$ is added to $C_{MPA}$ and $C_{FPA}$ on the overall passage running cost, the hypothesis of considering $C_{XPA}$ as a constant is reasonable.

$C_{PA}$ ($ cm$^{-2}$): corresponds to the total running cost per unit of area necessary to carry out one single passage. $C_{PA}$ can be estimated by the following equation:

# Methodology for Optimal In Vitro Cell Expansion in Tissue Engineering

$$C_{PA} = C_{MPA} + C_{FPA} + C_{XPA} \tag{21}$$

Since $C_{MPA}$, $C_{FPA}$, and $C_{XPA}$ can be considered as constants (for given culture conditions and under the hypotheses stated before), the value of $C_{PA}$ can also be considered as a constant.

$A_i$ (cm²): corresponds to culture area utilized during passage number $i$. For a given constant seeding density ($X0$), $A_i$ can be expressed as:

$$A_i = \frac{I_i}{X0} \tag{22}$$

$C_{Pi}$ ($): corresponds to the total running cost necessary to carry out passage number $i$. $C_{Pi}$ can be estimated by the following equation:

$$C_{Pi} = C_{PA} \cdot A_i \tag{23}$$

$C$ ($): corresponds to the total running cost of the expansion process. This parameter can be calculated by summing the cost of every single passage constituting the expansion process

$$C = \sum_{i=1}^{N} C_{Pi} = C_{PA} \sum_{i=1}^{N} A_i = \frac{C_{PA}}{X0} \sum_{i=1}^{N} I_i \tag{24}$$

$I_i$ can be expressed as a function of the expansion factors preceding passage number $i$ as follows:

$$I_i = I_{i-1} \times E_{i-1} = I_{i-2} \times E_{i-2} \times E_{i-1} = I_{i-3} \times E_{i-3} \times E_{i-2} \times E_{i-1} = \ldots = I \prod_{z=1}^{i-1} E_z \tag{25}$$

Therefore, Eq. (24) can be written as:

$$C = \frac{C_{PA}}{X0} \sum_{i=1}^{N} I_i = \frac{C_{PA} I}{X0} \sum_{i=1}^{N} \left( \prod_{z=1}^{i-1} E_z \right) \tag{26}$$

$\xi$ (–): corresponds to a dimensionless expression of the total running cost of the expansion process, $C$, and is defined by the following expression:

$$\xi = \frac{C}{\dfrac{C_{PA} I}{X0}} \tag{27}$$

Combining Eqs. (26) and (27), the dimensionless running cost of the expansion process can be expressed as follows:

$$\xi = \sum_{i=1}^{N} \left( \prod_{z=1}^{i-1} E_z \right) \tag{28}$$

From this equation, it is clear that the running cost of the expansion process is highly influenced by the values of the expansion factors ($E$) achieved in each passage. In other words, the selection of the length of each passage will have a tremendous effect on the running cost of the process, and this cost will be continuously decreased as we prolong the extent of each passage. The optimal culture conditions for cell expansion (based on growth kinetic) are not necessarily the most economical ones. Although the optimal passage length will always give the desired expansion factor in a minimal process time, longer values of the passage length could lead to the same desired expansion factor with a lower total process cost (in detriment of the process time) by reducing for example the number of passages required. The selection of culture conditions will be a compromise between optimal cell expansion (growth kinetics) and acceptable running cost. This compromise will normally translate into an increase of passage length further away from the optimal value dictated by the growth kinetic of the cells. This is the case, for instance, for chondroprogenitor cells, where the selection of longer passage lengths (120 h instead of the optimal 73 h) was shown to reduce the running cost of the expansion process by more than 60% at the expense of suboptimal proliferation and process time [29].

## 7 Preservation of Cell Phenotype

When selecting culture conditions for optimal cell expansion, it is crucial to ensure that the expanded cell population retains the required phenotype and functional characteristics for their final therapeutic use. Therefore, any expansion optimization approach should be accompanied by assays to evaluate the preservation of functional cell phenotypes. For instance, when dealing with chondroprogenitor cells, we need to ensure that the optimal expansion conditions do not compromise their ability to differentiate and produce cartilage-like extracellular matrix. After following the methodology proposed above, the chondroprogenitor cell population was grown in pellet cultures and was able to synthesize a cartilage-like matrix that stained strongly with safranin-O, indicating the presence of sulfated proteoglycans. In addition, immunohistochemistry analysis of the pellets showed the presence of collagen type II, another cartilage marker. These results proved the ability of chondroprogenitor cells to differentiate into chondrocytes and to synthesize cartilage-like matrix after serial expansion in monolayer using optimized culture conditions, an ability that ultimately validated the selection of those optimal conditions [15, 42].

In the case of cbEPCs, the endothelial phenotype of the expanded population was confirmed up to passage 15 [13]. cbEPCs consistently expressed endothelial markers CD34, VE-cadherin, VEGF-R2, CD146, CD31, eNOS, vWF, and CD105. In addition, cbEPCs were negative for mesenchymal marker CD90 and hematopoietic markers CD45 and CD14, confirming that the cells were not contaminated with either mesenchymal or hematopoietic cells at any stage of the expansion process.

Expanded cbEPCs were also able to up-regulate leukocyte adhesion molecules E-selectin, ICAM-1, and VCAM-1 in response to the inflammatory cytokine TNF-α. This response to an inflammatory cytokine is characteristic of endothelial cells and suggests that the use of cbEPCs in vivo could also provide physiologic proinflammatory properties.

The maintenance of an appropriate phenotype (e.g., expression of cellular surface markers), although essential, may not be sufficient depending on the application, and functional assays must be performed before establishing definitive expansion culture conditions. In this regard, despite their phenotypical stability, cbEPCs have been shown to undergo cellular changes during their expansion in vitro. In particular, the migratory capacity of cbEPCs in vitro decreased over time in culture [43]. Also, it has been shown that as cbEPCs are expanded in culture, their vasculogenic ability in vivo decreases [13]. Therefore, if cbEPCs are to be used for tissue vascularization, these functional changes can either impose a limitation on the extent to which cbEPCs could be expanded in vitro prior to implantation, or redefine the conditions for their final therapeutic use. The latter was actually the case of cbEPC at high passages, where the partial loss of vasculogenic ability in vivo was compensated by increasing the number of cells seeded into the implants [13].

## 8 Conclusions

Here, we provide a rational methodology for searching culture conditions that optimize the acquisition of large quantities of cells following a sequential expansion process. The methodology starts with the evaluation of growth kinetics in monolayer cultures, and leads to the selection of both optimal seeding density and passage length. The mathematical elucidation for the selection of optimal culture conditions has been specifically presented for those situations where the cell population presents either unaltered or altered growth kinetics during routine passaging. Additionally, a simplified set of equations was introduced to ease the analysis of cells that grow with exponential kinetics. This methodology also introduces additional considerations concerning the running cost of the expansion process, and shows that the selection of culture conditions should be a compromise between optimal cell expansion and acceptable running cost. This compromise will normally translate into an increase of passage length further away from the optimal value dictated by the growth kinetic of the cells. Finally, the importance of incorporating functional assays to validate the phenotypical and functional characteristics of the expanded cells has been highlighted. The optimization approach presented is expected to contribute to the development of feasible large-scale expansion of cells required by the tissue engineering industry.

**Acknowledgments** Part of this research was supported by funding from the US Army Medical Research and Material Command (W81XWH-05-1-0115).

# References

1. Eberli D, Atala A (2006) Tissue engineering using adult stem cells. Methods Enzymol 420: 287–302.
2. Langer R, Vacanti JP (1993) Tissue engineering. Science 260 (5110): 920–926.
3. Kaushal S, Amiel GE, Guleserian KJ, Shapira OM, Perry T, Sutherland FW, Rabkin E, Moran AM, Schoen FJ, Atala A, et al. (2001) Functional small-diameter neovessels created using endothelial progenitor cells expanded ex vivo. Nat Med 7(9): 1035–1040.
4. Hoerstrup SP, Sodian R, Daebritz S, Wang J, Bacha EA, Martin DP, Moran AM, Guleserian KJ, Sperling JS, Kaushal S, et al. (2000) Functional living trileaflet heart valves grown in vitro. Circulation 102 (19 Suppl 3): III44–III49.
5. Sutherland FW, Perry TE, Yu Y, Sherwood MC, Rabkin E, Masuda Y, Garcia GA, McLellan DL, Engelmayr GC, Jr, Sacks MS, et al. (2005) From stem cells to viable autologous semilunar heart valve. Circulation 111 (21): 2783–2791.
6. Oberpenning F, Meng J, Yoo JJ, Atala A (1999) De novo reconstitution of a functional mammalian urinary bladder by tissue engineering. Nat Biotechnol 17 (2): 149–155.
7. Vacanti JP, Langer R (1999) Tissue engineering: the design and fabrication of living replacement devices for surgical reconstruction and transplantation. Lancet 354 (Suppl 1): SI32–SI34.
8. Langer RS, Vacanti JP (1999) Tissue engineering: the challenges ahead. Sci Am 280 (4): 86–89.
9. Han Z, Vandevoort CA, Latham KE (2007) Therapeutic cloning: status and prospects. Curr Opin Mol Ther 9 (4): 392–397.
10. Petersen T, Niklason L (2007) Cellular lifespan and regenerative medicine. Biomaterials 28 (26): 3751–3756.
11. Holtzer H, Abbott J, Lash J, Holtzer A (1960) The loss of phenotypic traits by differentiated cells in vitro, I. Dedifferentiation of cartilage cells. Proc Natl Acad Sci U S A 46 (4): 1533–1542.
12. Melero-Martin JM, Al-Rubeai M (2007) In vitro expansion of chondrocytes. In: Ashammakhi N, Reis R, Chiellini E (eds.) Topics in tissue engineering, vol 3, Chap 2. University of Oulu, Finland, pp.1–37.
13. Melero-Martin JM, Khan ZA, Picard A, Wu X, Paruchuri S, Bischoff J (2007) In vivo vasculogenic potential of human blood-derived endothelial progenitor cells. Blood 109 (11): 4761–4768.
14. Dowthwaite GP, Bishop JC, Redman SN, Khan IM, Rooney P, Evans DJ, Haughton L, Bayram Z, Boyer S, Thomson B, et al.(2004) The surface of articular cartilage contains a progenitor cell population. J Cell Sci 117 (Pt 6): 889–897.
15. Martin JM, Smith M, Al-Rubeai M (2005) Cryopreservation and in vitro expansion of chondroprogenitor cells isolated from the superficial zone of articular cartilage. Biotechnol Prog 21 (1): 168–177.
16. Asahara T, Murohara T, Sullivan A, Silver M, van der Zee R, Li T, Witzenbichler B, Schatteman G, Isner JM (1997) Isolation of putative progenitor endothelial cells for angiogenesis. Science 275 (5302): 964–967.
17. Ingram DA, Mead LE, Tanaka H, Meade V, Fenoglio A, Mortell K, Pollok K, Ferkowicz MJ, Gilley D, Yoder MC (2004) Identification of a novel hierarchy of endothelial progenitor cells using human peripheral and umbilical cord blood. Blood 104 (9): 2752–2760.
18. Lin Y, Weisdorf DJ, Solovey A, Hebbel RP (2000) Origins of circulating endothelial cells and endothelial outgrowth from blood. J Clin Invest 105 (1): 71–77.
19. Yoo JJ, Meng J, Oberpenning F, Atala A (1998) Bladder augmentation using allogenic bladder submucosa seeded with cells. Urology 51 (2): 221–225.
20. Nielson LK, Smyth GK, Grrenfield PF (1991) Hemocytometer cell count distributions: implication of non poison behaviour. Biotechnol Prog 7: 560–563
21. Kell DB, Markx GH, Davey CL, Todd RW (1990) Real-time monitoring of cellular biomass: methods and applications. Trends Anal Chem 9: 190–194.
22. Konstantinov K, Chuppa S, Sajan E, Tsai Y, Yoon S, Golini F (1994) Real-time biomass-concentration monitoring in animal-cell cultures. Trends Biotechnol 12 (8): 324–333.

23. Olsson L, Nielsen J (1997) On-line and in situ monitoring of biomass in submerged cultivations. Trends Biotechnol 15: 517–522.
24. Sonnleitner B, Locher G, Fiechter A (1992) Biomass determination. J Biotechnol 25 (1–2): 5–22
25. Ducommun P, Ruffieux P-A, Furter M-P, Marison I, von Stockar U (2000) A new method for on-line measurement of the volumetric oxygen uptake rate in membrane aerated animal cell cultures. J Biotechnol 78 (2): 139–147.
26. Harding CL, Lloyd DR, McFarlane CM, Al-Rubeai M (2000) Using the Microcyte flow cytometer to monitor cell number, viability, and apoptosis in mammalian cell culture. Biotechnol Prog 16 (5): 800–802.
27. Shah D, Naciri M, Clee P, al-Rubeai M (2006) NucleoCounter – an efficient technique for the determination of cell number and viability in animal cell culture processes. Cytotechnology 51: 39–44.
28. Skehan P, Friedman SJ (1984) Non-exponential growth by mammalian cells in culture. 17 (4): 335–343.
29. Melero-Martin JM, Dowling MA, Smith M, Al-Rubeai M (2006b) Optimal in-vitro expansion of chondroprogenitor cells in monolayer culture. Biotechnol Bioeng 93 (3): 519–533.
30. Dietrich C, Wallenfang K, Oesch F, Wieser R (1997) Differences in the mechanisms of growth control in contact-inhibited and serum-deprived human fibroblasts. 15 (22): 2743–2747.
31. Buchanan RL, Whiting RC, Damert WC. (1997) When is simple good enough: a comparison of the Gompertz, Baranyi, and three-phase linear models for fitting bacterial growth curves. Food Microbiol 14: 613–626.
32. Deakin MA (1970) Gompertz curves, allometry and embryogenesis. Bull Math Biophys 32 (3): 445–452.
33. Skehan P, Friedman SJ (1982) Deceleratory growth by a rat glial tumor line in culture. 42 (5): 1636–1640.
34. Heegaard S, Spang-Thomsen M, Prause JU (2003) Establishment and characterization of human uveal malignant melanoma xenografts in nude mice. Melanoma Res 13 (3): 247–251.
35. Lloyd HH (1975) Estimation of tumor cell kill from Gompertz growth curves. Cancer Chemother Rep 59 (2 Pt 1): 267–277.
36. Spratt JA, von Fournier D, Spratt JS, Weber EE (1993) Decelerating growth and human breast cancer. Cancer 71 (6): 2013–2019.
37. Au P, Daheron LM, Duda DG, Cohen KS, Tyrrell JA, Lanning RM, Fukumura D, Scadden DT, Jain RK (2008) Differential in vivo potential of endothelial progenitor cells from human umbilical cord blood and adult peripheral blood to form functional long-lasting vessels. Blood 111 (3): 1302–1305.
38. Yoder MC, Mead LE, Prater D, Krier TR, Mroueh KN, Li F, Krasich R, Temm CJ, Prchal JT, Ingram DA (2007) Redefining endothelial progenitor cells via clonal analysis and hematopoietic stem/progenitor cell principals. Blood 109 (5): 1801–1809.
39. Santhalingam S (2007) Monitoring, optimisation and expansion issues of CHO cells producing gamma interferon in serum-supplemented and serum-free cultures [Ph.D. thesis]. University of Birmingham, Birmingham, UK.
40. Fox SR, Patel UA, Yap MG, Wang DI (2004) Maximizing interferon-gamma production by Chinese hamster ovary cells through temperature shift optimization: experimental and modeling. Biotechnol Bioeng 85 (2): 177–184.
41. Harada JJ, Porter CW, Morris DR (1981) Induction of polyamine limitation in Chinese hamster ovary cells by alpha-methylornithine. J Cell Physiol 107 (3): 413–426.
42. Melero-Martin JM, Dowling MA, Smith M, Al-Rubeai M (2006a) Expansion of chondroprogenitor cells on macroporous microcarriers as an alternative to conventional monolayer systems. Biomaterials 27 (15): 2970–2979.
43. Khan ZA, Melero-Martin JM, Wu X, Paruchuri S, Boscolo E, Mulliken JB, Bischoff J (2006) Endothelial progenitor cells from infantile hemangioma and umbilical cord blood display unique cellular responses to endostatin. Blood 108 (3): 915–921.

# Bioreactor Studies and Computational Fluid Dynamics

**H. Singh and D.W. Hutmacher**

**Abstract** The hydrodynamic environment "created" by bioreactors for the culture of a tissue engineered construct (TEC) is known to influence cell migration, proliferation and extra cellular matrix production. However, tissue engineers have looked at bioreactors as black boxes within which TECs are cultured mainly by trial and error, as the complex relationship between the hydrodynamic environment and tissue properties remains elusive, yet is critical to the production of clinically useful tissues. It is well known in the chemical and biotechnology field that a more detailed description of fluid mechanics and nutrient transport within process equipment can be achieved via the use of computational fluid dynamics (CFD) technology. Hence, the coupling of experimental methods and computational simulations forms a synergistic relationship that can potentially yield greater and yet, more cohesive data sets for bioreactor studies. This review aims at discussing the rationale of using CFD in bioreactor studies related to tissue engineering, as fluid flow processes and phenomena have direct implications on cellular response such as migration and/or proliferation. We conclude that CFD should be seen by tissue engineers as an invaluable tool allowing us to analyze and visualize the impact of fluidic forces and stresses on cells and TECs.

**Keywords** Bi-axial, Computational modelling, Fluid dynamics, Scaffolds, Tissue engineering

---

H. Singh
Division of Bioengineering, National University of Singapore, Engineering Drive 1, E2-04-01, Singapore 119260, Singapore

D.W. Hutmacher (✉)
QUT Chair in Regenerative Medicine, Institute of Health and Biomedical Innovation, Queensland University of Technology, 60 Musk Avenue, Kelvin Grove QLD, 4059 Australia

**Contents**

1  Introduction .................................................................................................... 232
2  Computational Fluid Dynamics and Modelling ............................................. 235
3  Future Directions ........................................................................................... 245
4  Conclusions .................................................................................................... 247
References ............................................................................................................ 247

# 1 Introduction

A bioreactor may be defined as a system that simulates physiological environments for the creation, physical conditioning, and testing of cells and tissues, support structures, and organs in vitro [1]—it provides a controlled environment [2]. Functions of such bioreactors include providing adequate nutrient supply to cells, waste removal, gaseous exchange, temperature regulation and mechanical force stimulation [2]. There are different types of bioreactors and they vary greatly in their size, complexity, and functional capabilities. Commonly utilized bioreactors used in tissue engineering applications include spinner flasks, rotating wall vessels, perfusion systems, hollow-fibre [3] systems as well as compression-loading systems. The most common operational modes of bioreactors include those of continuous, fed-batch and batch.

Tissue engineering-related bioreactors have long been thought of as black boxes within which cells are cultured mainly by trial and error. The science and technology involved in the design, functionality and application of such bioreactors clearly indicates otherwise. This review therefore presents various applications as well the biological implications of such dynamic culture systems on tissue growth, with a particular focus on convective transport. The application of computational fluid dynamics (CFD) with respect to tissue growth within such bioreactors is also discussed. We conclude that CFD has the potential to be regarded as invaluable in allowing us to analyze and visualize the impact of fluidic forces and stresses such as shear, in cell and tissue engineering.

Wendt et al. [4] aptly stated that "The optimal operating conditions of a bioreactor should not be determined through a trial and error approach, but should be instead defined by integrating experimental data and computational models". The synergy as a result of coupling the experimental and computational approaches can potentially yield data that will increase our understanding as well as increase the cohesiveness of data obtained from future bioreactor studies.

Dynamic culturing of cells and tissues has a direct impact on the composition, morphology and mechanical properties of engineered tissues grown in mechanically stimulated environments [5]. This is primarily due to effects of dynamic culture media transport, which often enhance the functions of dynamic flow-based bioreactors [6], as compared to diffusion-based static culture systems (see Figs. 1 and 2).

3D constructs cultured within a cell-culture well-plate for example, often exhibit tissue growth mainly along the external periphery of the scaffold, and not within

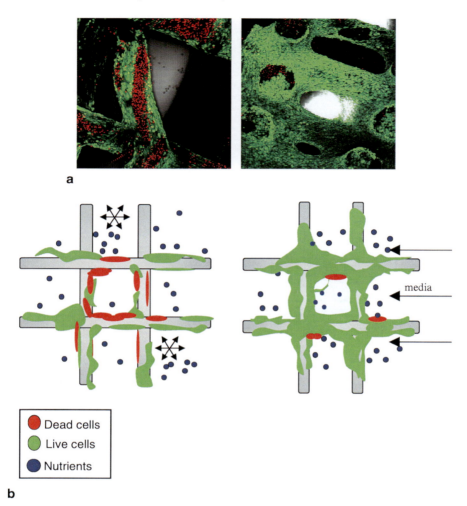

Fig. 1 (a) Confocal laser microscopy images of seeded scaffolds stained for live (green) and dead (red) cells reveal higher numbers of dead cells within the statically cultured scaffold (*left*) as compared to the dynamically cultured scaffold (*right*) and (b) graphical illustration showing that nutrient diffusion occurring within the statically cultured scaffold is driven by concentration (*left*) in contrast to convection driven (dynamic) flows where nutrients are uniformly distributed (*right*)

the innermost pores of the scaffold architecture. Where neo-tissue formation does occur within the innermost pores of scaffolds, it often becomes a matter of time before the neo-tissue becomes necrotic due to the lack of nutritional and gaseous transport. Carrier et al. [7] highlighted that an approximate 100-μm thick layer of cardiac cells can be supported via diffusion under static conditions, beyond which necrosis would likely occur.

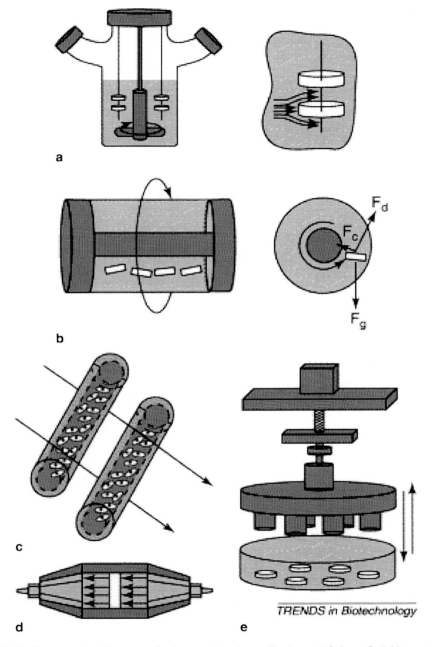

**Fig. 2** Representative bioreactors for tissue engineering applications. (a) Spinner-flask bioreactors have been used for the seeding of cells into 3D scaffolds and for subsequent culture of the constructs. During seeding, cells are transported to and into the scaffold by convection. During culture, medium stirring enhances external mass-transfer but also generates turbulent eddies, which could be detrimental for the development of the tissue. (b) Rotating-wall vessels provide a dynamic culture

## 2 Computational Fluid Dynamics and Modelling

Computational fluid dynamics is increasingly being seen as an excellent tool for analyzing biological systems as well as flows within bioreactors. For example, CFD work was carried out by Kimbell and Subramaniam [8] to investigate the uptake of gases via the nasal passageway for three species. Grotberg's [9] in-depth review of respiratory fluid mechanics further highlighted how fluid flow through the respiratory passages can be effectively modelled to understand and visualize flow characteristics under a variety of physiological conditions. Redaelli et al. [10] attempted to model pulsatile flow within arteries by means of interfacing a FORTRAN algorithm with FIDAP (Fluent, Inc.), a commercially available CFD package.

Other work performed using CFD include that of Varghese et al. [11], who modelled the influence of stenoses on pulsatile blood flow within blood vessels. The $k$-$\omega$ turbulence model, being more appropriate for lower Reynolds numbers (mild turbulence), was employed and results obtained indicated a higher level of accuracy as compared to the $k$-$\varepsilon$ turbulence model. The implications of heart valve design on blood flow, and the consequent downstream effects of blood flow within vessels, and the overall fluid dynamics involved in vascular diseases have been covered in detail [12, 13] within these reviews.

In order to better understand the role of the hydrodynamic environment and the factors that modulate it, computational fluid dynamics modelling is an invaluable tool which has only recently been applied to the area of tissue engineering. While traditionally applied to the chemical and mechanical engineering fields, CFD is now enabling tissue engineers to understand the implications of fluid flow and transport cell function and provide important insights into the design and optimization of 3D scaffolds suitable in bioreactors for in vitro tissue engineering.

Furthermore, CFD allows us to remove much of the trial-and-error involved in traditional culture methods [4]. CFD modelling is one of the most effective methods

**Fig. 2** (continued) environment to the constructs, with low shear stresses and high mass-transfer rates. The vessel walls are rotated at a rate that enables the drag force ($F_d$), centrifugal force ($F_c$) and net gravitational force ($F_g$) on the construct to be balanced; the construct thus remains in a state of free-fall through the culture medium. (c) Hollow-fibre bioreactors can be used to enhance mass transfer during the culture of highly metabolic and sensitive cell types such as hepatocytes. In one configuration, cells are embedded within a gel inside the lumen of permeable hollow fibres and medium is perfused over the exterior surface of the fibres. (d) Direct perfusion bioreactors in which medium flows directly through the pores of the scaffold can be used for seeding and/or culturing 3D constructs. During seeding, cells are transported directly into the scaffold pores, yielding a highly uniform cell distribution. During culture, medium flowing through the construct enhances mass transfer not only at the periphery but also within its internal pores. (e) Bioreactors that apply controlled mechanical forces, such as dynamic compression, to engineered constructs can be used as model systems of tissue development under physiological loading conditions, and to generate functional tissue grafts. Compressive deformation can be applied by a computer-controlled micro-stepper motor, and the stress on the constructs can be measured using a load cell. Reprinted with permission from [6]

to characterize flow fields, provided that the models are validated by experimental velocimetry techniques, such as laser-Doppler anemometry (LDA) and particle-image velocimetry (PIV) [14, 15].

While these experimental methods are reliable, they are too arduous to characterize the complete three-dimensional fluid flow in a bioreactor. CFD modelling provides a powerful means to overcome these limitations and enable the full characterization of three-dimensional flow fields in bioreactors with simple and complex geometries. Bioreactor designs and their fully characterized flows may thus be evaluated prior to fabrication. On the other hand, specific parameters such as fluid inlet velocities and shear stresses for example, may be varied to better predict their influence, leading to possible optimization of tissue growth. CFD codes generally enable visualization of flow phenomena as well, as it may be impractical to position probes within the fluid domains to measure parameters such as pressure and velocity.

Many CFD software packages are highly scaleable and allow for multiple processors or CPUs to be run in parallel to solve the flow simulation in a shorter amount of time. Savings in solution time may even be geometric, depending on the complexity of the problem and its set-up. Model meshing is carried out prior to solution and must be adequately refined to capture data at specific areas of interest.

For example, the mesh generation of boundary layers may be required in order to obtain accurate shear stresses along a specific surface. This however, depends on factors such as the type of flow (laminar or turbulent), fluid viscosity, and the level of detail required by the user. Unnecessary generation of refined mesh zones would increase the computational time and results in increased computational costs.

An important part of model mesh generation involves the selection of an "adequate" mesh that can meet the requirements of the anticipated study protocol. This is often known as performing a "grid independence" check. Grid independence involves the generation of meshes of varying degrees of refinement, followed by their solution under similar or identical conditions. While a highly refined mesh will provide for better solution accuracy, it may not be feasible to employ such a model (Fig. 3). Coarse, semi-refined and refined meshes are thus generated and their solutions compared.

The two key objectives for performing a grid independence study are: (a) to select an economical mesh that provides the user with the required degree of solution accuracy and (b) to check if the solutions obtained via the different meshes are consistent, thereby acting as a method of counter-checking the overall correctness of the simulation. Three examples are shown within Fig. 3, whereby a simple 2D rectangle was meshed with varying degrees of refinement. The first rectangle (bottom) was meshed with only nine elements, which is insufficient even for generic cases. The second example (middle) was meshed with 400 elements, which is likely to be adequate for general cases. The third rectangle (top) was meshed with 10,000 elements which would probably be excessive, requiring extra computational resources, time and money to solve, unless that extra degree of accuracy was required in the first place. An alternative would also be to refine particular zones of interest, rather than refining the entire mesh of the entire domain.

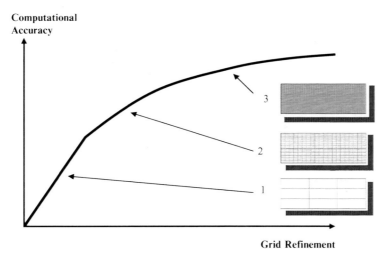

**Fig. 3** General relationship between solution accuracy and the degree of grid refinement. The linear range shows a proportional increment between both factors, whereas the range to the right highlights that any increase in refinement will not result in a proportional increase in accuracy

Many techniques and algorithms have been employed to solve for fluid-based problems. Both Porter et al. [16] and Raimondi et al. [17] modelled the effects of perfusion, to better comprehend the influence of perfusion and hence shear stresses, on 3D cultures. Porter et al. however, used a code based on the Lattice-Boltzmann principle to approximately obtain solutions to the Navier–Stokes equations (Fig. 4), whereas Raimondi et al. modelled their flows with the aid of the commercially available FLUENT package.

Lappa [18] on the other hand, developed equations based on the level-set principle to model fluid flow as well as to estimate soft tissue growth within a bioreactor. While the level-set method may provide higher levels of accuracy, it tends to be quite complex and might not always be feasible to apply.

CFD numerical techniques are highly capable of capturing the flow, pressure and concentration fields resolved down to the pore size of the scaffold. Simulations can show how the scaffold morphology influences the hydrodynamic shear and nutrient concentrations imposed on cells within constructs. We are therefore able to analyze the efficacy of the scaffold architecture with respect to nutrient and gaseous transport.

FIDAP was used to perform a flow analysis of culture media being perfused through scaffold models varying in porosity as well as pore size. Results showed that for a circular scaffold, pore size strongly influences wall stress levels, while porosity significantly affects the statistical distribution of the shear stresses, but not their magnitude [19]. The flow in a percolation porous structure by direct simulation was performed by Andrade et al. [20]. The impact of stagnant zones within an irregular scaffold structure was modelled and found to influence the overall flow field of the system.

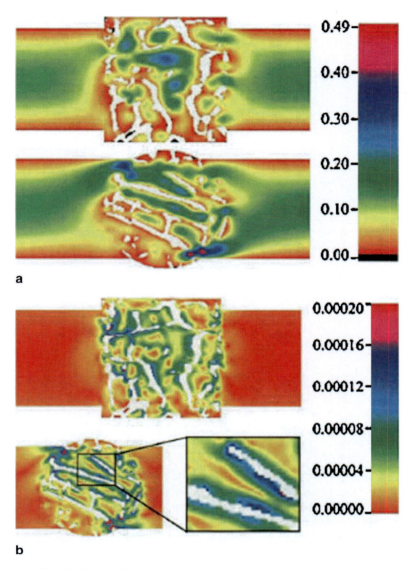

**Fig. 4** (a) Velocity flow field: speed of media flow (mm s$^{-1}$) through a transversely perfused cylindrical trabecular bone scaffold (*shown in white*) from side and top views. (b) Local shear stress field: map of shear stresses (Pa) in media transversely perfused through a 3D trabecular bone scaffold from side and top views. Reprinted with permission from [16]

Raimondi's group has an exceptional track record in the field of applying CFD modelling [19] to the area of tissue engineering (Fig. 5). They developed a CFD model of culture medium flowing through a 3D scaffold of homogeneous geometry, with the aim of predicting the shear stresses acting on cells as a function

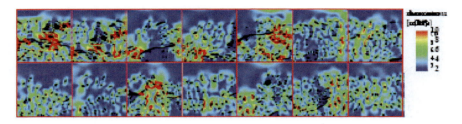

**Fig. 5** Results of the CFD simulations for the 14 images analyzed (model 1): the shear stress field is mapped for each fluid domain obtained from the histological images. Reprinted with permission from [19]

of other parameters, such as scaffold porosity and pore size, as well as medium flow rate and the diameter of the perfused scaffold section. Fig. 5: Results of the CFD simulations for the 14 images analysed (model 1): the shear stress field is mapped for each fluid domain obtained from the histological images. Reprinted with permission from Boschetti, F., M.T. Raimondi, F. Migliavacca, and G. Dubini, *Prediction of the micro-fluid dynamic environment imposed to three-dimensional engineered cell systems in bioreactors.* J Biomech, 2006. **39**(3): p. 418-25.

They modelled three scaffold groups corresponding to three pore sizes: 50, 100 and 150 µm. Each group was made of four sub-models corresponding to 59, 65, 77 and 89% porosities, respectively. A commercial finite-element code was used to set up and solve the problem. Results showed that the mode value of shear stress varied between 2 and 5 mPa, and was obtained for a circular scaffold of 15.5 mm diameter, perfused by a flow rate of 0.5 ml min$^{-1}$. The simulations showed that the pore size is a variable that strongly influences the predicted shear stress level, whereas the porosity is a variable that strongly affects the statistical distribution of the shear stresses, but not their magnitude.

Previous CFD simulations performed by Hutmacher's group demonstrated that bi-axial bioreactor (Fig. 6) rotation enhances fluid flow within the vessel and the scaffolds positioned within the vessel [21]. Comparison of the fluid velocities at specific locations within scaffolds revealed significant increments in velocity when the bi-axial vessel was compared with uni-axial rotation. These significant results highlighted that the problem of inadequate fluid and nutritional/waste transport to and from cells may be solved by rotating the vessel bi-axially. Other CFD work recently performed by the senior authors' group involved the flow modelling of fluid within and around scaffolds of 45° and 90° fibre laydown patterns, within a novel conical bioreactor system [22]. Results from the study showed that fluid is better able to permeate through the 90° fibre laydown scaffold due to its channel-like architecture (Fig. 7). This study therefore highlighted that the flow fields within scaffolds of varying architectures and porosities can play an important role in culturing cells, and such information may be used to meet the requirements of varying cell phenotypes and bioreactor systems to achieve the most out of them.

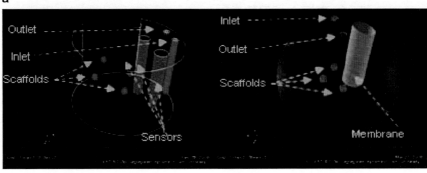

**Fig. 6** (**a**) Cylindrical chamber (*left*) and new Mk. 3 spherical chamber (*right*). The complex cylindrical chamber houses multiple probes which are likely to interfere with the flow dynamics of culture media to a greater extent than the spherical vessel and (**b**) model of cylindrical (*left*) and spherical vessels (*right*)

The objective of a study by Zhao et al. [23] was to quantitatively evaluate the effects of shear stress on hMSC proliferation, extracellular matrices (ECM) formation, and osteogenic differentiation in 3-D constructs using an in-house developed modular perfusion bioreactor system. The hMSCs seeded on a 3-D poly(ethylene terephthalate) (PET) scaffold were analyzed for two flow rates of 0.1 and 1.5 ml min$^{-1}$, respectively, over a 20-day culture period. Mathematical modelling indicated that shear stresses existed in the range of $1 \times 10^{-5}$ to $1 \times 10^{-4}$ Pa within the scaffold/cell constructs up to a depth of 70 μm. An analysis of oxygen transport

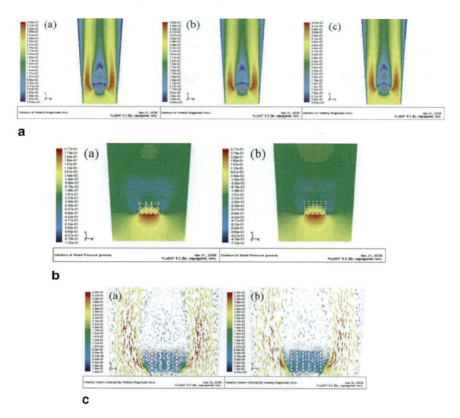

**Fig 7** (a) 90° scaffold—velocity contours at (a) $Re = 121$, (b) $Re = 170$, and (c) $Re = 218$; (b) (a) 45° and (b) 90° scaffolds—pressure contours at $Re = 121$, and (c) Cross-section of (a) 45° and (b) 90° scaffolds (note circular fibres)—velocity vector fields within scaffolds at $Re = 121$. Flow recirculation occurs downstream (*above*) of scaffolds, as indicated by the downward (recirculating) velocity vectors. Reprinted with permission from [22]

in the constructs for the two flow rates yielded oxygen levels significantly higher than those at which cell growth and metabolism for the range of flow rates were studied [24]. The authors concluded that differences in convective transport have no significant influence on cell growth and metabolism for the range of flow rates studied. However, the weakness of the study was that the computational models were 2D-based. In addition, the entire scaffold morphology was not fully modelled. Instead, only a depth of 200 mm from the bottom and top of the 2-mm thick scaffolds was considered in the model.

Toh et al. [25] hypothesized that the geometrical design of the micropillars would affect the fluid flow profile in a microfluidic channel, therefore affecting the cell immobilization efficiency of the micropillar array. Hence, a finite volume model of the microfluidic channel with different micropillar designs was constructed using the

ANSYS Workbench (version 9.0) and ANSYS CFX® (version 5.7.1) software packages (ANSYS, Inc., PA, USA). The model was used to simulate fluid flow in the microfluidic channel during cell immobilization using a Reynolds number of $Re = 0.1$. Microfluidic channels with various micropillar designs (designated P1–P4) were modelled by CFD to investigate their effects on flow characteristics in the channels. Using representative operating flow rates during the cell immobilization process, which were in the order of $Re = 0.1$, the corresponding velocity profiles for P1–P4 at the proximal, middle and distal volumes of the microfluidic channels were discussed.

In the CFD models, the micropillars would not disrupt the laminar flow in the microfluidic channel during the cell immobilization process, except at volumes spanning approximately 0.4 mm at the proximal end of the micropillar arrays. The flow profiles of the bulk volume of the microfluidic channels were similar for all four micropillar designs, with minimal cross flow from the centre cell channel to the peripheral perfusion channels. Significant cross flow between these two compartments was only observed at the proximal volume of the microfluidic channel, with P1 having the largest cross flow followed by P4, P3 and P2 in descending order. The magnitude of this cross flow may affect the propensity for clogging when a cell suspension is directed into the micropillar array. A clogging phenomenon indicated that the rate of cell accumulation (dependent on cross flow velocity) was larger than the rate of cell removal (dependent on tangential flow velocity). The velocity profile around the micropillars was much more uniform for P2, P3 and P4 micropillars, as compared to the P1 micropillar, with P4 having the most uniform flow. On the basis of the CFD analysis, the performance of parallelogram micropillars was expected to be better than semi-circular micropillars because the former had more desirable flow profiles for minimizing the risk of clogging. Since the simulation was carried out without the presence of cells, these findings were verified experimentally by testing the cell immobilization efficiency of the different micropillar designs.

Ye [26] utilized the more recent FEMLAB code to model the rate of nutrient uptake in the culture of bone tissue with respect to their hollow-fibre membrane bioreactor. Pollack et al. [27] performed numerical simulations in conjunction with their experiments to verify the trajectory of microcarrier beads within a rotating bioreactor vessel. They also managed to establish a basic relationship between the density of the microcarrier beads, the surrounding fluid's density, and the resulting trajectories of the beads. A similar application involved human SaOS-2 line cells seeded onto degradable poly(lactic-co-glycolic acid) micro-carriers for bone tissue engineering. Experimental results showed that these cells retained their osteoblastic phenotype and cells expressed increased amounts alkaline phosphatase. The maximum shear stresses exerted on attached cells were numerically modelled and calculated to be 3.9 dyn $cm^{-2}$ [28].

Sucosky et al. [14] were among the first groups that focused on the experimental and numerical characterization of the flow field within a spinner flask operating under conditions used in cartilage tissue engineering. Laboratory experiments carried out in a scaled-up model bioreactor employed PIV to determine velocity and shear-rate fields in the vicinity of the construct closest to the stir bar. Numerical computations calculated using FLUENT, a commercial software package, simulate the flow field in the same model bioreactor under similar operating conditions. In the computations, scaffolds modelled as both solid and porous media with different permeabilities and flow rates through various faces of the construct nearest the stir bar were examined.

Bueno et al. [29] verified the efficacy of an in-house developed wavy-walled spinner flask using CFD by investigating and characterizing the flow regimes within their system, which were found to support cell proliferation by enhancing fluidic transport and stimulating matrix deposition of chondrocytes. The wavy profile was found to "dampen" the effects of turbulence as compared to a cylindrical spinner flask [30].

Building on their previous work, Bilgen and Barabino [31] characterized the complex hydrodynamic environment and examined the changes in the flow field due to the different positions of tissue-engineered cartilage constructs. The geometries for the spinner flask and two different configurations of the wavy-walled bioreactor, where the constructs were located in the lobe or in the centre of the bioreactor, were created via the GAMBIT mesh generation software (Fluent Inc, Lebanon, NH). The meshes created in GAMBIT were then exported to FLUENT 6.1 for the unsteady simulation of 3D turbulent flow.

A turbulence model was necessary to modify the instantaneous transport equations to remove the small scales in order to produce a set of equations that were computationally less expensive to solve. The CFD modelling showed that the flow-induced shear stress range experienced by engineered constructs cultivated in the wavy-walled bioreactor (0–0.67 dyn cm$^{-2}$) was found to be significantly lower than that within the spinner flask (0–1.2 dyn cm$^{-2}$), and was modulated by the radial or axial position of the constructs. The authors concluded that results from this recent study indicated that the location of constructs in the bioreactor not only affected the magnitude and distribution of the shear stresses on the constructs, but also other hydrodynamic parameters. These parameters include the directional distribution of the fluid velocity and the degree of fluid recirculation, all of which may differentially influence the development of tissue-engineered constructs.

Dusting et al. [32] presented a different perspective on the fluid-mixing phenomena within a spinner flask, stating that vortex breakdown flows may enhance the mixing of culture media. The authors concluded that size, profile and intensity of these fluid recirculation regions are highly Reynolds number dependent. However, further studies need to be designed and executed to support these conclusions.

Williams et al. [33] applied Fluent CFD models to calculate flow fields, shear stresses, and oxygen profiles around nonporous constructs which simulated cartilage development in a concentric cylinder bioreactor (Fig. 8). Their findings showed that shear stress distributions ranged from 1.5 to 12 dyn cm$^{-2}$ across the exposed construct surfaces and varied little with the relative number or placement of constructs in the bioreactor. They concluded that approximately 80% of the construct surface exposed to flow experiences shear stresses between 1.5 and 4 dyn cm$^{-2}$. Species mass transport modelling for oxygen demonstrated that fluid-phase oxygen

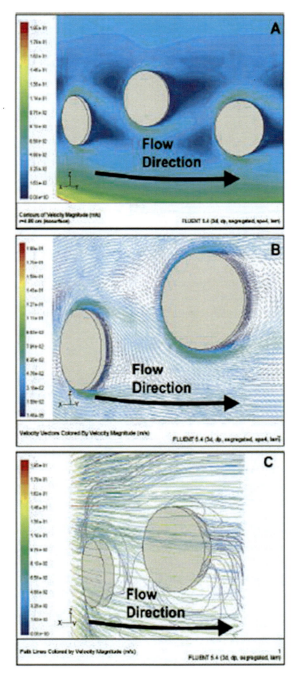

**Fig. 8** Bioreactor velocity profiles and cell trajectories. (**a**) Velocity magnitude contour plot showing overall flow pattern for radial isosurface bisecting the middle of the construct. (**b**) Velocity vectors for radial isosurface bisecting the middle of the construct showing velocity magnitude (*arrow colour*) and direction. (**c**) Pathlines of neutrally buoyant particles around constructs simulating cell trajectories during bioreactor seeding. Velocity magnitudes are colour coded according to the *scale on the left*. Reprinted with permission from [33]

transport to constructs is rather uniform, such that $O_2$ depletion near the downstream edge of constructs was noted with minimum $pO_2$ values near the constructs. From these computational modelling studies they concluded that these values are above oxygen concentrations in cartilage in vivo, suggesting that bioreactor oxygen concentrations likely do not affect chondrocyte growth.

## 3 Future Directions

Further progress in creating suitable environments for the growth of tissues requires an understanding of how chemical, mechanical and other environment factors influence cell proliferation, differentiation and production of ECM. Within this context, it is worthwhile to note how the available "mechanical" theories have focused on "what happens inside the tissue"; on the contrary numerical methods for simulating the biomechanical laws that govern so-called neo-tissue (cells and ECM) formation in terms of "surface incorporation/conversion conditions" (interface kinetics of the growth) still remain poorly developed. In practice from a numerical point of view, the growing biological specimen gives rise to a moving boundary problem. Moving boundary problems remain a challenging task for numerical simulation, prompting much research and leading to many different solutions.

As discussed above, CFD modelling is able to capture flow, pressure and concentration fields resolved at the scaffold's pore level. The simulations show how the scaffold micro-architecture influences hydrodynamic shear and nutrient concentrations imposed on cells attached to the surface of a scaffold. These studies provide a foundation for exploring the effects of dynamic flow on cell function and provide important insights into the design and optimization of 3D scaffolds suitable for use in bioreactors for in vitro tissue engineering. However, a major weakness of these models is that it is assumed that either no neo-tissue is within the interconnected pore space or that cells are attached only along the surfaces of the scaffold. Because of the complexity of the biological processes involved and a relative lack of experimental data needed for validation, a comprehensive tissue-growth model has not yet been presented. Some preliminary models are currently available in the literature. Cheng et al. [34] presented a 3D model based on cellular automata able to simulate the dynamics of a cell population taking into account cell division, migration, different seeding modes and contact inhibition. The model can be tuned by modifying the values of the parameters which govern the population dynamics. The work is a development of the 2D model by Lee et al. [35], for which preliminary experimental validation was performed. Nevertheless, some limitations common to the two papers have been recognized by the authors. Nutrients and growth factors are assumed to be uniform in space and time and only very simple geometries of the spatial domains are considered.

Chung et al. [36] adopted a differential formulation of the problem. An equation modelling cell mass conservation was defined and this equation included a source term representing cell proliferation and a diffusive term representing cell migration. Although the authors stated that the diffusive term is able to reproduce cell random walk, this approach is not as sophisticated and comprehensive as the cellular automata model (random walks, cell division and contact inhibition can be effectively accounted for). However, the differential formulation of the problem allowed for the inclusion of nutrient and oxygen transport and consumption. In a follow-up paper [37], cellular interaction, proliferation, nutrient consumption and culture medium circulation were investigated. The model incorporated modified Contois cell-growth kinetics that included effects of nutrient saturation and limited cell growth. Nutrient uptake was depicted through the Michaelis–Menton kinetics. To describe the culture medium convection, the fluid flow outside the cell-scaffold construct was described by the Navier–Stokes equations, while the fluid dynamics within the construct was modelled by Brinkman's equation for porous media flow. Effects of the media perfusion were examined by including time-dependant porosity and permeability changes due to cell growth. The overall cell volume was considered to consist of cells and extracellular matrices as a whole without treating the ECM separately. Numerical simulations showed that when cells were cultured subjected to direct perfusion, they penetrated deeper into scaffolds, which resulted in a more uniform spatial distribution. Cell numbers were increased by perfusion and ultimately approached an asymptotic value as the perfusion rates increased in terms of the dimensionless Peclet number that accounts for the ratio of nutrient perfusion to diffusion. In addition to enhancing the nutrient delivery, perfusion simultaneously imposes flow-mediated shear stress to the engineered cells. Shear stresses were found to increase with cell growth as the scaffold void space was occupied by the cell and ECM volumes. The macro average stresses increased from 0.2 mPa to 1 mPa at a perfusion rate of 20 µm s$^{-1}$ with the overall cell volume fraction growing from 0.4 to 0.7. This made the overall permeability value decrease from $1.35 \times 10^{-2}$ cm$^2$ to $5.51 \times 10^{-4}$ cm$^2$. The authors concluded that when relating the simulation results with perfusion experiments in the literature, the average shear stresses were found to be below the critical value that would induce the cell death of chondrocytes.

Lemon et al. [38] combined partial differential equations and the mixture theory, which was originally derived from the general theory of multiphase porous flow, where tissue, water and solid scaffold are modelled as different phases. Such a model is able to account for tissue–tissue, tissue–water and tissue–scaffold interactions. Hence, it can be hypothesized that a more realistic calculation of stresses and flow field can be performed. It can be argued that this approach constitutes a viable alternative to tissue-growth models based on cellular automata, however nutrient transport and consumption were not included in the model. It can be argued that this approach constitutes a viable alternative to tissue-growth models based on cellular automata, although nutrient transport and consumption were not included within the model.

Galbusera et al. [37] therefore developed a CFD model coupled with the cellular automation algorithm based on this background. This model studies in vitro tissue growth and is able to account for both cell migration and the cellular microenvironment, as "created" by the perfusion bioreactor system (e.g. nutrient and oxygen transport, mechanical stimuli). The group published a preliminary study in which conditions of static culture and of culture in a perfused bioreactor were simulated and compared. Oxygen diffusion, convective transport and consumption in combination with a simplified scaffold were incorporated into the model.

# 4 Conclusions

CFD modelling is one of the most effective methods used to characterize flow fields, provided that the models are validated by experimental velocimetry techniques, such as LDA and PIV. While these experimental methods are reliable, they are too arduous to characterize the complete three-dimensional fluid flow in a bioreactor. CFD modelling provides a powerful means to overcome these limitations and enable the full characterization of three-dimensional flow fields in bioreactors with simple and complex geometries.

CFD allow us to predict and possibly optimize design, flow, nutritional and metabolic requirements without having to perform numerous and expensive bioreactor experiments, potentially saving significant amounts of time and resources. The coupling of experimental methods and computational simulations forms a synergistic relationship that can potentially yield greater and yet, more cohesive data for future bioreactor studies. The next steps would likely involve simulating fluid–solid interactions, where cells and ECM, for example, may be modelled to deform or detach under fluidic forces under real-time conditions. A primary difficulty lies with the fact that cell membranes are visco-elastic in nature. With the further advancement of computing and software technologies, such challenging simulations may indeed become possible, and consequently increase our understanding of the impact of fluidic forces on tissue growth in bioreactors.

## References

1. Barron V, Lyons E, Stenson-Cox C, McHugh PE, Pandit A (2003) Bioreactors for cardiovascular cell and tissue growth: a review. Ann Biomed Eng 31(9):1017–1030
2. Portner R, Nagel-Heyer S, Goepfert C, Adamietz P, Meenen NM (2005) Bioreactor design for tissue engineering. J Biosci Bioeng 100(3):235–245
3. Kumar S, Wittmann C, Heinzle E (2004) Minibioreactors. Biotechnol Lett 26(1):1–10
4. Wendt D, Jakob M, Martin I (2005) Bioreactor-based engineering of osteochondral grafts: from model systems to tissue manufacturing. J Biosci Bioeng 100(5):489–494
5. Vunjak-Novakovic G, Meinel L, Altman G, Kaplan D (2005) Bioreactor cultivation of osteochondral grafts. Orthod Craniofac Res 8(3):209–218
6. Martin I., Wendt D, Heberer M (2004) The role of bioreactors in tissue engineering. Trends Biotechnol 22(2):80–86

7. Carrier RL, Rupnick M, Langer R, Schoen FJ, Freed LE, Vunjak-Novakovic G (2002) Perfusion improves tissue architecture of engineered cardiac muscle. Tissue Eng 8(2):175–188
8. Kimbell JS, Subramaniam RP (2001) Use of computational fluid dynamics models for dosimetry of inhaled gases in the nasal passages. Inhal Toxicol 13(5):325–334
9. Grotberg JB (2001) Respiratory fluid mechanics and transport processes. Annu Rev Biomed Eng 3:421–457
10. Redaelli A, Boschetti F, Inzoli F (1997) The assignment of velocity profiles in finite element simulations of pulsatile flow in arteries. Comput Biol Med 27(3):233–247
11. Varghese SS, Frankel SH (2003) Numerical modeling of pulsatile turbulent flow in stenotic vessels. J Biomech Eng 125(4):445–460
12. Yoganathan AP, He Z, Casey Jones S (2004) Fluid mechanics of heart valves. Annu Rev Biomed Eng 6:331–362
13. Wootton DM, Ku DN (1999) Fluid mechanics of vascular systems, diseases, and thrombosis. Annu Rev Biomed Eng 1:299–329
14. Sucosky P, Osorio DF, Brown JB, Neitzel GP (2004) Fluid mechanics of a spinner-flask bioreactor. Biotechnol Bioeng 85(1):34–46
15. Venkat RV, Stock LR, Chalmers JJ (1996) Study of hydrodynamics in microcarrier culture spinner vessels: a particle tracking velocimetry approach. Biotechnol Bioeng 49:456–466
16. Porter B, Zauel R, Stockman H, Guldberg R, Fyhrie D (2005) 3-D computational modeling of media flow through scaffolds in a perfusion bioreactor. J Biomech 38(3):543–549
17. Raimondi MT, Boschetti F, Falcone L, Migliavacca F, Remuzzi A, Dubini G (2004) The effect of media perfusion on three-dimensional cultures of human chondrocytes: integration of experimental and computational approaches. Biorheology 41(3–4):401–410
18. Lappa M (2005) A CFD level-set method for soft tissue growth: theory and fundamental equations. J Biomech 38(1):185–190
19. Boschetti F, Raimondi MT, Migliavacca F, Dubini G (2006) Prediction of the micro-fluid dynamic environment imposed to three-dimensional engineered cell systems in bioreactors. J Biomech 39(3):418–425
20. Andrade JS, Almeida MP, Filho JM, Havlin S, Suki B, Stanley HE (1997) Fluid flow through porous media: the role of stagnant zones. Phys Rev Lett 79(20):3901–3904
21. Singh H, Teoh SH, Low HT, Hutmacher DW (2005) Flow modelling within a scaffold under the influence of uni-axial and bi-axial bioreactor rotation. J Biotechnol 119(2):181–196
22. Singh H, Ang ES, Lim TT, Hutmacher DW (2007) Flow modeling in a novel non-perfusion conical bioreactor. Biotechnol Bioeng 97(5):1291–1299
23. Zhao F, Chella R, Ma T (2007) Effects of shear stress on 3-D human mesenchymal stem cell construct development in a perfusion bioreactor system: experiments and hydrodynamic modeling. Biotechnol Bioeng 96(3):584–595
24. Jiang BH, Semenza GL, Bauer C, Marti HH (1996) Hypoxia-inducible factor 1 levels vary exponentially over a physiologically relevant range of O2 tension. Am J Physiol 271(4 Pt 1):C1172–C1180
25. Toh YC, Zhang C, Zhang J, Khong YM, Chang S, Samper VD, van Noort D, Hutmacher DW, Yu H (2007) A novel 3D mammalian cell perfusion-culture system in microfluidic channels. Lab Chip 7(3):302–309
26. Ye H (2006) Modelling nutrient transport in hollow fibre membrane bioreactors for growing three-dimensional bone tissue. J Membr Sci 272:169–178
27. Pollack SR, Meaney DF, Levine EM, Litt M, Johnston ED (2000) Numerical model and experimental validation of microcarrier motion in a rotating bioreactor. Tissue Eng 6(5):519–530
28. Botchwey EA, Pollack SR, Levine EM, Laurencin CT (2001) Bone tissue engineering in a rotating bioreactor using a microcarrier matrix system. J Biomed Mater Res 55(2):242–253
29. Bueno EM, Bilgen B, Barabino GA (2005) Wavy-walled bioreactor supports increased cell proliferation and matrix deposition in engineered cartilage constructs. Tissue Eng 11(11–12):1699–1709

30. Bilgen B, Chang-Mateu IM, Barabino GA (2005) Characterization of mixing in a novel wavy-walled bioreactor for tissue engineering. Biotechnol Bioeng 92(7):907–919
31. Bilgen B, Barabino GA (2007) Location of scaffolds in bioreactors modulates the hydrodynamic environment experienced by engineered tissues. Biotechnol Bioeng 98(1):282–294
32. Dusting J, Sheridan J, Hourigan K (2006) A fluid dynamics approach to bioreactor design for cell and tissue culture. Biotechnol Bioeng 94(6):1196–1208
33. Williams KA, Saini S, Wick TM (2002) Computational fluid dynamics modeling of steady-state momentum and mass transport in a bioreactor for cartilage tissue engineering. Biotechnol Prog 18(5):951–963
34. Cheng G, Youssef BB, Markenscoff P, Zygourakis K (2006) Cell population dynamics modulate the rates of tissue growth processes. Biophys J 90(3):713–724
35. Lee Y, Kouvroukoglou S, McIntire LV, Zygourakis K (1995) A cellular automaton model for the proliferation of migrating contact-inhibited cells. Biophys J 69(4):1284–1298
36. Chung CA, Yang CW, Chen CW (2006) Analysis of cell growth and diffusion in a scaffold for cartilage tissue engineering. Biotechnol Bioeng 94(6):1138–1146
37. Galbusera F, Cioffi M, Raimondi MT, Pietrabissa R (2007) Computational modeling of combined cell population dynamics and oxygen transport in engineered tissue subject to interstitial perfusion. Comput Methods Biomech Biomed Eng 10(4):279–287
38. Lemon G, King JR (2007) Multiphase modelling of cell behaviour on artificial scaffolds: effects of nutrient depletion and spatially nonuniform porosity. Math Med Biol 24(1):57–83

# Fluid Dynamics in Bioreactor Design: Considerations for the Theoretical and Practical Approach

**B. Weyand, M. Israelowitz, H.P. von Schroeder, and P.M. Vogt**

**Abstract** The following chapter summarizes principles of fluid dynamics in bioreactor design with a focus on mammalian cell-culture systems.

**Keywords** Bioreactor, Computational fluid dynamics, Flow pattern, Fluid dynamics, Laminar flow, Shear stress, Perfusion system, Pulsatile flow, Rotating vessel, Stirring system

## Contents

1  Introduction ............................................................................................ 252
2  Fundamentals of Fluid Mechanics ........................................................ 252
3  Effects of Fluid Dynamics on Mammalian Cells ................................. 254
4  Types of Bioreactor and Their Underlying Fluid Mechanics .............. 256
    4.1  Stirring Systems ............................................................................. 256
    4.2  Perfusion Systems .......................................................................... 257
    4.3  Rotating Systems ........................................................................... 258
    4.4  Pulsatile Systems ........................................................................... 263
    4.5  Other Systems ................................................................................ 263
5  Measurement and Calculation of Fluid Dynamics in Bioreactors ....... 264
    5.1  Specific Manometer Methods ........................................................ 264
    5.2  Laser Doppler Anemometry .......................................................... 264
    5.3  Particle Dynamics Analysis ........................................................... 265

---

B. Weyand(✉), P.M. Vogt
Department of Plastic, Hand and Reconstructive Surgery OE 6260, Hannover Medical School, Carl-Neubergstr. 1, 30625, Hannover, Germany

M. Israelowitz, H.P. von Schroeder
Biomimetics Technologies Inc., 191 Ellis Avenue, Toronto, Ontario ON M6S 2X4 Canada

H.P. von Schroeder
Department of Surgery, University Hand Program and Bone Lab, University of Toronto, Toronto Western Hospital, 399 Bathurst Street 2E, Toronto, ON M5T 2S8, Canada

5.4 Particle Image Velocimetry (PIV) .................................................................. 265
5.5 Planar Laser-Induced Fluorescence (PLIF) ................................................. 265
5.6 Computational Fluid Dynamics .................................................................. 265
6 Perspective ............................................................................................................ 266
References .................................................................................................................. 266

# 1 Introduction

The development of bioreactors began with the introduction of vessel shakers for the cultivation of microorganisms that where used primarily for antibiotic screening tests and also antibiotic production by bacteria in the 1940s and 1950s. With the observation that oxygen transfer of the bacterial cultures depended on the speed of the shaker, the flask angle and the orbit of the vessel, a basis was established for the development of the first fermenters. Fermentation plants were developed for biochemical engineering and industrial microbiology, and these still comprise the major contingent of bioreactor designs. But as biotechnology expands, together with the evolving field of tissue engineering, different vessels have been designed and built for the special demands of mammalian cell and tissue culture including culturing mammalian cells in two- and three-dimensional structures. Important factors in bioreactor design include enzyme kinetics to keep pH equilibrium and temperature constant, mass transfer at interfaces, and substrate and product interrelationships for cell growth and fluid dynamics.

Herein we will address the important role of fluid dynamics in bioreactor design: starting with the fundamental principles, we will explain the importance of fluid dynamics in cell proliferation and differentiation, describe general bioreactor types and their underlying fluid mechanics, and discuss methods to measure and calculate flow in bioreactor systems. We cannot cover every single bioreactor model ever invented, patented or currently on the market, but intend to point out the importance of fluid dynamics in modern bioreactor design. Given the rapid expansion of the tissue engineering field and the focus on mammalian cells, we will not include bioreactor types and designs such as fermentation plants and photo-bioreactors for plant cell cultures in detail. Instead we will focus on specific bioreactor models for mammalian cell culturing and tissue engineering with special attention to the fundamentals of fluid mechanics.

# 2 Fundamentals of Fluid Mechanics

The term "fluid" comprises fluids (liquids) and gases. Fluid mechanics consists of two fields: fluid statics (such as hydrostatics and aerostatics) and fluid dynamics (such as hydrodynamics and aerodynamics). The scientific basis of fluid dynamics is derived from the conservation laws of classical mechanics and thermodynamics,

# Fluid Dynamics in Bioreactor Design

such as conservation of mass (Lomonosov–Lavoisier law), conservation of energy (first law of thermodynamics) and conservation of momentum (Newton's second law of motion). Further aspects are derived from the second law of thermodynamics, the equation and moment of momentum, as well as similarity laws and laws of friction.

Fluids have certain properties which are defined by temperature, specific volume and pressure. The temperature ($T$) is a unit independent from mass. The specific density ($\rho$) of any fluid in a certain phase (e.g., liquid phase, gas phase) can be derived from the mass ($m$) and the volume ($V$):

$$\rho = \frac{m}{V}$$

The reciprocal of the specific density gives the specific volume (v):

$$v = \frac{1}{\rho}$$

The pressure ($p$) is defined as the normal force ($\Delta F$) acting per area ($\Delta A$):

$$p = \frac{\Delta F}{\Delta A}$$

In fluid mechanics, fluids can be divided into Newtonian and non-Newtonian fluids. Newtonian fluids are incompressible. They follow the law of friction and begin to flow when a minimal force acts on them tangentially.

In order to have flow movements, a shear force has to overcome the frictional resistance. When a fluid flows through a pipe, the shear forces are stronger at the boundary layers where the fluid movement is slower than in the middle. The different layers of the fluid have different velocities, and the shear forces are affected by the viscosity. The viscosity is a transport property of fluid; it is a measure of the fluid characteristic to transfer impulses.

In fluid mechanics, the dynamic viscosity is independent of the pressure, but dependent on the temperature: in liquid fluids the dynamic viscosity decreases by increasing temperature. The dynamic viscosity is simply a proportional constant of the specific fluid and its phase.

$$\eta = \frac{\tau \cdot s}{A \cdot c}$$

($\tau$=shear stress, $A$=area, $s$=distance, $c$=velocity) The kinematic viscosity is defined as

$$v = \frac{\eta}{\rho}$$

In addition, in hydrostatics, a surface tension is acting at any liquid–gas or liquid–liquid interface if both liquids possess different densities.

On the basis of the three conservation laws (conservation of energy, conservation of mass and conservation of momentum), the Navier—Stokes equations have been developed as differential equations describing the state and the motion in time of fluids (Claude-Luis Navier, 1822; George Gabriel Stokes, 1845). For more detailed description of these differential equations and derivates, please see standard textbooks on fluid mechanics and computational fluid dynamics [1, 2].

Further properties of fluids are the so-called fluid patterns. Here we distinguish between laminar flow and turbulent flow. In laminar flow, the fluid particles move parallel to the movement axis, and there is no movement in the right angle to the axis. In turbulent flow, fluid particles have a main movement axis, which is disturbed by further components moving in different directions. The ratio of inert forces to shear forces is described by the Reynolds number, which can be used to distinguish between laminar and turbulent flow. A Reynolds number lower than 2,320 describes a laminar flow; if the Reynolds number is greater than 2,320, the flow is turbulent [2, 3]. Under certain conditions, laminar flow is only considered up to the Reynolds number 1,000 [4].

The turbulent flow pattern is difficult to define: it includes characteristics such as irregularity, diffusivity, instability, three-dimensional features and dissipation of turbulent kinetics energy. Turbulent flow has been defined by Hinze (1975) as: "Turbulent fluid motion is an irregular condition of flow in which the various quantities show a random variation with time and space coordinates, so that statistically distinct average values can be discerned" [5]. The three hypotheses by Kolmogorov (hypothesis of local isotropy of small scale motions and Kolmogorov's first and second similarity hypothesis) describe a theoretical concept on which modelling theories of turbulent fluids are built [5].

The task of describing fluid mechanics of porous media was first reported by Darcy (1856) who discovered that the fluid velocity per area through a column of porous material is proportional to the pressure gradient maintained along the column [6]. More advanced models have been developed for the modelling of transition from laminar to turbulence in porous media in recent years [6].

## 3 Effects of Fluid Dynamics on Mammalian Cells

Cellular physiology is influenced by external stimuli, such as extracellular matrix properties, mechanical stresses, microgravity and nutritional and oxygen availability. Mammalian cells respond to external stresses by changes in proliferation, migration, differentiation, metabolism or even cell death [7–9]. Animal cells have lower metabolic rates and oxygen demands compared to yeast and bacteria, but their tolerance to fluid forces is also much lower. This has implications for bioreactor design since the high stirring and agitation rates used for microbial cultures to provide sufficient oxygenation will lead to apoptosis and cell death in mammalian cell cultures due to fluid turbulences. Cell exposure to fluid shear stresses of 10–100 dyne cm$^{-2}$ leads to a death rate of 20–80% after 10min [8]. Hybridoma cells are more resistant to fluid shear stresses compared to cells grown on surfaces of microcarriers. To reduce wall shear stresses, macromolecules such as serum proteins or polymeric surfactant molecules (Pluronics F68) can be added to the medium in order to increase survival rates of cells [8]. To improve oxygen supply to the cells, some approaches utilize aeration of the culture medium with gas, which may create new problems: gas bubbles rupturing at liquid surfaces lead to increased cell death due to release of surface energy from the bubbles: this also can be reduced in the presence of the non-ionic surfactants mentioned above [8, 10, 11].

The biological effects of external fluid forces also depend on the cell type. Cells that are exposed to significant haemodynamic forces in vivo, such as endothelial cells or smooth muscle cells, respond differently to shear stresses in vitro than hepatocytes or haematopoietic stem cells for example. Endothelial cells that line our blood vessels are exposed to changing blood pressure and shear stresses by blood flow and as a result have developed protective feedback mechanisms in order to maintain cellular homeostasis [12]. Mechanical forces activate sensors in the cell membrane, directly or through receptor ligands on intracellular signalling pathways to affect transcription factors in the nucleus that further regulate gene expression and protein translation [12]. Mechanosensors of the endothelial cell membrane are membrane proteins such as receptor tyrosine kinases (e.g., the receptor for vascular endothelial growth factor: Flt-1), integrins, G proteins and G-coupled receptors, $Ca^{2+}$ channels and intercellular junction proteins, membrane lipids and also the glycocalix [12]. Studies of endothelial cells have demonstrated differing cellular responses depending on the nature of the flow: endothelial cell proliferation is down-regulated under steady shear stress conditions ("laminar flow") by activation of factors such as the tumour suppressor protein p53 leading consecutively to cell cycle arrest. In contrast, endothelial cells at branching points where flow disturbances are present, show increased proliferation and cell turnover [12].

Shear stresses with a significant forward direction lead to activation of monocyte chemotactic protein-1 in endothelial cells (a protector against atherogenesis). Here both constant laminar shear stresses and pulsating flow with an oscillating component show similar effects. These effects are transient but time-dependant, and include for example the sequential activation of factors of the Ras-mitogen activated protein kinases pathway with a short-term up-regulation followed by a long-term down-regulation [12]. In contrast, survival factors, such as Krüppel-like factor-2, are significantly up-regulated in endothelial cells exposed to both pulsating and reciprocating oscillating flow, but not when exposed to steady laminar flow without oscillations [12].

Fluid dynamics in bioreactors can be used in order to maintain a certain cell phenotype either by enhancing proliferation and expansion or by initiating differentiation processes. Suspension cultures have been under widespread use for a variety of cells: non-anchorage-dependent cells such as haematopoietic stem cells and mesenchymal stem cells can be cultured in free suspension; whereas anchorage-dependent cells such as chondrocytes require to be cultured as aggregates or seeded on microspheres or microcarriers to enhance the growth-surface-to-reactor-volume [13–17].

Studies on cellular differentiation stimulated by external forces have mainly focused on the effects of mechanical stresses, such as stretch or strain application, but in recent bioreactor models the influence of fluid shear stresses on the cell phenotype has been recognized. Chrondrogenic cell-matrix constructs have been shown to preserve their differentiation state under laminar flow, whereas they develop a fibrous capsule when exposed to turbulent flow conditions and dedifferentiate in a hydrostatic environment [18]. Osteoblastic cells respond to short-term exposure to fluid shear with changes in intracellular calcium and release of nitric oxide, alteration in the cell cytoskeleton and up-regulation of several transcription and intracellular messenger factors [19–22]. Long-term exposure to shear stress can initiate and enforce differentiation processes in osteoblastic cells, resulting in elevation of early and late osteoblast differentiation markers and an increased amount of mineralized matrix deposition [19–22].

## 4 Types of Bioreactor and Their Underlying Fluid Mechanics

For tissue engineering, several different types of bioreactors have been developed and some are already commercially available. These can be roughly categorized under the main headings: stirring systems, perfusion systems, rotating systems, pulsatile flow systems and others.

### *4.1 Stirring Systems*

#### 4.1.1 Spinner Flask

The basic stirring system is the spinner flask. A spinner flask consists of a cylindrical glass container with a stirring element (stirring rod/propeller/impeller) at the bottom of the vessel. It demonstrates advantages to static cell culture by mixing the culture environment around the cells and enhancing oxygen and nutrition supply. Here concentration gradients can occur between the feeding zone in the upper part of the vessel, where the main nutrition consumption occurs, and the impeller zone, where the oxygen mass transfer predominantly takes place [23]. Constant rotating of the rod is required to keep cells in suspension. Two general types of impellers exist: Impellers that generate axial flow (which goes parallel to the impeller shaft) and impellers that generate radial flow (e.g., a Rushton turbine) [5]. Fluid dynamics also depend on the size of the impeller and the rotating speed of the stirring rod, from which the Reynolds number can be calculated [24].

$$\mathrm{Re} = \frac{\rho N L^2}{\mu}$$

(Re=Reynolds number; $\rho$=medium density; $N$=rotating speed of the stirrer rod; $L$=rod length; $\mu$=dynamic viscosity)

If the Reynolds number is less than the critical size of 1,000, the flow is assumed to be laminar, if it is higher than 1,000, the flow is turbulent [4].

Since the fluid movement goes from the rotating rod towards the wall, then upwards and returns down near the axis, a recirculating flow is created above the stirring rod, which becomes stronger at higher Reynolds numbers [4, 25]. This recirculating flow pattern can enhance medium mixing and therefore improve oxygenation, but may also exert centrifugal forces driving suspended cells against the vessel wall. The hydrodynamic forces that damage the cells arise from velocity gradients, which can be expressed in the energy dissipation rate. The maximal local energy dissipation rate is located close to the tip of the rod.

The requirement for aeration of the medium by addition of gas (oxygen) can become inevitable for larger-sized spinner flasks. Here fluid dynamics may be affected by gas bubbles, which can damage cells by cell membrane rupture due to bubble-cell attachment [10, 53]. The airlift bioreactors even work without impellers, relying on the buoyancy of the circulating air bubbles for medium mixing, but coalescence or collapse of air bubbles produce high shear stresses as outlined above.

### 4.1.2 Wavy-Walled Bioreactor (WWB)

To reduce turbulent stresses that may damage the cells, a modified form of the spinner flask, the wavy-walled bioreactor has been developed [26, 27]. The cross-section of the wavy-walled bioreactor resembles a six-pointed star with smoothed edges. Mixing of the medium is carried out by means of a magnetic stir bar at the bottom of the flask. Cell-matrix-constructs can be placed within the lobes or centrally. The flow field differs significantly from traditional spinner flasks. By calculation of computational fluid dynamics of the different models, the velocity vectors around the constructs change in their magnitude from spinner flasks to the wavy-walled bioreactor. The spinner flasks show the highest numbers for tangential velocity and the lowest numbers for axial velocity, which is reversed in the WWB with lobe-placed scaffolds. The maximum shear stress on the constructs is lower in the wavy-walled bioreactor compared to the spinner flasks at the same rotation speed. Here, there is also a difference in shear stress distribution with lower numbers at constructs in the upper part of the wavy-walled bioreactor compared to constructs situated close to the bottom of the vessel [26].

## 4.2 Perfusion Systems

Perfusion cultures have been developed to enhance cell growth within three-dimensional matrices. Here, a matrix is often held in chambers, columns or special cartridges [28]. Flow patterns depend on matrix material, size and distribution of pores, shape of the vessel and mode of perfusion. Perfusion systems have been found to be superior to hydrostatic systems in that they improve cell distribution (especially within the inner core of the chosen matrix), cell proliferation and cell differentiation [28]. Cellular alignment appears along the perfusion axis. Some reactor models allow dynamic cell seeding which showed a higher efficiency rate and a more homogeneous cell distribution within the scaffold compared to static seeding in a dish prior to cultivation [29].

We have recently developed a novel perfusion bioreactor for culturing tissue-engineering bone constructs under laminar flow conditions [30]. The design of the model is based on computational fluid dynamics (Figs. 1–6), and different porosities of the scaffold have been integrated into the calculations.

### 4.2.1 Estimation of Fluid Dynamics within Porous Scaffolds in Perfusion Bioreactors

Although many perfusion bioreactors intend to achieve laminar flow patterns, local shear stresses and turbulences may arise depending on bioreactor configuration, medium flow rate, dynamic viscosity and porosity and micro-architecture of the scaffolds. New computational models have demonstrated highly non-uniform flow through porous scaffolds with varying porosity where highest speeds appear at the centre of small orifices [31, 32]. Cioffi calculated fluid dynamics in a micro-computed tomography (CT)-based model and found similar numbers for shear stresses when using the CT-based model or a simple estimation for scaffolds with interconnected pores and at low Reynolds numbers [33]. Others have modelled perfusion through a porous scaffold using the Brinkman and continuity equation for describing flow through porous media and the Navier–Stokes equation for the flow in the fluid layers [34].

### 4.2.2 Hollow Fibre Bioreactor

A specialized perfusion system is the hollow fibre bioreactor for artificial liver or kidney replacement. Here fluid turbulences occur mainly at branching points but other issues such as hypo-perfusion of cell channels and pressure drop in the shell region between cell-packed and cell-free cases can also occur [35, 36]. An increase in flow rate or fluid viscosity, or doubling the number of capillaries all lead to increased shear stress along the channels [37]. A new model of computer-controlled microfluid cell culture has been recently described, where fluid through a capillary-like network is being controlled by piezoelectric, movable pins using a Braille display for positioning and controlling the pins [38].

## 4.3 Rotating Systems

### 4.3.1 Rotating-Wall Perfused-Vessel Bioreactor

The rotating wall bioreactor has been developed by NASA/JSC from the viscous pump bioreactor of the University of Houston [39, 40]. The design should provide

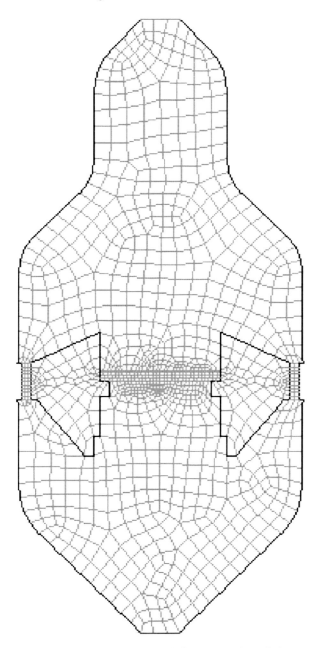

**Fig. 1** Generating a suitable grid model is the basis for computational fluid dynamics

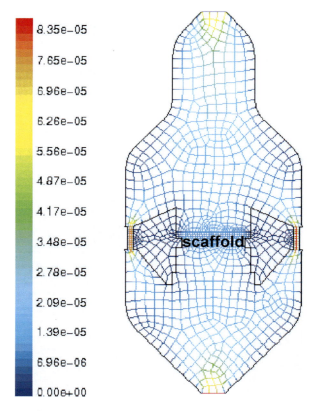

**Fig. 2** Distribution of different velocity magnitudes (m s$^{-1}$) calculated for the model

anchorage-dependent mammalian cells a shear-reduced, microgravity environment. Cells are grown on microcarriers suspended in the medium. The vessel consists of a concentric outer and inner cylinder, which can be independently rotated. The flow comes from the cylinder base and exits through pores into the inner cylinder. Primary flow movement is in the azimuthal direction; secondary flow patterns are more complex (three-fluid-cell pattern). Depending on the rotation speed and direction, a laminar Couette flow can change into turbulent Taylor-vortex flow regimes [41]. To create a microgravity environment, differential rotation is required: the outer cylinder rotates at a slow speed, whereas the inner cylinder spins at a faster speed [40]. Thereby, a radial-axial flow is created, which increases by higher differential rotation rates. The centrifugal forces are increased by higher rotation rates, driving cells towards the outer wall. On the other hand, as cells grow and form aggregates, rotation rate has to be increased to maintain cells in suspension and microgravity conditions. If the bioreactor is used in space, rotation is not required for cell suspension, but some fluid motion for mass transport is still necessary.

Fluid Dynamics in Bioreactor Design 261

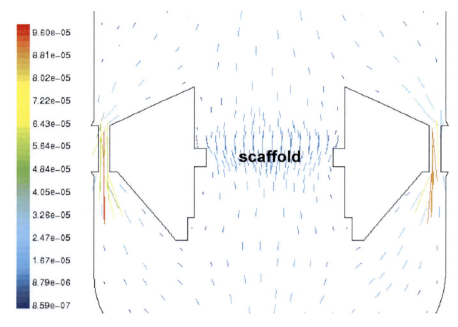

**Fig. 3** Vector distributions describing the different velocity magnitudes (in m s$^{-1}$) in the core vessel

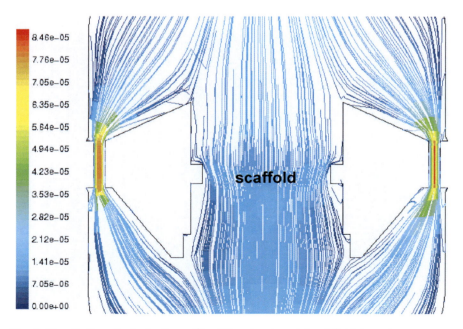

**Fig. 4** Visualization of velocity (in m s$^{-1}$) pathlines by computational fluid dynamics

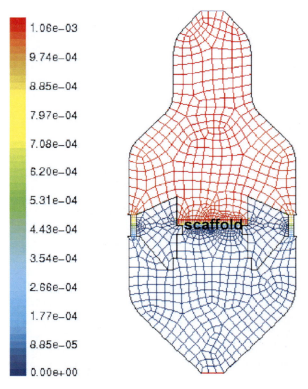

**Fig. 5** Pressure distribution (in Pascal) in the CFD model during tissue growth in the scaffold

**Fig. 6** Prototype of a new laminar flow perfusion bioreactor during its testing phase

Fluid Dynamics in Bioreactor Design

Various modified forms of the rotating-wall perfused-vessel bioreactor, such as the slow lateral turning vessel and the high aspect ratio vessel, also exist.

### 4.3.2 Rotating Shaft Bioreactor

In the rotation shaft bioreactors, the scaffolds are attached to an inner shaft that rotates around its axis within a vessel that is generally a cylindrical shape. The cylinder can be either completely filled with medium (Biostat® RBS, B. Braun Biotech.) [42, 43] or can consist of a liquid and a gas phase [44]. From the latter, a computational fluid dynamic model revealed quite low maximal and mean shear stresses acting on the matrix at 10rpm, which are significantly lower than forces at a comparable speed in any of the rotating-wall vessel bioreactors, the spinner flask or the concentric cylinder bioreactor [44].

## 4.4 Pulsatile Systems

### 4.4.1 Pulsatile Flow Reactor

Pulsatile flow bioreactors have been developed to mimic cardiovascular conditions in vitro for the culture of vascular grafts or heart valves in vitro. Systems are either run by a pulsatile pump [45], or pulsatile flow waves that are created by an elastic membrane which is being inflated and deflated by an air pump or a ventilator [46, 47]. Newer developments contain sophisticated feedback control systems to control changes in pressure and flow waveforms during sterile cultivation [48]. Flow patterns are laminar to turbulent, depending on parameters, pulsation rate and pressure control.

## 4.5 Other Systems

### 4.5.1 Cell Cube Bioreactor

The cell cube bioreactor is a disposable bioreactor that has been developed for the production of vaccine and other gene therapeutic tools [49]. The cell cube bioreactor has a square cross-section with medium inflow at the bottom and medium outflow at the top in the opposite corner. Cells are grown on stacked parallel plates with 1mm spacing. Fluid flow is laminar throughout the reactor with Reynolds numbers from 1 to 160. Inside the reactor there is a high heterogeneity of flow velocity, and high shear stresses are found in certain regions such as the top of the channels which corresponds with low cell growth.

### 4.5.2 Hydrostatic Pressure/Perfusion Culture System

Wanatabe introduced a pressure-proof chamber with a V-shaped hanger for cultivation of cells under specific high-frequency (<0.5Hz) hydrostatic pressure conditions and additionally constant laminar flow with minimal fluid velocity for three-dimensional tissue engineering [50].

### 4.5.3 Bottom-Driven Cylinders with Free-Surface

Dusting described a rotating bottom-driven cylinder with free-surface as a new geometrical approach for bioreactor design based on fluid dynamic measurements that combines advantages of the spinner flask and the rotating-wall perfused vessel [51].

## 5 Measurement and Calculation of Fluid Dynamics in Bioreactors

### 5.1 Specific Manometer Methods

Different methods have been used to calculate and measure fluid flows. Henri de Pitot (1695–1771) developed the Pitot tube to measure pressure differences in the flow of rivers and calculated the water velocities at single points. Ludwig Prandtl (1875–1953) constructed the so-called Prandtl tube (Pitot static tube), a manometer that measures pressure differences within a flow field from which the flow speed also can be calculated. Another method is the Venturi flowmeter, which was designed to measure the speed of fluids in a pipe. The pipe narrows from a larger to a smaller constricted cross-sectional area: the difference in pressures between the two regions is determined by differences in heights of the marker in the U-shaped manometer that connects the two regions. From these data the speed of the fluid in the pipe can be calculated using the Bernoulli equation [52].

### 5.2 Laser Doppler Anemometry

Laser Doppler Anemometry (LDA) is an optical measuring technique to determine velocities of fluids at a single point. The underlying physical principle of this method is the Doppler effect: Laser light waves interact with a moving observer or the modulation of waves is perceived by a stationary observer from a moving emitter. An advantage of the LDA is that the system is non-intrusive and therefore does not affect the flow field. The data obtained by LDA consist of a time series of instantaneous velocity measurements at each measuring position, which gives more data points for fast moving particles than for slow particles. For correction, the arithmetic weighting factor can be used under certain conditions [5].

## 5.3 Particle Dynamics Analysis

Particle Dynamics Analysis is an extension of LDA and based on phase Doppler principles. Here, two or more detectors capture the optical interference pattern generated by light scattering by the single particles moving through the measured volume.

## 5.4 Particle Image Velocimetry (PIV)

Particle Image Velocimetry (PIV) is another optical measuring technique, where a whole velocity field can be measured at one time point. The fluid flow is seeded with particles or droplets, and their movement is recorded by cameras taking images of a light sheet of the region of interest at known time intervals. For illumination often a high-energy pulsed laser is used (e.g., a Nd:YAG laser with a pulse duration of 5–10ns). The scattered light is detected by CCD (charged coupled device) cameras. Seeding particles can be metal-coated particles (12μm diameter), metal-coated hollow-glass spheres (10μm), or silica particles (3.5, 10, 12μm).

## 5.5 Planar Laser-Induced Fluorescence (PLIF)

Planar Laser-Induced Fluorescence (PLIF) is an optical measuring technique used to obtain instant whole-field concentration or temperature maps in liquid flows. A fluorescent dye is added to the liquid fluid, excited, and the fluorescent light intensity is measured by the camera and further processed. A dye for temperature measurement is rhodamine B, for concentration measurement rhodamine 6G can be used, and fluorescin disodium is suitable for both, temperature and concentration measurements.

## 5.6 Computational Fluid Dynamics

Computational fluid dynamics is a computer-based method that brings together methods of fluid dynamics and numerical analysis to simulate flow patterns, velocities and other aspects of fluid mechanics and to solve complex differential equations of mathematical fluid models. The calculations are based on the generation of a suitable numerical grid model, which can be very difficult to obtain when dealing with complex geometries. The validation of results derived from computational fluid dynamics depends on the analysis of discretization, iteration and modelling errors, the quality of the numerical grid and the detection of programming and user errors [1].

# 6 Perspective

Increasing knowledge about the effects of fluid dynamics on physiology, proliferation and differentiation of mammalian cells has provided tremendous impact on the development of bioreactors for tissue engineering. New advances in the theoretical background and application technologies of fluid dynamics such as computational fluid dynamics can help to improve design and construction of modern bioreactors. By computer simulation of flow vectors, velocity distribution and pressure gradients, problems of bioreactor design such as turbulent flow patterns can be identified prior to the experimental testing phase of the prototype. Computational fluid dynamics allows adjusting the bioreactor design for optimal culture conditions of specific mammalian cells and tissue-engineering constructs. Modern software for computational fluid dynamics can achieve three-dimensional fluid dynamics models, which help to design even more precise bioreactor models for tissue engineering.

## References

1. Ferzinger JH, Peri M (2002) Computational methods for fluid dynamics, 3rd edn. Springer-Verlag, Berlin Heidelberg New York
2. Von Böckh P (2004) Fluidmechanik. 2. Auflage, Springer-Verlag, Berlin Heidelberg New York
3. Gersten K (1991) Einführung in die Strömungsmechanik, 6. überarbeitete Auflage. Herausgeber Th. Lehmann, Verlag Vieweg, Braunschweig, Wiesbaden
4. Nagata S (1975) Mixing-principles and applications. Wiley-Verlag, New York
5. Svensson FJE (2005) Fluid dynamics in stirred vessels — experiments and simulations of single-phase and liquid-liquid systems. Chalmers University of Technology, Chalmers Reproservice, Göteburg, Sweden
6. Bejan A, Dincer I, Lorente S, Miguel AF, Reis AH (2004) Porous and complex flow structures in modern technologies. Springer-Verlag, Berlin Heidelberg New York
7. Ludwig A, Kretzmer G (1993) Shear stress induced variation of cell condition and productivity. J Biotechnol 27:217–223
8. Cherry RS (1993) Animal cells in turbulent fluids: details of the physical stimulus and the biological response. Biotechnol Adv 11:279–299
9. Olivier LA, Yen J, Reichert WM, Truskey GA (1999) Short-term cell/substrate contact dynamics of subconfluent endothelial cells following exposure to laminar flow. Biotechnol Prog 15:33–42
10. Meier SJ, Hatton A, Wang DIC (1999) Cell death from bursting bubbles: role of cell attachment to rising bubbles in sparged reactors. Biotechnol Bioeng 62(4):468–478
11. Dey D, Emery AN (1999) Problems in predicting cell damage from bubble bursting. Biotechnol Bioeng. 65(2):240–245
12. Chien S (2007) Mechanotransduction and endothelial cell homeostasis: the wisdom of the cell. Am J Physiol Heart Circ Physiol 292:H1209–H1224

13. Baksh D, Davies JE, Zandstra PW (2003) Adult human bone marrow-derived mesenchymal progenitor cells are capable of adhesion-independent survival and expansion. Exp Haematol 31(8):723–732
14. Croughan MS, Hamel J-F, Wang DIC (1987) Hydrodynamic effects on animal cells grown in microcarrier cultures. Biotechnol Bioeng 29:130–141
15. Nielsen LK (1999) Bioreactors for hematopoietic cell culture. Annu Rev Biomed Eng 01:129–152
16. Van Wezel AL (1967) Growth of cell-strains and primary cells on micro-carriers in homogenous culture. Nature 216(5110):64–65
17. Youn BS, Sen A, Behie LA, Girgis-Garbado A, Hassell JA (2006) Scale-up of breast cancer stem cell aggregate cultures to suspension bioreactors. Biotechnol Prog 22(3):801–810
18. Vunjak-Novakovic G, Meinel L, Altman G, Kaplan D (2005) Bioreactor cultivation of osteochondral grafts. Orthod Craniofacial Res 8:209–218
19. Kreke MR, Huckle WR, Goldstein AS (2005) Fluid flow stimulates expression of osteopontin and bone sialoprotein by bone marrow stromal cells in a temporally dependent manner. Bone 36:1047–1055
20. Sikavitas VI, Bancroft GN, Holtorf HL, Jansen JA, Mikos AG (2003) Mineralized matrix deposition by marrow stromal osteoblasts in 3D perfusion culture increases with increasing fluid shear forces. Proc Natl Acad Sci U S A 100(25):14683–14688
21. Wang Y, Uemura T, Dong J, Kojima H, Tanaka J, Tateishi T (2003) Application of perfusion culture system improves in vitro and in vivo osteogenesis of bone marrow-derived osteoblastic cells in porous ceramic materials. Tissue Eng 9(6):1205–1214
22. Yu X, Botchwey EA, Levine EM, Pollack SR, Laurecin CT (2004) Bioreactor-based bone tissue engineering: the influence of dynamic flow on osteoblast phenotypic expression and matrix mineralization. Proc Natl Acad Sci U S A 101(31):11203–11208
23. Davidson KM, Sushil S, Eggleton CD, Marten MR (2003) Using computational fluid dynamics software to estimate circulation time distributions in bioreactors. Biotechnol Prog 19:1480–1486
24. Sucosky P, Osorio DF, Brown JB, Neitzel P (2004) Fluid mechanics of a spinner-flask bioreactor. Biotechnol Bioeng 85(1):34–46
25. Yu P, Lee TS, Zeng Y, Low HT (2005) Fluid dynamics of a micro-bioreactor for tissue engineering. Fluid Dya Mater Process 1(3):235–246
26. Bilgen B, Barabion GA (2007) Location of scaffolds in bioreactors modulates the hydrodynamic environment experienced by engineered tissues. Biotechnol Bioeng 98(1):282–294
27. Bueno EM, Bilgen B, Carrier RL, Barabino GA (2004) Increased rate of chondrocyte aggregation in a wavy-walled bioreactor. Biotechnol Bioeng 88(6):767–777
28. Chen H-C, Hu Y-C (2006) Bioreactors for tissue engineering. Biotechnol Lett 28:1415–1423
29. Wiliams C, Wick TM (2004) Perfusion bioreactor for small diameter tissue-engineered arteries. Tissue Eng 10(5/6):930–941
30. Israelowitz M, Rizvi S, von Schroeder HP, Holmes C, Gille C (2007) Laminar flow reactor. United States Patent Application, 11/895645
31. Porter B, Zauel R, Stockman H, Guldberg R, Fyhrie D (2005) 3-D computational modelling of media flow through scaffolds in a perfusion bioreactor. J Biomech 38:543–549
32. Singh H, Ang ES, Lim TT, Hutmacher DW (2007) Flow modelling in a novel non-perfusion conical bioreactor. Biotechnol Bioeng 97(5):1291–1299
33. Cioffi M, Boschetti F, Raimondi MT, Dubini G (2006) Modelling evaluation of the fluid-dynamic microenvironment in tissue engineered constructs: a micro-CT based model. Biotechnol Bioeng 93(3):500–510
34. Chung CA, Chen CW, Chen CP, Tseng CS (2007) Enhancement of cell growth in tissue-engineering constructs under direct perfusion: modelling and simulation. Biotechnol Bioeng 97(6):1603–1616
35. Moussy Y (2003) Convective flow through a hollow fiber bioartificial liver. Artif Organs 27(11):1041–1049
36. Wolfe SP, Hsu E, Reid LM, Macdonald JM (2002) A novel multi-coaxial hollow fiber bioreactor for adherent cell types. Part 1: hydrodynamic studies. Biotechnol Bioeng 77(1):83–90
37. Marfels G, Poyck PPC, Eloot S, Chamuleau RAFM, Verdonck PR (2006) Three-dimensional numerical modelling and computational fluid dynamics simulations to analyze and improve oxygen availability in the AMC bioartificial liver. Ann Biomed Eng 34(11):1729–1744

38. Gu W, Zhu X, Futai N, Cho BS, Takayama S (2004) Computerized microfluidic cell culture using elastomeric channels and Braille display. Proc Natl Acad Sci U S A 101(45):15861–15866
39. Kleis SJ, Shreck S, Nerem RM (1990) A viscous pump bioreactor. Biotechnol Bioeng 36:771–777
40. Begley CM, Kleis SJ (2000) The fluid dynamic and shear environment in the NASA/JSC rotating-wall perfused-vessel bioreactor. Biotechnol Bioeng 70(1):32–40
41. Curran SJ, Black RA (2005) Oxygen transport and cell viability in an annular flow bioreactor: comparison of laminar Coquette and Taylor-Vortex flow regimes. Biotechnol Bioeng 89(7):766–774
42. Märkl H, Pörtner R (2003) Bioreaktoren. Chemie Ingenieur Technik 75(12):1888–1889
43. Suck K, Behr L, Fischer M, Hoffmeister H, van Griensven M, Stahl F, Scheper T, Kasper C (2007) Cultivation of MC3T3-E1 cells on a newly developed material (Sponcram) using a rotating bed system bioreactor. J Biomed Mat Res A 80(2):268–275
44. Chen H-C, Lee H-P, Sung M-L, Liao C-J, Hu Y-Cr(2004) A novel rotating-shaft bioreactor for two-phase cultivation of tissue-engineered cartilage. Biotechnol Prog 20:1802–1809
45. Niklason LE, Gao J, Abbott WM, Hirschi KK, Houser S, Marini R, Langer R (1999) Functional arteries grown in vitro. Science 284:489–493
46. Hoerstrup SP, Zund G, Sodian R, Schnell AM, Grunenfelder J, Turina MJ (2001) Tissue engineering of small calibre vascular grafts. Eur J Cardiothorac Surg 20:164–169
47. Thompson CA, Colon-Hernandez P, Pomerantseva I, MacNeil BD, Nasseri B, Vacanti JP, Oesterle SN (2002) A novel pulsatile, laminar flow bioreactor for the development of tissue-engineered vascular structures. Tissue Eng 8(6):1083–1088
48. Hildebrand DK, Wu ZJJ, Mayer JE, Sacks MS (2004) Design and hydrodynamic evaluation of a novel pulsatile bioreactor for biologically active heart valves. Ann Biomed Eng 32:1039–1049
49. Aunins JG, Brader B, Caola A, Griffiths J, Katz M, Licari P, Ram K, Ranucci CS, Zhou W (2003) Fluid mechanics, cell distribution, and environment in CellCube bioreactors. Biotechnol Prog 19(1):2–8
50. Watanabe S, Inagaki S, Kinouchi I, Takai H, Masuda Y, Mizuno S (2005) Hydrostatic pressure/perfusion culture system designed and validated for tissue engineering. J Biosci Bioeng 100(1):105–111
51. Dusting J, Sheridan J, Hourigan K (2006) A fluid dynamics approach to bioreactor design for cell and tissue culture. Biotechnol Bioeng 94(6):1196–1208
52. Fishbane PM, Gasiorowicz S, Thornton ST (1993) Physics for scientists and engineers, 2nd edn. Prentice-Hall, Inc., Upper Saddle River, New Jersey
53. Koynov A, Tryggvason G, Khinast JG (2007) Characterization of the localized hydrodynamic shear forces and dissolved oxygen distribution in sparged bioreactors. Biotechnol Bioeng 97(2):317–331
54. Navier CLMH (1822) Memoire sur les lois du mouvement des fluids. Mem. Acad. Sci. Inst. France 6: 389–440
55. Stokes GG (1845) On the theories of the internal friction of fluids in motion, and of the equilibrium and motion of elastic solids. Transactions of the Cambridge Philosophical Society Vol VIII, p287
56. Stokes GG (1880) Mathematical and Physical Papers (reprinted from the original journals and transactions, with additional notes by the author). Cambridge at the university press, Vol I, p75–129
57. Hinze JO (1975) Turbulence. 2nd Edition, McGraw-Hill, New York, p537–566
58. Darcy HPG (1856) Les fontaines publiques de la ville de Dijon. Dalmont, Paris (647p)
59. De Pitot H (1732) Description d'une machine pour mesurer la vitesse des eaux et la sillage des vaisseaux. Memoires de l'Academie des Sciences, Paris
60. Prandtl L (1942) Führer durch die Strömungslehre. 1. Auflage Verlag Vieweg & Sohn, Braunschweig
61. Prandtl L, Oertel H jr (2002) Kapitel 4: Dynamik der Flüssigkeiten und Gase, p57-176 in: Prandtl - Führer durch die Strömungslehre. Grundlagen und Phänomene. 11. überarbeitete und erweiterte Auflage, Herausgeber Oertel H jr., Verlag Vieweg, Braunschweig

# Index

**A**
Adipose-derived mesenchymal stem cells (AdMSC) 106
Alkaline phosphatase (AP) 100
Antibodies, mAb production, stable hybridomas/established mammalian cells 201
Apoptosis 103
Articular cartilage 127
Autologous cells, in vitro expansion 211

**B**
Bag bioreactors, biomanufacturing 196
   disposable 186
   static 189
   stirred 189
   vibromixer 192
BEV/insect cell-culture systems, r-proteins 199
Biomaterials 95, 100
Bioreactor optimization 11
Bioreactor systems 130
Bioreactors, 3D in vitro model systems 7
   clinical perspective 20
   dynamic compression 131
   fluid mechanics/dynamics 256, 264
   hollow fibre 258
   hybrid bag 195
   in vivo/in vitro 83
   magnetic force 86
   rotating bed 95
   rotating shaft 263
   rotating-wall perfused-vessel 258
   wave-mixed 192
   wavy-walled (WWB) 257

ZRP 113
BMSC 101
Bone 95
Bone engineering 100
Bone formation, mechanical forces 97
Bone morphogenetic proteins (BMPs) 100
Bone regeneration 15
Bone sialoprotein 100

**C**
Capillarised matrix 34
Cardiac muscle tissue 9, 29, 36
Cardiovascular tissue engineering 29
Cartilage 8, 125, 145
   articular 127
      biomechanics 129
      permeability 129
Cartilage formation, hydrostatic/mechanical load 162
   in vitro, calcium phosphate carriers 150
   reduced oxygen tension/optimized growth factor combination 153
Cartilage generation, perfusion systems 134
Cell cube bioreactor 263
Cell cultivation, ZRP-bioreactor 113
Cell death 68
Cell expansion 183, 209, 218
Cell phenotype, preservation 226
Cell seeding, three-dimensional matrices 3
Cell sources 98, 211
Cell/tissue development, computational models 13
Cell-scaffold constructs 21
Cellular necrosis 68
Ceramics 100

269

Chitosan 125
Chondrocytes 145
Collagen 10, 100, 127
Computational fluid dynamics 6, 251, 265
    modelling 235
Computational modelling 1, 231
    bioreactor systems 10
Connective tissues 81
Costs 224
Culture, 2D 119
Culture environment, controlled 5
Culture models, in vitro 82

**D**
Disposable bag bioreactor 183
Doppler optical coherence tomography
    (DOCT) 18
Dynamic compression bioreactors 131
Dynamic mechanical stimulation 29

**E**
Electric cell–substrate impedance sensing
    (ECIS) 112
Engineered heart tissue 35

**B**
Flow pattern 251
Flow velocity calculations 138
Fluid dynamics 231, 251, 264
    computational 251
Fluid mechanics 252
Force production 39
Functional tissue 183

**G**
Gene therapy, viruses 202
Glucose assay 114
Glycosaminoglycans 14, 128, 148
Graft manufacturing processes,
    streamlining 21
Growth curves 215
Growth factors 100, 145, 160
Growth kinetics 219
    exponential 222

**H**
Haematopoietic cell expansion 198
Heart valves 29, 33
Histology 138

Hollow-fiber bioreactors 131, 258
Human cell sources 16
Human chondrocytes 125
Hybrid bag bioreactor 195
Hydrostatic pressure/perfusion culture
    system 264

**I**
Intraoperative engineering 23

**L**
Laminar flow 251
Laser Doppler anemometry (LDA) 264

**M**
Magnetic force bioreactor 86
Mammalian cells, fluid dynamics 254
    transient transfection 200
Manometer methods 264
Matrix mineralization 115
Mechanical load 98, 145
Mechanical strain 95
Mechanics 81
Mesenchymal stem cells 83
Metabolic failure 67
Metabolic parameters 138
Microelectrode dishes,
    flexible 111
Milieu parameters 16
Mineralization 104
Monitoring, on-line 88
Muscle 41
    chemical stimulation 63
    electrical stimulation 60
    mechanical stimulation 57
    multiple-mode stimulation 64
    perfusion 62
Muscle bioreactor 41
Muscle development 39
    dynamic feedback control 43
Muscle organ maintenance 46
Muscle tissue, cardiac 29
    / organ development, guided 55
Muscle tissue bioreactors, failure
    modes 65
    mechanical failure within tissue 66
    septic contamination 65

**N**
Necrosis 104

# Index

## O
On-line monitoring 81
Operator error 69
Optical coherence tomography (OCT) 19, 88
Organ bioreactor system 41
Organ culture 39
Osteocalcin 100
Osteochondral implants 145

## P
Particle dynamics analysis 265
Particle image velocimetry (PIV) 11, 265
Perfusion bioreactors, porous scaffolds, fluid dynamics 258
Perfusion compression bioreactor 85
Perfusion systems 32, 131, 251, 257
Planar laser-induced fluorescence (PLIF) 265
Porcine cartilage, cartilage-carrier-constructs 171
Procollagen 127
Production facilities, centralized/de-centralized 23
Progenitor cells 209
Proteins, therapeutic 199
Proteoglycans 10, 127
Pulsatile flow 251
    reactor 263

## R
Re-differentiation 145
Regenerative medicine 125, 209
    tissue engineering 210
Reverse transcriptase-polymerase chain reaction (RT-PCR) 107, 117
Rotating bed bioreactor 95, 113, 121
Rotating shaft bioreactor 263
Rotating systems 258
Rotating vessel 251
Rotating-wall perfused-vessel bioreactor 258
Rotating-wall vessels 101, 131

## S
Safety 22

Scaffolds 4, 231
    design, micro-scale models 11
Scanning electron microscopy 115
Sensors 16
Sequential monolayer expansion 214
Shear stress 251
Skeletal muscle construct 55
Small intestine scaffold 34
Spinner flask 130, 256
Sponceram 95, 113
Stem cells 183, 209
Stirring systems 251, 256
Strain 98, 109
    cellular signals 99

## T
Therapeutic agents 183
Tissue damage, electrochemical 68
Tissue engineering, bioreactors, sensing 16
    cardiovascular 30
Tissue fluid 128
Tissue-to-tissue interfaces, mechanical failure 66
Tissues, bioengineered functional 197
    in vitro vascularised 29, 34
    muscle bioreactor 41
    physical conditioning 6
Toxic contamination 68
Traceability 22

## V
Vascular grafts 29, 30
Viruses, therapeutic 199
    vaccines 202
Wall shear stresses 13

## W
Wave-mixed bioreactors 192
Wavy-walled bioreactor (WWB) 257
Western blotting 105

## Z
ZRP-bioreactor, cell cultivation 113